ELECTROMAGNETIC and ACOUSTIC SCATTERING: DETECTION and INVERSE PROBLEM

ELECTROMAGNETIC and ACOUSTIC SCATTERING: DETECTION and INVERSE PROBLEM

21 May – 3 June, 1988,
Marseille-Luminy, France

Editors: **C Bourrely**
P Chiappetta
B Torresani
Centre de Physique Theorique,
CNRS, Marseille, France

World Scientific
Singapore • New Jersey • London • Hung Kong

Published by

World Scientific Publishing Co Pte Ltd
5 Toh Tuck Link, Singapore 596224
USA office: 27 Warren Street, Suite 401-402, Hackensack, NJ 07601
UK office: 57 Shelton Street, Covent Garden, London WC2H 9HE

British Library Cataloguing-in-Publication Data
A catalogue record for this book is available from the British Library.

ELECTROMAGNETIC AND ACOUSTIC SCATTERING: TIC DETECTION AND INVERSE PROBLEM

ISBN-13 9789971507480
ISBN-10 997150748X

Advisory Committee

C. BARDOS	Laboratoire de Mathématiques Appliquées, E.N.S., Paris
R. DELEUIL	Laboratoire de Radioélectricité, Univ. de Provence, Marseille
Ph. LAMY	Laboratoire d'Astronomie Spatiale, Marseille
B. POIREE	D.R.E.T., Paris
A. ROGER	Elf-Aquitaine, Pau
D. SORNETTE	Laboratoire de Physique de la Matière Condensée, Nice
P. TOURNOIS	Thomson-CSF, Boulogne-Billancourt

Organizing Committee

C. BOURRELY	Centre de Physique Théorique, CNRS, Marseille
P. CHIAPPETTA	Centre de Physique Théorique, CNRS, Marseille
B. TORRESANI	Centre de Physique Théorique, CNRS, Marseille

List of Chairmen

SCATTERING - ELECTROMAGNETIC AND ACOUSTIC DIFFRACTION

A. BERTHON, P. CHIAPPETTA, B. POIREE, J.P. SESSAREGO

IMAGING

J.P. LEFEBVRE

PROPAGATION IN RANDOM AND INHOMOGENEOUS MEDIA

F. BENTOSELA, A. ROUET, B. SOUILLARD

INVERSE PROBLEM

M. BERTERO, A. ROGER, P. SABATIER

FOREWORD

The Colloquium *Electromagnetic and Acoustic Scattering, Detection and Inverse Scattering Problem* was held at Marseille from 31 Mai to 3 June 1988. This colloquium is the first organized on this subject at the University of Luminy, the program covers a large variety of fields namely : scattering and diffraction in electromagnetism and acoustics, imaging, waves propagation in disordered and inhomogeneous media, inverse scattering problem.

By making this choice of topics we were aware that each subject by itself provided enough matter to make a specialized colloquium. However we adopted the idea that it would be of great interest to gather physicists and engineers working in these different domains with the hope that they communicate not only their results but also their methods and techniques used in their own field as a mutual benefit. We thought that one possibility to facilitate this cross disciplinary exchange was to organize round tables at the end of each topical sessions with a panel of specialists animating discussions.

A more local interest of the colloquium was to stimulate a dynamical interaction amongst the Marseille laboratories working in acoustics, optics, electromagnetic scattering and theoretical physics. Let us mention that a serie of seminars have already taken place, our hope is that the colloquium will strengthen their relationship.

It is a pleasure to thank the speakers who made the effort to make their results accessible to their non specialist colleagues. It should be mentioned that searchers from the industry have attended the meeting and brought interesting contributions on very practical problems they have to deal with.

The colloquium being the first organized locally in this field, it seemed to us reasonable to make an anouncement mainly to the national laboratories in order to estimate its impact. In fact the interest stimulated among the physics community was very encouraging and from the proposals made by numerous participants we propose to transform this meeting in a regular biennal event extended to an international audience.

We would like to thank the members of the advisory committee, C. Bardos, R. Deleuil, Ph. Lamy, B. Poirée, A. Roger, D. Sornette, P. Tournois, for providing very helpful advice on the choice of the subjects and their related speakers.

The colloquium secretary Mrs. D. Escalier has done an excellent job which has contributed to the smooth running of the colloquium.

We would like to thank the following organizations for providing the necessary financial support : le Centre National de la Recherche Scientifique, l'Université d'Aix-Marseille II, l'Université de Provence, le Conseil Regional PACA, le Conseil Général des Bouches du Rhône and la Société OCE.

The Organizing Committee
November 1988

CONTENTS

* No written contribution received

IMAGING

PROPAGATION IN RANDOM AND INHOMOGENEOUS MEDIA

SCATTERING -
ELECTROMAGNETIC AND ACOUSTIC DIFFRACTION

LIGHT SCATTERING PROBLEMS IN ASTROPHYSICS

Jean-Marie PERRIN
Laboratoire d'Astronomie Spatiale, CNRS

As is the case for various physical topics, astrophysics is concerned in the study of interactions between light and dust particles. Such interactions reveal various components of universe e.g. the interplanetary and interstellar media.

The cosmogonic interest of interplanetary grains has triggered intensive research on this subject, following two main approaches. The first one is concerned with studies of the brightness of the solar light scattered by interplanetary grains (zodiacal light). These studies are extended to interplanetary sources of dust such as comets, asteroïdal and planetary surfaces. The second approach is concerned with laboratory studies of meteorites, lunar rocks and dust particles collected in the upper atmosphere.

To obtain physical properties of interplanetary grains from caracteristics of the light they scatter (the zodiacal light), various methods are used. Classically the brightness of the zodiacal light as observed along a line-of-sight is described by a double integral, one over the size distribution of the interplaneary grains and one over their spatial distribution. With a limited set of assumptions it is possible to get rid of the integral over the spatial distribution using so-called inversion or reverse scattering technics; a function, the volume scattring function (VSF), containing the integral over the size distribution is obtained: it

characterizes the light scattered by a unit volume of interplanetary dust versus the scattering angle. As the size distribution can be obtained from space measurements and studies of microcraters grooved on lunar rocks, the VSF can be calculated using models describing interactions between light and various sorts of dust grains. This process must be used over a wide range of wavelength. When employed with polarization measurements of the zodiacal light, it permits to obtain information about the polarization of the scattered light.
The models describing the interaction between the solar light and the interplanetary dust cannot always be used for the study of the interstellar grains, especially models based on eikonal methods. The population which contributes mainly to the zodiacal light in composed of grains sizes are much larger than the wavelength in the ultraviolet and visible spectral domains. On the contrary the interstellar medium of our Galaxy is composed of submicronic dust particles.

The nature of the interstellar dust is still controversed. None of the various models of grains is able to explain the totality of the diffuse spectral features observed in the interstellar medium. Nevertheless some of them permit interesting confrontation between theoretical predictions and observational data. Some spectral features can be explained by homogeneous grains; other need coated or heterogenous particles of various shapes. In the framework of the classical electrodynamics these particles can be decomposed in a lattice of elementary dipoles whose sizes are compatible with the description of material given by a complex index of refraction (i.e. a macroscopic parameter).

This method can also be used for the study of the interstellar medium in other galaxies. More generally it can be used everywhere in the universe, each time a problem of interaction of electromagnetic field with dust particles is encountred.

REFERENCES

Dodd, R.T., "Meteorites", Cambridge University Press, (1981).

"Ices in the Solar System" (J. Klinger, D. Benest, A. Dollfus, R. Smoluchoswski, Editors), D. Reidel Publisbing Company (1985).

"Comets" (L.L. Wilkening, Editor), The University of Arizona Press, (1982).

"Properties and Interactions of Interplanetary Dust", (R.H. Giese, P. Lamy, Editors), D. Reidel Publishing Company, (1985).

"Cosmic Dust" (J.A.M. McDonnell, Editor), A. Wiley Interscience Publication (1978).

"Interstellar Processes" (D.J. Hollenbach, H.A. Thronson, Editors), D. Reidel Publishing Company (1987).

6

SURFACE INTEGRAL OPERATORS AND THEIR USE IN ACOUSTICS AND ELECTROMAGNETICS

André BERTHON
Société AERO - 3 avenue de l'Opéra - 75001 PARIS

ABSTRACT

We review the method of integral equations, its formalism, numerical implementation and domain of applicability, with emphasis on the properties of the singular surface integral operators which are central to the method. In the first part, after having defined these operators we discuss the ways of deriving a set of integral equations in order to solve a problem of scattering or radiation. The second part deals with the numerical aspects of the use of such operators. In the third part the practical limitations are discussed, and a few possible ways for future extensions are indicated.

1.- OPERATORS AND EQUATIONS

1.1 Equivalence Principle

Quite generally in a homogeneous medium the propagation of a harmonic field u -not necessarily scalar- is governed by an equation $Pu = 0$, where P is a second-onder elliptic partial differential operator with constant coefficients, and has no singularities except at the boundary. Now if there are some (harmonic) sources given by some distribution S the equation becomes $Pu = S$ and its solution is given in integral form by :

$$u(x) = \int G(x,y)\, S(y)\, dy \tag{1}$$

provided one knows the GREEN function $G(x,y)$ associated with P and satisfying suitable boundary conditions. The above integral makes sense at least if S has a compact support to which x does not belong.

By applying this to the exterior domain of a closed surface Σ, a GREEN function of free space which is translation-invariant and a freely propagating u, we obtain an integral representation for the field outside of Σ, which vanishes everywhere in the interior. For, if χ is the characteristic function of the interior domain we may write :

$$P(\chi u) = \chi P u + U_\Sigma \tag{2}$$

$$\chi u = G * U_\Sigma \tag{3}$$

where the distribution U_Σ has support Σ, and depends on the values of u and its derivatives on the surface.

This may be viewed as an expression of HUYGENS's principle or a KIRCHHOFF's formula. It states anyway an equivalence : in the exterior domain the sources of the field u , which are interior to Σ , are equivalent to a distribution of sources on the surface.

1.2 Integral representations

Let us take as examples the scalar field of acoustics and the electromagnetic field. To derive U_Σ we introduce δ_Σ, the DIRAC measure on Σ, the exterior normal vector of length unity $\hat{n}$, and a few formulas :

$$\text{grad}\,(\chi\, u) = \chi\,\text{grad}\, u + u\,\hat{n}\,\delta_\Sigma \tag{4}$$

$$\text{rot}\,(\chi\, v) = \chi\,\text{rot}\, v + \hat{n} \times v\,\delta_\Sigma \tag{5}$$

$$\text{div}\,(\chi\, v) = \chi\,\text{div}\, v + \hat{n} . v\,\delta_\Sigma \tag{6}$$

For the HELMHOLTZ operator $\Delta + k^2$ they give :

$$\begin{aligned} \chi u &= G * \frac{\partial u}{\partial n}\,\delta_\Sigma + \text{grad}\, G * \hat{n}\, u\,\delta_\Sigma \\ &= \int_\Sigma \left(G \frac{\partial u}{\partial n} - \frac{\partial G}{\partial n}\, u \right) dS \end{aligned} \tag{7}$$

where G, the GREEN function, is $\exp(-ikr) / 4\pi r = \varphi(r)$ for $r = /x\text{-}y/$ if u is to obey SOMMERFELD's asymptotic condition at infinity, the implicit harmonic time dependence of the field beeing $\exp(i\omega t)$.

For electromagnetics both the electric field E and the magnetic field H obey the double-rotational equation (P = rot rot - k^2) derived from the MAXWELL equations. It is integrated by means of the so-called dyadic GREEN function : for a vector distribution V the solution of Pu = V reads :

$$u = (\varphi + 1/k^2 \text{ grad} \otimes \text{grad } \varphi) * V \tag{8}$$

Applying the above formulae, and using MAXWELL's equation in the result one obtains expressions like :

$$\chi E = \int_{\Sigma} (-i\omega\mu (\hat{n} \times H) \varphi + (\hat{n} \times E) \times \text{grad } \varphi + (\hat{n}.E) \text{ grad } \varphi) \, dS \tag{9}$$

and a similar one for H. By using the relations between the fields and their rotationals other expressions are obtained, see for instance STRATTON [1]). In particular everything may be expressed in terms of the equivalent current sheets of surface density $J = \hat{n} \times H$ (electric current) and $K = \hat{n} \times E$ (magnetic current). If we set :

$$A_{\Sigma} (J) = \varphi * J \, \delta_{\Sigma} \tag{10}$$

$$H_{\Sigma} (J) = \text{rot } A_{\Sigma} (J) \tag{11}$$

$$E_{\Sigma} (J) = A_{\Sigma} (J) + \text{grad div } A_{\Sigma} (J)/k^2 \tag{12}$$

the integral representation of the fields may be written as :

$$\chi E = -i\omega\mu \, E_{\Sigma} (J) + H_{\Sigma} (K) \tag{13}$$

$$\chi H = H_{\Sigma} (J) + i\omega\varepsilon \, E_{\Sigma} (K) \tag{14}$$

Furthermore for any tangent vector field J on Σ the distributions $E_\Sigma(J)$ and $H_\Sigma(J)$ are related to each other by MAXWELL's equations. This remains true when Σ is not a closed surface but has a boundary Γ, which still allows to define $A_\Sigma(J)$. But in that case the expression of E_Σ as a surface integral is no more valid ; as an application of Stoke's theorem shows, the correct expression is :

$$E_\Sigma(J) = A_\Sigma(J) + (\operatorname{div} J \ \delta_\Sigma * \operatorname{grad}\varphi - (\hat{n} \times J).\hat{\tau}\ \delta_\Gamma * \operatorname{grad}\varphi)/k^2 \quad (15)$$

thus a line integral (distribution of support Γ, where $\hat{\tau}$ is the unit tangent vector) shows up.

This is KOTTLER's formula, which was devised to restore MAXWELL's equations (but not proper boundary conditions) when an approximate equivalent current is used on an aperture.

1.3 Integral operators

Since the field in the exterior domain is given by an integral extending over the surface Σ, this integral expression must have a limit when approaching the surface from the exterior. Thus for a point $x_0 \in \Sigma$ with normal n_0, the integral over a small patch surrounding x_0, evaluated at $x = x_0 + \varepsilon\, n_0$, has a limit for $\varepsilon \to 0$, $\varepsilon > 0$, although it may be divergent for $\varepsilon = 0$, whereas the integral over the rest of Σ is obviously continuous. But by the same reasoning there will be a limit, perhaps different, for $\varepsilon \to 0$, $\varepsilon < 0$, and this holds also for the derivatives of the field, and also for non-closed surfaces, as long as x_0 is not on the boundary. These limits define singular integral operators operating on suitable linear spaces of functions supported by Σ.

If the surface Σ coïncides with a physical boundary between homogeneous media, with some linear and local boundary conditions on it, the above-mentioned operators attached to Σ describe completely the reflection from or transmission through the surface of any incident field.

Specifically for the scalar field, as shown by formula (7), two integral operators on scalar functions defined on Σ are introduced, with kernels respectively φ and $\partial\varphi/\partial n$. When the integral (7) is evaluated at $x = x_o + \varepsilon\, n_o$, $x_o \in \Sigma$, the first part is continuous whereas the second is responsible for the jump :

$$\lim_{\varepsilon \to 0} \int_\Sigma u(y) \frac{\partial \varphi}{\partial n}(x,y)\, dy = \mathrm{sgn}(\varepsilon)\, u(x_o)/2 + \int_\Sigma u(y) \frac{\partial \varphi}{\partial n}(x_o,y)\, dy \quad (16)$$

The singularities of these operators are independent of the wavenumber k , thus their mathematical properties are essentially the same as for the corresponding operators of stationary problems, which are studied extensively, for instance, in the book of DAUTRAY and LIONS [2].

In electromagnetics, according to formula (13-14), we take the limits of the projections on Σ of both the electric and the magnetic field radiated by a current sheet J, and obtain two integral operators on the space of tangent vector fields having a continuous surface divergence div J. If they are denoted $\mathbf{H}_\Sigma$ and $\mathbf{E}_\Sigma$, we have the following limits for $x = x_o + \varepsilon\, \hat{n}_o$, $x_o \in \Sigma$:

$$\lim_{\varepsilon \to 0} \hat{n}_o \times H_\Sigma(J) = \mathrm{sgn}(\varepsilon)\, J(x_o)/2 + \hat{n}_o \times \mathbf{H}_\Sigma(J) \quad (17)$$

$$\lim_{\varepsilon \to 0} \hat{n}_o \times E_\Sigma(J) = \hat{n}_o \times \mathbf{E}_\Sigma(J) \quad (18)$$

Now from a computational point of view H_Σ is given, like in the scalar case, by a surface integral :

$$\hat{n}_o \times H_\Sigma(J) = \hat{n}_o \times \int_\Sigma \operatorname{grad} \varphi(x_o,y) \times J(y)\; dy$$

whereas the analog expression for E_Σ is not defined due to the appearance of second derivatives of φ in (12). But consider a patch Σ_η surrounding x_o, symmetric around that point and shrinking towards x_o when η goes to zero. Applying formula (15) to Σ_η and taking the limit we observe that the term with div J vanishes due to the symmetry, whereas the boundary integral may be computed up to terms vanishing with η . Finally we obtain the following principal-value integral expression for E_Σ :

$$\hat{n}_o \times E_\Sigma(J) = \lim_{\eta \to 0} n_o \times \left\{ \int_{\Sigma \setminus \Sigma_\eta} \left(\varphi(x_o,y)\, J(y) + 1./k^2 \operatorname{grad} \otimes \operatorname{grad} \varphi(x_o,y).J(y) \right) dy - J(x_o) / 4k^2 \eta \right\} \tag{19}$$

Now let us derive some relations between both operators. It is convenient to introduce the operator **R** which rotates a tangent vector field by $\pi/2$ around the normal. Let J and K be two currents on the closed surface Σ, E and **H** the fields given by formulae (13-14). By taking the limit of their projections on the surface we can write, in matrix operator form :

$$\begin{pmatrix} \hat{n} \times E \\ \hat{n} \times H \end{pmatrix} = \begin{pmatrix} -i\omega\mu\, R\, E_\Sigma & 1/2 + R\, H_\Sigma \\ 1/2 + R\, H_\Sigma & i\omega\varepsilon\, R\, E_\Sigma \end{pmatrix} \begin{pmatrix} J \\ K \end{pmatrix} \tag{20}$$

But by the equivalence principle we may replace J and K on the right-hand side by the equivalent currents $\hat{n} \times H$, $\hat{n} \times E$ respectively and express them again as functions of J and K, so we obtain equalities valid for every pair of currents, which imply the following relations between operators :

$$(\mathbf{R}\,\mathbf{E}_\Sigma)(\mathbf{R}\,\mathbf{H}_\Sigma) + (\mathbf{R}\,\mathbf{H}_\Sigma)(\mathbf{R}\,\mathbf{E}_\Sigma) = 0 \qquad (21)$$

$$k^2(\mathbf{R}\,\mathbf{E}_\Sigma)^2 + (\mathbf{R}\,\mathbf{H}_\Sigma)^2 = 1/4 \qquad (22)$$

1.4 Integral equations

The simplest problem to address is the scattering of an incident field by a non penetrable object, for instance a perfect conductor in the electromagnetic field. The boundary condition may be expressed as $\hat{n} \times E = 0$; the unknown of the problem is $\hat{n} \times H = J$, not just an equivalent current but the true surface density of electric current, since the field actually vanishes in the interior domain. Thus J is constrained by two different equations, the well-known electric field and magnetic field integral equations :

$$\text{EFIE} \qquad -i\omega\mu\,\mathbf{R}\,\mathbf{E}_\Sigma(J) = -\hat{n} \times E_{inc} \qquad (23)$$

$$\text{MFIE} \qquad (1/2 - \mathbf{R}\,\mathbf{H}_\Sigma)(J) = \hat{n} \times H_{inc} \qquad (24)$$

Using equation (22) and the fact that the incident fields are related to each other it can be shown easily that both equations are equivalent as long as the solution is unique, i.e. away from resonances of the interior domain ; at a resonant frequency non vanishing solutions exist for the homogeneous EFIE, which fulfill also the interior MFIE $(1/2 + \mathbf{R}\,\mathbf{H}_\Sigma)(J) = 0$, but not the exterior one.

Turning to penetrable objects, we consider the boundary between two homogeneous media with an incident field in medium 1. Now the unknowns are $J = \hat{n} \times H$ and $K = \hat{n} \times E$ on the boundary Σ, since both tangential fields are continuous. Again the fields in each medium are expressed in terms of J and K and, by taking the limit on Σ, we obtain two expressions for each unknown, where 1, 2 as indices for the constants and exponents for the operators refer to medium 1 and 2 respectively :

$$K = \hat{n} \times E_{inc} - i\omega\mu_1 \mathbf{RE}_\Sigma{}^1(J) + (1/2 + \mathbf{RH}_\Sigma{}^1)(K) \qquad (25)$$
$$= \quad i\omega\mu_2 \mathbf{RE}_\Sigma{}^2(J) + (1/2 + \mathbf{RH}_\Sigma{}^2)(K)$$

$$J = \hat{n} \times H_{inc} + i\omega\varepsilon_1 \mathbf{RE}_\Sigma{}^1(K) + (1/2 + \mathbf{RH}_\Sigma{}^1)(J) \qquad (26)$$
$$= \quad -i\omega\varepsilon_2 \mathbf{RE}_\Sigma{}^2(K) + (1/2 + \mathbf{RH}_\Sigma{}^2)(J)$$

In order to fulfill the boundary conditions we have to write down two equations for J and K that combine the relations valid on each side for any free field. This may be achieved in different ways, all equivalent by relation (22). As the operators $\mathbf{H}_\Sigma$ are compact -then the $\mathbf{E}_\Sigma$'s, by (22), are not- by eliminating the strongest singularities of $\mathbf{E}_\Sigma$ one obtains a system of two integral equations to which the theory of FREDHOLM applies and it is shown by MUELLER [3)] that it has a unique solution. This is achieved by the following linear combinations :

$$(\varepsilon_1+\varepsilon_2)K/2 = \varepsilon_1 \hat{n} \times E_{inc} - i(k_1{}^2\mathbf{RE}_\Sigma{}^1 - k_2{}^2\mathbf{RE}_\Sigma{}^2)(J) + (\varepsilon_1\mathbf{RH}_\Sigma{}^1 + \varepsilon_2\mathbf{RH}_\Sigma{}^2)(K) \qquad (27)$$

$$(\mu_1+\mu_2)J/2 = \mu_1 \hat{n} \times H_{inc} + i(k_1{}^2\mathbf{RE}_\Sigma{}^1 - k_2{}^2\mathbf{RE}_\Sigma{}^2)(K) + (\mu_1\mathbf{RH}_\Sigma{}^1 + \mu_2\mathbf{RH}_\Sigma{}^2)(J) \qquad (28)$$

Along these lines we are able to handle any configuration of homogeneous media, that is to produce a set of integral equations where the unknowns are the tangential fields on every interface between two different media. For each side of such an interface S there is a closed surface Σ bounding the corresponding medium ; the tangential limiting values of E and H are expressed as a sum of contributions from the unknown fields on the different pieces of Σ , including S, and for each S two equations are chosen that ensure the equality of tangential fields on both sides ; thus they relate together all the unknown equivalent currents that interact with S, i.e. that are not separated from S by a closed boundary, and the structure of the resulting system of integral equations reflects the topological setting of the different media. The field of the incident wave or any impressed sources gives of course the non homogeneous part of the system. It is worth mentioning that no assumption is made except homogeneity ; the media may well be of complex propagation constant k, for instance imperfect conductors.

Another example of the use of integral operators can be given, where the incident field is not completely known : consider a radiating structure powered via a waveguide, S beeing one of its cross-sections; let us define the antenna as the set of all structures downstream of S, including any reflector. The contributions to the radiated field are those of the (true or equivalent) currents on the external surfaces of the antenna ; this includes the currents on the terminal part of the waveguide unless it is separated from free space by some medium. We know how to write down integral equations for the unknown tangential fields on the different interfaces ; however the incident field, since it comes through the waveguide, must be expressed, in the antenna region, via the equivalent currents $\hat{z} \times H$ and $\hat{z} \times E$ on S,taking for the z axis the axis of the guide, which is normal to S, directed outwards. These currents are also unknown. But at the given frequency a waveguide propagates a finite number of modes ; in each mode we can have only a linear combination of two solutions of MAXWELL's homogeneous equations inside the waveguide, namely an incoming and an outgoing wave, which are known up to a complex constant.

Thus if there are N modes the incident field belongs to a 2N-dimensional complex linear space, and by solving the system of integral equations we obtain a linear space of the same dimension for the unknown currents, in other words we know the solution up to a finite number of constants. But the fields are continuous on the cross section S, where on the waveguide side they are just given by a linear combination of mode fields with the unknown parameters, whereas on the antenna side they have an integral expression in terms of the structure currents, including $z \times H$ and $z \times E$ on S. So the equality of, say, the transverse electric field on both sides of S gives two relations between our unknown parameters for each point of the section ; taking N such points suffices to solve the problem completely. The solution is of course unique up to an overall complex constant, and provides the standing wave ratios of the antenna simultaneously with the currents, from which the near field as well as the far field are easily calculated. Taking more than N points should not change the rank of the linear system for the mode coefficients, otherwise there is evidence that the field on S cannot be expanded on the N modes only ; this assumption requires obviously that the cross section be far enough from the waveguide ends, so that higher, vanishing modes may actually be neglected at that place. For more detail we refer to a previous and to a forthcoming paper [4,5)]

2.- NUMERICAL METHODS

Although the systems of integral equations discussed above solve the problems of radiation and scattering, they are not helpful unless they can be efficiently implemented on a computer. One is faced with the two classical problems of the discretization of a linear equation in a Banach space, and of the numerical treatment of singularities, since we are dealing with singular operators.

2.1 Discretization

In order to reduce an integral equation to a linear system of finite dimension on looks for an approximate solution in a finite dimensional space spanned by suitably chosen functions ; equations for the unknown coefficients are obtained by projecting the original equation on each vector of the basis (the method of moments) or with the help of any set of independent linear functionals having the right dimension , for instance just by matching at given points of the surface.

Discretization occurs also necessarily in the computation of the matrix elements, which are integrals. Consider for example the EFIE, equation (23), with the unknown current J expanded as :

$$J = \sum_{i=1}^{m} J_i \ \psi_i \tag{29}$$

and a set of functionals w_j . The finite system is :

$$\sum_{j=1}^{m} M_{ij} \ J_j \quad = V_i \tag{30}$$

$$V_i \ = \ - < w_i \, , \hat{n} \times E_{inc} \ > \tag{31}$$

$$M_{ij} \ = \ -i\omega\mu \ < w_j \, , \hat{n} \times E_{\Sigma}(\psi_j) \ > \tag{32}$$

The computation of M_{ij} involves a double integration on Σ , if the w_i's are scalar products with continuous functions like the ψ_j themselves, and a single integration if they are of the Dirac type like in point matching ; therefore this second possibility, though less secure in principle, is often preferred in practice. One may remark that quite generally the numerical quadrature formulae replace an integral by some weighted sum of Dirac measures.

2.2 Singularities

Moment methods have been extensively used and are discussed in many books [6,7] ; usually however, equations of the EFIE type are treated in integrodifferential form. The pure integral form (23) requires more work since an integrable kernel is not available. But it has the advantage not to introduce derivatives of the unknown function, which reduces the number of basis functions to introduce since derivatives involve higher spatial frequencies. A further advantage comes from the difficulty itself. Indeed the diagonal matrix elements of formula (32) is equal to the value of the singular operator RE_{Σ} applied to ψ_j

at some point that belongs to the support,which we assume limited, of this basis function (assuming point - matching ; otherwise it is a sum of such values). This is defined by the limiting procedure of equation (19), which shows that the matrix element diverges when the size of the domain of integration vanishes. When the discretization is refined the diagonal or self-patch interaction terms increase indefinitly, which is of great advantage for the conditioning of the system. For FREDHOLM integral equations of the first kind with regular kernels the discretized system becomes unavoidably more and more instable.

It is true that according to (19), the diagonal element must be calculated as the difference of two diverging quantities ; this cannot be safe unless the integration of the singular part of the kernel is performed analitycally, which is always possible in some neighborhood of the point x_0, say $\Sigma_\varepsilon \supset \Sigma_\eta$ if the surface is regular ; then the $1/\eta$ term appears in the integration over $\Sigma_\varepsilon \setminus \Sigma_\eta$ and cancels the second term of (19) ; we are left with analytic terms depending on ε, plus the nonsingular terms,which may be written as a limited development around x_0, to be computed and integrated numerically.

2.3 The particular case of axisymmetric problems

Assume that all surfaces of the problem have a common axis of revolution. This is an often-treated case because the azimuthal angle θ becomes more or less a dummy variable [8,9]. However the complete reformulation of the integral equation in terms of one variable -the arc length s- has not been studied so much. It leads to one-dimensional singular integral operators whith we briefly introduce now.

Since MAXWELL's equations are rotationnally invariant, E is a polar and H an axial vector, operators E_Σ and H_Σ commute with rotations and have parity + and - respectively. If we FOURIER-analyze all tangent vector fields :

$$J_\alpha(s,\theta) = \sum_{n=-\infty}^{\infty} J_{\alpha n}(s)\exp(in\theta) \qquad \alpha=1,2 \tag{33}$$

where $\alpha = 1$ for the longitudinal component and $\alpha = 2$ for the azimuthal one, and if we call $\Gamma_n^{\pm}$, $n > 0$, the subsets with azimuthal number n and $J_{1n} = \pm J_{1-n}$, $J_{2n} = \mp J_{2-n}$, the result is :

$$E_\Sigma(\Gamma_n^{\pm}) \subset \Gamma_n^{\pm} \qquad H_\Sigma(\Gamma_n^{\pm}) \subset \Gamma_n^{\mp} \tag{34}$$

So every problem of radiation and scattering splits into independent problems for each value of n, where the operators are given by 2 x 2 matrices of one-dimensional kernels $E_{\alpha\beta}{}^n(s,t)$, $H_{\alpha\beta}{}^n(s,t)$.

For example the EFIE becomes the set of following equations :

$$\lim_{\eta \to 0} \left\{ \int_{|t-s|<\eta} -i\omega\mu \sum_{\beta=1}^{2} E_{\alpha\beta}{}^n(s,t)\, J_{\beta n}(t)\, \rho(t)\, dt \right. \\ \left. -i\omega\mu/(\pi k^2 \eta)\; J_{\alpha n}(s)\, \delta_{\alpha 1} \right\} = -E^{inc}{}_{\alpha n}(s) \tag{35}$$

with the FOURIER coefficients of the incident field on the r.h.s. ; $\rho(s)$ is the distance to the axis of the point with abscissa s, the KRONECKER symbol shows that only the longitudinal-to-longitudinal kernel has a strong singularity. Complete expressions for the kernels, efficient methods for calculating them and examples of applications are given in [4,5].

3.- TRENDS AND OPEN QUESTIONS

We have sketched the properties of singular integral operators and shown their interest for stating and solving problems of radiation or scattering. It is interesting to examine more generally their possibilities and limitations compared with other methods.

3.1 Domain of usefulness

The method of integral equations (MIE) is commonly said to be limited to low-frequency problems. This means $D \lesssim N\lambda$, where D stands for some characteristic dimension of the surfaces in the problem, λ is the wavelength and N a small number. Typically the number n of discretization points night be 25 to 100 times N^2 - 5 to 10 points per wavelength. The achievable N depends on the size, also strongly on the accuracy of the computer. As time goes on and the capacities of computers increase, the limit of low frequencies shifts upwards. It seems fair to say that N may presently reach a few units for general 3D problems, whereas for bodies of revolution where n increases like N, N could even be a few hundreds. At least in the latter case, this matches the validity domain of asymptotic methods.

As compared with differential methods the MIE has the drawback of providing linear systems with full matrices ; but this is compensated by having to discretize only the boundaries and not to calculate fields at any volume points ; futhermore, since the radiation condition is built-in, the integral method is particularly well suited for external problems.

3.2 Further problems

Among the assumptions made was the homogeneity of each propagating medium ; of course it may be relaxed if one considers inhomogeneities as secondary sources proportional to the local fields - the medium remaining linear. The formalism stays the same, the GREEN function to use beeing that of an average homogeneous medium. What is lost is that now on must discretize the volume also.

The boundaries between media have been supposed to be regular ; in fact piecewise regular is enough since the singular integral operators are defined for non closed surfaces - However it is known that at an edge the unknown current densities have definite weak singularities or definite vanishing behaviours. It would be useful to incorporate this knowledge into the choice of basis functions and dicretization points.

It was mentioned that there is nothing particular with complex propagation constants. However this means absorption, and the skin depth may be much smaller than the width of the medium. Again computation time could be saved by taking properly into account the fact that far spaced points do not influence each other.

Finally the most important open question is that of the relations between integral and asymptotic methods. The concept of rays does not emerge naturally from the integral formalism and there seems to be no bridge between these two descriptions of the electromagnetic or acoustic field. However it would be extremely useful to use both methods simultaneously in many problems where objects are present of very different sizes and curvatures. This is beeing tried with the development of the so-called hybrid methods, but much progress has still to be made.

REFERENCES

{ 1 } STRATTON (J.A). Electromagnetic Theory, McGraw-Hill, New-York, 1941.

{ 2 } DAUTRAY (R), LIONS (J.L). Analyse mathématique et calcul numérique pour les Sciences et les Techniques, Masson, 1985.

{ 3 } MUELLER (C). Foundations of the Mathematical Theory of Electromagnetic Waves, Springer-Verlag, New-York, 1969.

{ 4 } BERTHON (A). Equations intégrales pour les problèmes à symétrie de révolution. Traitement du Signal, vol. 4, n° 2, 1987.

{ 5 } BERTHON (A), BILLS (R). Integral Equations Analysis of Radiating Structures of Revolution. To appear in IEEE Trans. on AP, February 1989.

{ 6 } MITTRA (R) editor. Numerical and Asymptotic Techniques in Electromagnetics, Springer-Verlag, 1975.

{ 7 } USLENGHI (P.L). Electromagnetic Scattering, Academic Press, 1978.

{ 8 } MEDGYESI-MITSCHANG (L.N), PUTNAM (J.M). Electromagnetic Scattering from Axially Inhomogeneous Bodies of Revolution. IEEE Trans. Antennas Propagat., vol. AP-32, n° 8, pp. 797-806, August 1984.

{ 9 } GOVIND (S), WILTON (D.R), GLISSON (A.W). Scattering from Inhomogeneous Penetrable Bodies of Revolution. IEEE Trans. Antennas Propagat., vol. AP-32, n° 11, pp. 1163-1173, november, 1984.

ACOUSTIC DIFFRACTION BY SLENDER BODIES OF ARBITRARY SHAPE

Michel TRAN-VAN-NHIEU
THOMSON-SINTRA
1, Av. Aristide Briand
94117 ARCUEIL

No written contribution received.

HIGH AND LOW ENERGY APPROXIMATIONS FOR THE ELECTROMAGNETIC SCATTERING BY IRREGULAR OBJECTS

B. Torrésani

Centre de Physique Théorique
C.N.R.S. Luminy Case 907
F-13288 Marseille Cedex 09
FRANCE

Abstract: We discuss two models of electromagnetic scattering by arbitrarily-shaped targets, respectively in the high and low energy approximations.Numerical results are compared to exact theories in simple cases, and to experimental data.

I INTRODUCTION

The study of the electromagnetic scattering by arbitrarily-shaped targets is currently recognized as a field of crucial importance, both from the theoretical as well as experimental points of view. Experimental measurements of scattered intensities by irregularly-shaped objects have been performed in the optical as well as microwave domains, and have exhibited large deviations from the results obtained in the case of scattering by simple objects, like for instance spheres or spheroids. Moreover, applications of such studies to particular situations are of great importance, for instance in astrophysics (see J.M. Perrin's lecture [1] for a review), and also in solid state physics, metrology, chemistry, medicine...

Generically, the problem is to solve vector Helmholtz equations for the Magnetic and Electric fields, with the continuity of their tangential components as boundary condition. But such a program is only tractable in the simple cases (i.e. in the cases where the geometry of the scatterer is simple), using for instance the Stratton method [2]. However, up to now, this method has been successfully applied only to spheres [3], spheroids [4] and infinitely long cylinders. Moreover, for large values of the reduced frequency λ/a (i.e. when the size of the scattering object is large compared to the incident wavelength), one is led to important numerical convergence problems.

The aim of this contribution is to report on joint works with C. Bourrely, P. Chiappetta, T. Lemaire (C.P.T. Marseille), J.M. Perrin (L.A.S. Marseille) and R. Deleuil (Laboratoire de Radioélectricité, Marseille), on the modelisation of the electromagnetic scattering by arbitrarily-shaped objects, at high and low energies, and comparison with experimental results.

II HIGH ENERGY APPROXIMATION: THE EIKONAL MODEL:

The intensity scattered by large particles is the superposition of three components: the first one due to the reflection at the surface of the scatterer, the second one due to to absorption and third one due to diffraction. We will discuss here only the diffractive component, refering for instance to the Wolff[5] and Perrin-Lamy[6] methods to treat the reflective part. Due to the energy domain we are concerned with (high energy, i.e. wavelength much greater than the size of the scatterer), it is relevant to work in the framework of a scalar approximation. Let us then represent the electromagnetic field by a scalar wavefunction :

$$\Psi(\vec{r}) \exp\{-i\omega t\} = \phi(\vec{r})\; e^{i\,\vec{k}.\vec{r}} \exp\{-i\omega t\}$$

satisfying the scalar Helmholtz equation:

$$[\Delta + n^2(\vec{r})\, k^2]\, \Psi(\vec{r}) = 0$$

Here $n(\vec{r})$ is the complex index of refraction and $\vec{k}$ the wave vector of the incident field. Asymptotically, $\Psi(\vec{r})$ fulfills the Sommerfeld radiation condition:

$$\Psi(\vec{r}) \underset{r\to\infty}{\longrightarrow} e^{i\,\vec{k}.\vec{r}} + f(\theta)\, \frac{e^{ikr}}{r}$$

In the scalar approximation, the scalar Helmholtz equation has the general solution:

$$\Psi(\vec{r}) = e^{i\,\vec{k}.\vec{r}} - \frac{k^2}{4\pi} \int_{\mathcal{V}} \frac{e^{ik\,|\vec{r}-\vec{r}\,'|}}{|\vec{r}-\vec{r}\,'|}\, [1 - n^2(\vec{r}\,')]\, \Psi(\vec{r}\,')\, d\vec{r}\,'$$

$\mathcal{V}$ being the volume inside which $1 - n^2(\vec{r}) \neq 0$.

For $r \to \infty$, the scattering amplitude reads:

$$f(\vec{k}_r,\vec{k}) = -\frac{k^2}{4\pi} \int_{\mathcal{V}} d\vec{r}\,' \exp\{-i\,k\,\vec{n}.\vec{r}\,'\}\; V(\vec{r}\,')\; \Psi(\vec{r}\,')$$

where $V(\vec{r}) = [1 - n^2(\vec{r})]$, $\vec{n} = \vec{r}/r$ is the direction of observation, and $\vec{k}_r = k\,\vec{n}$.

Such a solution requires the knowledge of the scalar wave function $\Psi(\vec{r})$ at each point within $\mathcal{V}$. To obtain it, one uses the Glauber assumptions[7]:

G1: Let a be the mean range of the scattering potential $V(\vec{r})$ (which can be defined for instance by: $\mathcal{V} = \frac{4}{3}\pi a^3$); then $k.a \gg 1$.

G2: For a point $\vec{r}$ within $\mathcal{V}$, the product $V(\vec{r})\, \phi(\vec{r})$ varies slowly on distances less or equal than λ .

G3: For some point $\vec{r}$ within $\mathcal{V}$, the product $V(\vec{r})\, \phi(\vec{r})$ varies appreciably only within a distance much larger than λ.

The last two assumptions impose severe restrictions on the scattering potential and on its variations inside the scattering body (note that there is not any geometrical constraint on the surface of the scatterer). They also indicate that the largest contribution to the integral is given by the domain $(\vec{r}-\vec{r}') = \vec{r}''$ parallel to the $\vec{k}$ axis.

Setting $\vec{r} = \vec{b} + z.\vec{e_z}$, where $\vec{e_z}$ is the unit vector along the z axis and $b = |\vec{b}|$ the impact parameter, and using the stationary phase principle, one then gets in spherical coordinates:

$$f(\vec{k_r},\vec{k}) = -\frac{k^2}{4\pi}\int_{\mathcal{V}} dz\, d\varphi\, b\, db\ [1 - n^2(\vec{b},z)]\ \exp\{-i\,(kb\sin\theta\cos\varphi - 2kz\sin^2\frac{\theta}{2})\}$$
$$\cdot\ \exp\{-i\frac{k}{2}\int_{-\infty}^{z} dz'\,[1 - n^2(\vec{b},z)]\}$$

(Here $\theta = (\vec{k_r},\vec{k})$ is the scattering angle)

The integration over the z variable at a fixed value (b_0,φ_0) of the b and φ coordinates can straightforwardly be carried out as follows: introduce the intersecting points $z_j(b_0,\varphi_0)$ of the scatterer surface with the straigth line defined by b_0 and φ_0. Then [8]

$$f(\vec{k_r},\vec{k}) = -\frac{k^2}{4\pi}\ \Sigma_j\int_0^{2\pi} d\varphi\int_{b_{inf}(\varphi)}^{b_{sup}(\varphi)} b\,db\int_{z_{2j-1}(b,\varphi)}^{z_{2j}(b,\varphi)} dz\ \ G(b,z,\varphi)$$
$$G(b,z,\varphi) = \exp\{-i\,(\,kb\sin\theta\cos\varphi\ -\ 2kz\sin^2\frac{\theta}{2}\,)\}\ \ [1 - n^2(\vec{b},z)]$$
$$\exp\{-\frac{ik}{2}\int_{z_{2j-1}(b,\varphi)}^{z_{2j}(b,\varphi)}\ [1 - n^2(\vec{b},z)]\}$$

This result is valid for the whole angular domain, but does not include variations of the refractive index over distances smaller than the wavelength at the surface of the particle (or inside in the inhomogeneous case). Moreover, it cannot take into account surface irregularities over scales less than λ.

In the cases where the object has a symmetry with respect to the z axis, this expression takes a simpler form [9]:

$$f(\vec{k_r},\vec{k}) = -\frac{k^2}{2}\int_0^{b_{sup}} b\,db\ \ J_0(kb\sin\theta)\ g(b)$$
$$g(b) = \int_{-\infty}^{\infty} dz\ \ [1 - n^2(b,z)]\ \ \exp\{2ik\sin^2(\frac{\theta}{2})\,z\}$$
$$\cdot\ \exp\{-i\frac{k}{2}\int_{-\infty}^{z} dz'\,[1 - n^2(b,z')]\}$$

The asymptotic scattered intensity is finally derived, according to:

$$I(\theta,\varphi) = k^2\,|\,f(\vec{k_r},\vec{k})\,|^2$$

One is therefore led to an expression of the diffracted intensity, valid in the whole angular domain, and taking exactly into acount the shape of the scattering object. When applied to the case of scattering by perfect spheres, this model gives results very close to those obtained in the framework of Mie theory [10]; moreover, the convergence problems at very high energies (i.e. in the domain of frequencies very much larger than the size of the scatterer) are absent in our case. But its most interesting predictions are those obtained in the case of rough scatterers. Numerical computations have been performed in the case where the rough particle (assumed to possess a cylindrical symmetry with respect to the direction of the incident beam, in order to allow faster numerical computations) is modelled by a fractal function[9], the fractal dimension being considered as a roughness parameter. More precisely, the scattering object is obtained by rotating the shape given at the top of the figures 1 and 2 around the symmetry axis (note that the term fractal has not to be taken here in its proper sense, since a scale-cutoff has been introduced to eliminate fluctuations of size smaller than λ, because of G2 assumption). The numerical computations exhibit two strong qualitative features: the presence of strong oscillations, and an important enhancement of the backscattered intensity, increasing when the fractal dimension (i.e. the roughness) increases. It is important to note that these are important deviations from Mie theory, and that these two features are confirmed by experiments (see [11] for instance, for the observation of the enhancement of backscattered intensity).

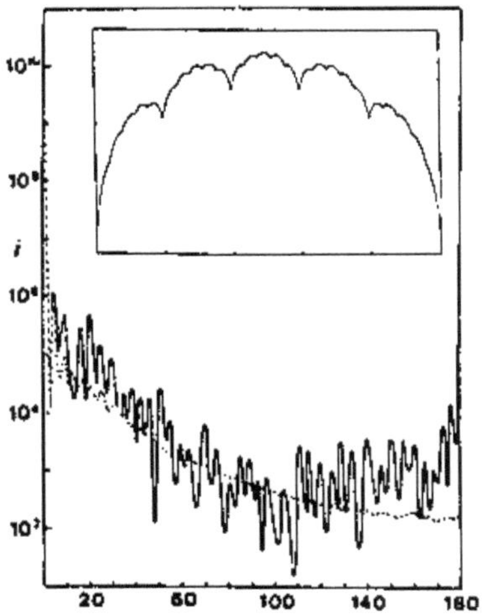

Fig. 1

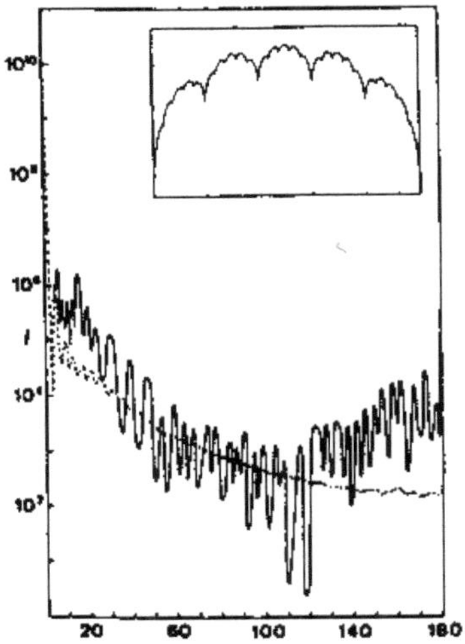

Fig. 2

Differential cross section $i(\theta)$ for a particle of mean radius 50 µm and refractive index 1.1+.005i at the wavelength .6328 µm (Fig 1: fractal dimension = 1.2; Fig 2: fractal dimension = 1.3). Dashed curve: perfect sphere of the same radius and refractive index.

A complete check of the Eikonal model being actually difficult from current optical experimental data, an experimental study of scattering from various types of irregular objects was realized in the microwave range [12] (A more complete description, together with the first experimental results on backscattering can be found in [13]), and is currently in progress. Typically, all objects (of index 1.89+.01i have axial symmetry and

the same length along this axis (a=8 cm), and the section b=S(z) of their boundary is made of quarter circles; the incident wavelength is set to .86 cm.
Hence the hypotheses of the Eikonal Model are satisfied. Numerical computations of the diffracted intensities have been performed, in the conditions of axial symmetry. Fig. 3 and 4 show the intensity diffracted by two of these objects, as a function of the scattering angle θ. As usual in the case of irregular objects, one sees the presence of strong oscillations, together with an increase of backscattered intensity. Moreover one gets a very interesting correlation between the number of quarter-circles appearing in the shape b=S(z) of the scatterer) and the number of maxima of intensity in the angular domain $85° \leq \theta \leq 180°$: for 16 quarter circles (Fig.4) one gets 3 maximas, while for 8 quarter circles (Fig.3) one gets two additional maximas at medium angles. Other numerical computations not presented here with respectively 4 and 6 quarter circles also lead to additional peaks of intensity at medium angles. This can be easily interpreted as interference phenomena, with a frequency of oscillations proportional to the distance between two consecutive elementary patterns (quarter circles here); for more details, we refer to [14].

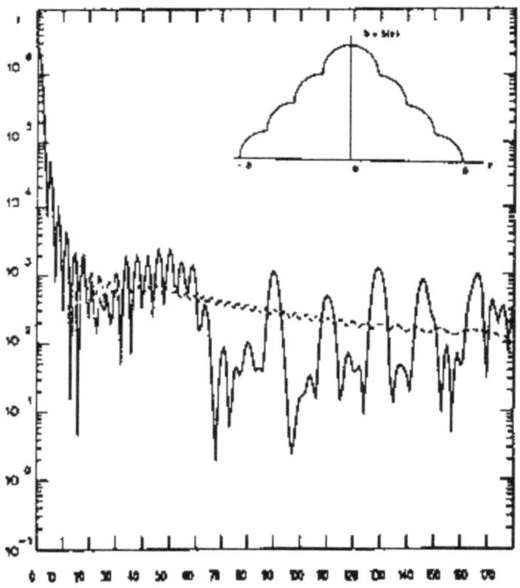

Fig. 3

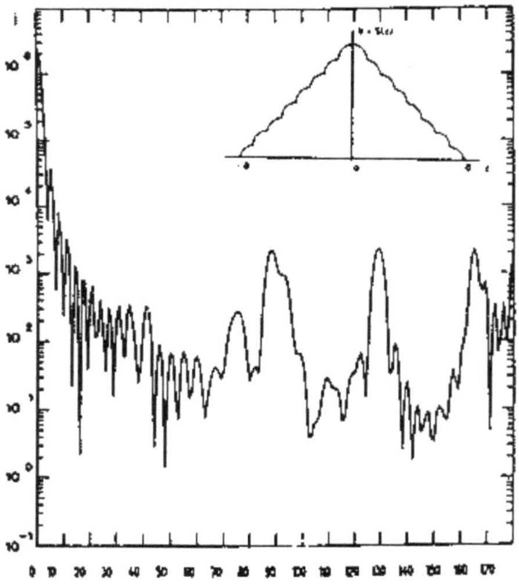

Fig. 4

Differential cross sections for objects whose geometry is described at the top of the figures, of mean radius 8 cm, refractive index 1.89+.01i at the wavelength .86 cm

Up to now, the Eikonal model gives only a scalar description of the electromagnetic scattering by dielectric objets at high energy. For lower energies, the polarisation effects have clearly to be taken into account, but it is far from clear how to incorporate them in the Eikonal model, since Stratton method seems to be uncompatible with the model. A possibility seems to be to adapt the Wolf method[5] (work in progress, by T. Lemaire).

III LOW ENERGY APPROXIMATION:

At low energies, it is fairly natural to expect that the scattered intensities do not depend on slight fluctuations of the scattering potential (i.e. slight variations of the refractive index, or slight irregularities of the surface of the target). However, quantum scattering theory predicts that, at low energies, the scattering function f can be quite well approximated by a scattering function obtained from a discretized version of the potential, the accuracy of the approximation depending on how low the energy is[15]. The model presented here, in which the scatterer is in some sense discretized, is an electromagnetic application of this general idea (note that the scalar approximation can no longer be used now, since the polarisation effects will be here very sensitive).

The dielectric scattering object is modelled by an array of polarizable elements (dipoles), located on the sites of a cubic lattice (following an original idea of Purcell and Pennypacker[16]). The Lorentz-Lorenz relation expresses the mean complex electric polarizability α_e in terms of the complex refractive index n:

$$\alpha_e = \frac{3}{4\pi\rho}\,\frac{n^2 - 1}{n^2 + 2}$$

ρ being the density of polarizable elements (here $\rho = d_o^{-3}$, the inverse cube of the lattice spacing).The scatterer is acted on by an incident beam, linearly polarized electromagnetic field:

$$\vec{E}_{inc}(\vec{r}) = \vec{E}_o \;\exp\{i\,\vec{k}.\vec{r}\}$$

By standard results, the dipole moment of the ith polarizable element is proportional to the effective electric field at the ith site:

$$\vec{d}_i = \alpha_e\,\vec{E}_{eff}(\vec{r}_i)$$

where $\vec{E}_{eff}(\vec{r}_i)$ is the superposition of the incident electric field $\vec{E}_{inc}(\vec{r}_i)$ and the electric field radiated by the other sites:

$$\vec{E}_{rad}(\vec{r}_i) = \sum_{\substack{j=1\\ j\neq i}}^{N} \mathbb{T}(\vec{r}_{ij})\,.\,\vec{d}_j$$

the $\vec{r}$-dependent $\mathbb{T}$ matrix being expressed as follows (see e.g. [17]):

$$\boxed{\mathbb{T}(\vec{r}) = \frac{e^{ikr}}{r^3}\,\left[\,k^2r^2\left(1 - \frac{\vec{r}\otimes\vec{r}}{r^2}\right) + (1 - i\,kr)\left(3\,\frac{\vec{r}\otimes\vec{r}}{r^2} - 1\right)\right]}$$

where one has denoted $\vec{r}_{ij} = \vec{r}_i - \vec{r}_j$

Taking together the expressions of the radiated electric fields and the $\mathbb{T}$ matrices for all the dipoles leads to a system of 6N complex linear equations, which can be solved by various methods (for instance, one can follow Purcell-Pennypacker's original proposition, or use numerical matrix inversion procedures). The method we present here is based on an approximation of each dipolar moment by a multiple scattering development[18][19], and presents suitable convergence properties: the dipolar moment at the ith site is decomposed into the contributions of all the other sites, after a given number of scatterings. It then takes the form:

$$\vec{d_i} = \alpha_e \vec{E_o} \exp\{i\vec{k}.\vec{r_i}\} + \alpha_e^2 \sum_{\substack{j=1\\ j\neq i}}^{N} \mathbb{T}(\vec{r_{ij}}) . \vec{E_o} \exp\{i\vec{k}.\vec{r_j}\}$$
$$+ \alpha_e^3 \sum_{\substack{j=1\\ j\neq i}}^{N} \sum_{\substack{m=1\\ m\neq j}}^{N} \mathbb{T}(\vec{r_{ij}}) . \mathbb{T}(\vec{r_{jm}}) . \vec{E_o} \exp\{i\vec{k}.\vec{r_m}\} + \ldots$$

which physically can be thought as a development on successive Born approximations (Numerically, this is to a resolution of the system of 6N equations by the iterative method). Note that the value of the refractive index n has to be small enough to ensure the numerical convergence of such a development.

From the dipolar moments $\vec{d_i}$, one deduces the expression of the integrated cross sections:

$$\sigma_{abs} = 4\pi k \frac{Im(\alpha_e)}{|\alpha_e|^2} \sum_{i=1}^{N} \frac{|\vec{d_i}|^2}{|\vec{E_o}|^2}$$

$$\sigma_{ext} = -4\pi k \sum_{i=1}^{N} \frac{Im[\exp\{i\vec{k}.\vec{r_i}\}\, \vec{d_i}^{\,*} . \vec{E_o}\,]}{|\vec{E_o}|^2}$$

$$\sigma_{sca} = \sigma_{ext} - \sigma_{abs}$$

from which one easily deduce the efficiency factors : $Q_i = \frac{\sigma_i}{A}$ (A being the area of the geometrical cross section of the scattering object, and i standing for sca, ext or abs).

The differential cross sections are then easily computed by applying to the total dipolar moment $\sum \vec{d_i}$ of the scatterer the matrix $\mathbb{T}(\vec{r})$, $\vec{r}$ being the location of the observer. Since we are currently interested on the asymptotic problem, let us first note that asymptotically, the $\mathbb{T}(\vec{r})$ matrix reads:

$$\mathbb{T}(\vec{r}) \underset{r\to\infty}{\longrightarrow} k^2 \frac{e^{ikr}}{r} [\, \mathbb{1} - \vec{n}\otimes\vec{n}\,]$$

The scattered field in the asymptotic $\vec{r}$ direction is then given by:

$$\vec{E}(\vec{r}) = k^2 \frac{e^{ikr}}{r} \sum_{i=1}^{N} \exp\{-ik\,\vec{r_i}.\vec{n}\}\, [\, \mathbb{1} - \vec{n}\otimes\vec{n}\,].\vec{d_i}$$

The differential scattering cross section reads:

$$\frac{d\sigma}{d\Omega} = \frac{\vec{P}.\vec{n}}{|\vec{P_0}|}\, r^2$$

where $\vec{P}$ (resp. $\vec{P}_0$) is the time-averaged Poynting vector of the incident (resp. scattered) wave. The intensity functions, corresponding to different incident and scattered polarization states, can then be computed by the formulas:

$$i_{\varepsilon_i \varepsilon_s}(\theta, \varphi) = \lim_{r \to \infty} \; 2\, k^6 \left| \sum_{i=1}^{N} \exp\{-ik\, \vec{r_i}.\vec{n}\}\; \vec{d_i} \cdot \vec{e}_{\varepsilon_s} \right|^2$$

ε_i (resp. ε_s) corresponding to incident (resp. scattered) polarization state perpendicular (ε=1) or parallel (ε=2) to the scattering plane. The angles (θ,φ) are the spherical coordinates of the unit $\vec{n}$-vector. One can also be interested in the φ-integrated intensity functions; one then computes:

$$i_{\varepsilon_s}(\theta) = \frac{1}{2\pi} \int_0^{2\pi} d\varphi \;\; i_{1\varepsilon_s}(\theta,\varphi)$$

Like in the case of the Eikonal model, the predictions of the Low Energy Model have been compared with the results of exact theories and experimental data.The first test of the accuracy of the Low Energy Model is the comparison to exact theories for spheres and spheroids (a more detailed comparison can be found in [19]).

In the case of spheres, we modeled a sphere of refractive index 1.33, with k.a = 1.5 by a regular array of 305 dipolar subunits, and computed the dipolar moment by a 20th order multiple scattering development. The plot of the efficiency factors[19] exhibits a remarkable agreement with the exact theory (Mie theory) in the range $0 \leq k.a \leq 1.9$ (discrepancies smaller than 3%), and are slightly better than those obtained by Purcell and Pennypacker [16]. Moreover the plot of the intensity functions i_1 and i_2 (Fig.5) and the polarization exhibits a very close agreement to Mie Theory, in the whole angular domain.

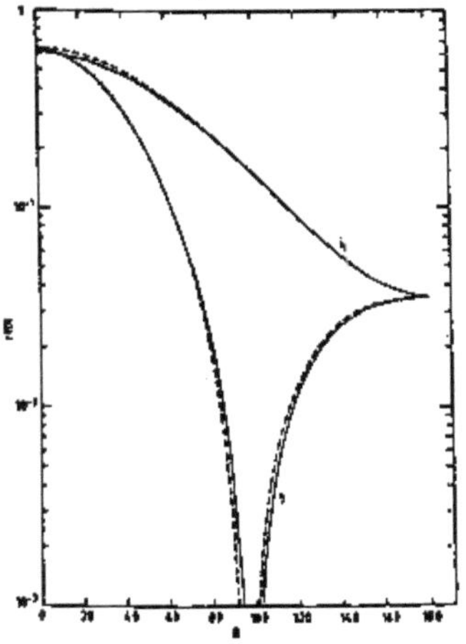

Fig. 5

Scattering intensities $i_1(\theta)$ and $i_2(\theta)$ for a sphere of refractive index 1.33+.1i, with size parameter k.a = 1.5. Full curve: Mie Theory; Dashed curve: Low Energy Model.

The Low Energy Model has also been tested by comparison to Asano-Yamamoto's results on scattering by prolate spheroïds (of equation $(x^2+y^2)/b^2 + z^2/a^2 = 1$, with a>b) and oblate spheroïds ($x^2/b^2 + (y^2+z^2)/a^2 = 1$, with a>b) for different values of the size parameter $c=k[a^2-b^2]^{1/2}$. The numerical results, to be compared to those of [4], are once more fairly interesting.

More precisely, the integrated efficiency factors for prolate and oblate spheroïds for a given refraction index and for many values of the ratio a/b have been numerically computed[19]; these results are in close agreement with those of [4], in the range of c values between 0 and 5 (more precisely, the discrepancies are smaller than 5%). For higher c-values, the agreement is less good, and this seems to be the limit of the validity domain of the Low Energy Model.

Fig.6 shows the plots of the intensity functions $i_1(\theta)$ and $i_2(\theta)$ for prolate spheroïds of refraction index 1.33, with a/b=2, and values of the c parameter between 1 and 5, computed in the framework of the Low Energy Model. This plot has to be compared with the corresponding one of [4] (more exactly to Fig.3 in this reference). The numerical results of the Low Energy Model have been obtained with a modelisation of the spheroïds as regular arrays of about 500 dipolar subunits, and with a number of iterations of about 20. The agreement between the Low Energy Model and Asano-Yamamoto's theory is rather impressive, and the Low Energy Model reproduces the general structures as well as the numerical values of Asano-Yamamoto's plot. Note that the polarisation effects are very sensible for low c-values, and tend to disappear when this size parameter grows.

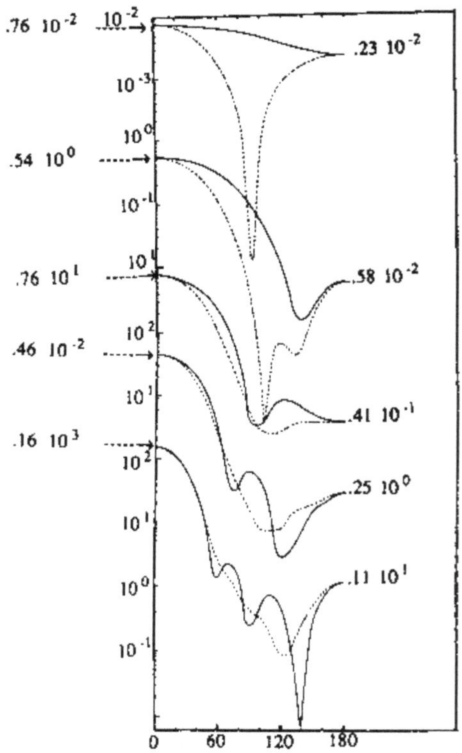

Fig. 6

Intensity functions for prolate spheroids of refractive index 1.33, with a/b=2, and respectively c=1, c=2, c=3, c=4, c=5.
Low Energy Model.

Finally, the accuracy of the Low Energy Model can be tested[20] by comparison with experimental data on microwave scattering by a dielectric helix[21]. This author has studied the scattering by a dielectric helical rod, acted on by a linearly polarised electromagnetic field, with different orientations of the helix in the beam.
We study here the particular case of an array of about 600 elements modelling a right-handed seven turns helix (of radius 1.83 cm, wire radius 0.24 cm and pitch 0.553 cm), with refractive index $n = 1.626 + i\, 0.012$ at the wavelength $\lambda = 3.18$ cm, with axis perpendicular to the wavevector, in the scattering plane.
The numerical results, taken from [20] are shown in Fig.7. They exhibit an interesting agreement vith experimental data, both for the $i_{11}(\theta)$ and the $i_{22}(\theta)$ curves. The maximal discrepancies are of about one order of magnitude. Other orientations of the scattering helix (helical axis parallel to the wavevector, or perpendicular to the scattering plane) have also been considered in [20]; the accuracy of the corresponding numerical results is comparable to that of the ones presented here. The cross polarized intensities i_{12} and i_{21} have also been studied, and have been found to be completely negligible, except in the case where the helical axis is in the scattering plane, perpendicular to the scattering vector. Finally, the predictions given by the Low Energy Model are closer to data than those of Ref.[22] (in which the authors used only a first order approximation of Shifrin's theory[23]).

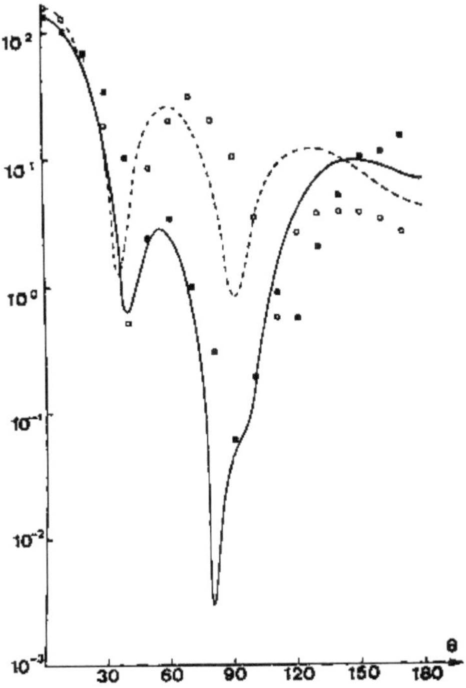

Fig. 7

Intensity functions $i_{11}(\theta)$ (full curve) and $i_{22}(\theta)$ (dashed curve), for a right-handed seven turns helix of radius 1.83 cm, wire radius 0.24 cm and pitch 0.553 cm, with refractive index $n = 1.626 + .012\, i$ at the wavelength $\lambda = 3.18$ cm, placed in the scattering plane, perpendicular to the incident beam. Comparison to experimental results

However, we have to wait for more improved experimental data before drawing definite conclusions.

IV CONCLUSIONS:

In this paper, we have reviewed two models of electromagnetic scattering, taking exactly into account the shape of the scatterer, and thus well adapted to the scattering by irregular objects.

Essentially, the Eikonal model is very well adapted to the very high energy domain, in which the scalar approximation it involves can be considered valid. Moreover, it does not present numerical convergence problems at such energies. Unfortunately, when applied to scatterers without radial symmetry, it leads to very time consuming numerical computations. The other restriction that can be made is an intrinsic characteristic of the model, i.e. its scalar character: at the limits of its validity domain, it is able to describe the diffracted intensity, but not the polarisation effects, which begin to be very sensible at such energies. And up to now, there is no other way of describing these polarisation effects in the framework of the Eikonal model, apart from Perrin-Lamy's ansatz[6].

The Low Energy Model, consisting in a discretization of the scattering potential, is very well adapted to targets much smaller than the incident wavelength (the basic point of the approximation involved is that the variations of the electric field between two neighbour lattice points inside the target are small enough). Taking into account the contribution of magnetic multipoles would lead to more precise results. Moreover, the method can easily be generalised to magnetic scatterers.

REFERENCES:

[1] **J.M. Perrin**: These proceedings.
[2] **J. Stratton**: *'Electromagnetic Theory'* (Mc Graw-Hill, 1941)
H.C. Van De Hulst: *'Light Scattering by small particles'* (Wiley, 1957).
[3] **G. Mie**: *Ann. Phys.* 25 (1908) 377.
[4] **S. Asano, G. Yamamoto**: *Appl. Opt.* 14 (1975) 29.
S. Asano: *Appl. Opt.* 18 (1979) 712.
[5] **M. Wolff**: *Appl. Opt.* 14 (1975) 1395, *Appl. Opt.* 20 (1981) 2493, *Icarus* 44 (1980) 780.
[6] **J.M. Perrin, P.L. Lamy**: *Optica Acta* 33 (1986) 1001.
[7] **R.J. Glauber**: *'Lectures in Theoretical Physics'* Vol. 1, Brittin and Durham Ed., New-York, Interscience, 315.
[8] **C. Bourrely, P. Chiappetta, T. Lemaire**: Preprint CPT-88/P.2139
[9] **C. Bourrely, P. Chiappetta, B. Torrésani**: *J.O.S.A.* A3 (1986) 250.
[10] **P. Chiappetta, J.M. Perrin**: *Optica Acta* 32 (1985) 907.
[11] **K. Weiss-Wrana**: *Astron. Astrophys.* 126 (1983) 240.
[12] **R. Deleuil**: *Optica Acta* 16 (1969) 23.
[13] **C. Bourrely, P. Chiappetta, R. Deleuil, B. Torrésani**: Preprint CPT-88/P.2086 to appear in the special issue *'Radar Cross Sections of Complex Objects'* of the "Proceedings of the I.E.E.E.".
[14] **C. Bourrely, P. Chiappetta, B. Torrésani**: *Opt. Comm.* 58 (1986) 365.
[15] **S. Albeverio, F. Gesztesy, R. Høegh-Krohn, H. Holden**: *'Solvable Models in Quantum Mechanics'* Springer Verlag (1988)
[16] **E.M. Purcell, C.R. Pennypacker**: *Astrophys. J.* 186 (1973) 705.
[17] **M. Born, E. Wolf**: *'Principles of Optics'* (Pergamon, Oxford, 1970).
[18] **P. Chiappetta**: *J. of Phys.* A13 (1980) 2201.
[19] **P. Chiappetta, J.M. Perrin, B. Torrésani**: *Nuovo Cimento* 9D (1987) 717.
[20] **P. Chiappetta, B. Torrésani**: *'Electromagnetic Scattering by a dielectric Helix'*, Preprint CPT-88/P.2086 (1988), to appear in *Appl. Opt.*.
[21] **R.T. Wang**: Preprint CRDC-CR 86071 University of Florida
[22] **R.D. Haracz, L.D. Cohen, A. Cohen, R.T. Wang**: *Appl. Opt.* 26 (1987) 4632.
[23] **K.S. Shifrin**: *'Scattering of light in a turbid medium'* Moscow 1951.
C. Aquista: *Appl. Opt.* 15 (1976) 2932.

ELECTROMAGNETIC SCATTERING AND MUTUAL INTERACTIONS BETWEEN CLOSELY SPACED SPHEROIDS

by

Jeannine DALMAS et Roger DELEUIL

Département de Radioélectricité, Université de Provence, Centre de St Jérôme, Avenue Escadrille Normandie-Niemen, 13397 - MARSEILLE CEDEX 13, FRANCE

ABSTRACT.

The problem of multiple scattering from two prolate spheroids with arbitrary orientation of their axes of symmetry has been theoretically solved. For particular geometries, involving more convenient computations (axes of symmetry mutually parallel and perpendicular and the wave vector of the linearly polarized incident plane wave perpendicular to the line joining the centers of the two scatterers), multiple scattering polar patterns are shown.

The study of electromagnetic scattering from one or more scatterers can be approached in two ways :

1) the scatterers are sufficiently separated in order that each scatter effectively receives only the incident field and so each of them scatters as if it were alone (single scattering by one or more scatterers),

2) the scatterers are closer to each other and hence the mutual interactions between scatterers must be taken into account (multiple scattering).

In Electromagnetic Theory, it can be shown that each divergenceless field vector (**E** for example) satisfies the vector wave equation $\Delta \mathbf{E} + K^2 \mathbf{E} = 0$ and, consequently, can be expressed by means of two vector wave functions [1] $\mathbf{M}^a = \text{curl}\,(\psi \mathbf{a})$ and $\mathbf{N}^a = (1/K)\,\text{curl}\,\mathbf{M}^a$. As it can be established, these vector wave functions are obtained from a vector **a** (such as curl **a** = o and div **a** = const.) and a scalar function ψ (with $\Delta\psi + K^2\psi = 0$, where Δ is the scalar Laplacian). Consequently, expressing the solution to the vector wave equation in terms of vector wave functions requires firstly the solution of the scalar wave equation satisfied by ψ. This is possible for curvilinear coordinate systems in which the scalar wave equation is separable such as spherical coordinates, prolate and oblate spheroidal coordinates, ...). In addition, the vector **a** is chosen in practrice as the position vector **r** or one of the three Cartesian unit vectors.

We consider a perfectly conducting prolate spheroid with major axis lying along the z - axis and receiving an electromagnetic plane wave (E^i, H^i), propagating in the x, z plane (Fig. 1). This plane wave is linearly polarized and its direction of polarization is either in the x,z plane (denoted //) or coincides with the y-axis (denoted ⊥).

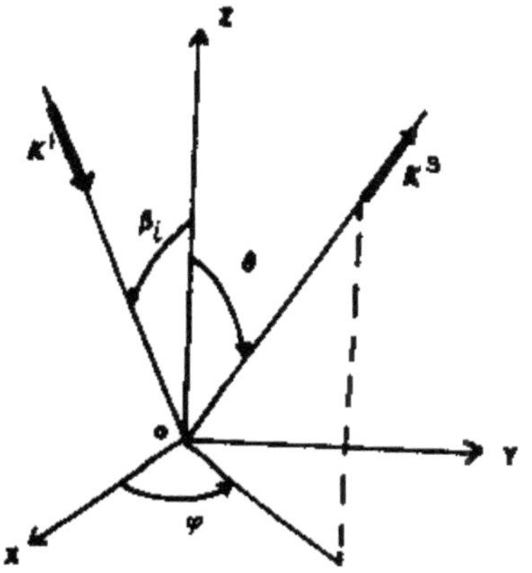

Fig. 1

As in the work by Flammer [2], we write :

$$E^{i}(\eta,\xi,\phi) = \sum_{m=0}^{\infty} \sum_{n=m}^{\infty} B\,(c_{mn}\, M^{(1)r}_{emn}(c;\eta,\xi,\phi) + d_{mn}\, N^{(1)r}_{omn}(c;\eta,\xi,\phi))$$

$$E^{i}(\eta,\xi,\phi) = \sum_{m=0}^{\infty} \sum_{n=m}^{\infty} jA\,(-d_{mn}\, M^{(1)r}_{omn}(c;\eta,\xi,\phi) + c_{mn}\, N^{(1)r}_{emn}(c;\eta,\xi,\phi)) \tag{1}$$

with

$$\varphi^{(h)}_{e,omn}(\eta,\xi,\phi) = R^{(h)}_{mn}(c,\xi)\; S^{(1)}_{mn}(c,\eta) \begin{matrix} \cos \\ \sin \end{matrix}\, m\phi$$

and $\mathbf{a} = \mathbf{r}$, $c = kd$, $k = \|\mathbf{k}^{i}\|$ and $d = \sqrt{a^2 - b^2}$ where d, a, b denote the semi-interfocal distance, the semi-major axis and the semi-minor axis, respectively.

The scattered field is written as a series-expansion similar to that of the incident field :

$$E^{s}_{\perp}(\eta,\xi,\phi) = \sum_{m=0}^{\infty} \sum_{n=m}^{\infty} B\,(-a_{\perp mn}\, M^{(3)r}_{emn}(c;\eta,\xi,\phi) - b_{\perp mn}\, N^{(3)r}_{omn}(c;\eta,\xi,\phi))$$

$$E^{s}_{/\!/}(\eta,\xi,\phi) = \sum_{m=0}^{\infty} \sum_{n=m}^{\infty} jA\,(+b_{/\!/ mn}\, M^{(3)r}_{omn}(c;\eta,\xi,\phi) - a_{/\!/ mn}\, N^{(3)r}_{emn}(c;\eta,\xi,\phi)) \tag{2}$$

where the $e^{-j\omega t}$ time-dependence of the field vectors is selected. In Eq. (2), the superscripts   and   are used to indicate the polarization of the incident field.

The scattered coefficients $a_{\perp,/\!/mn}$ and $b_{\perp,/\!/mn}$ are determined by satisfying the boundary conditions over the surface of the perfectly conducting scatterer. We obtain equations containing the variables η and ϕ. The variable ϕ is easily eliminated using the orthogonal properties of cos $m\phi$ and sin $m\phi$. The elimination of the variable η is more difficult to carry out and requires the calculation of nine definite integrals [3, 4].

Problems requiring multiple scattering to be taken into consideration are complicated because the field striking each scatterer is composed of the sum of the incident field and all the secondary fields coming from the other scatterers. Consequently, it is necessary to establish *addition theorems* for the spheroidal vector wave functions **M** and **N** [5, 6], in order to transform the scattered field coming from each spheroid (outgoing wave) into an ingoing wave striking the spheroid for which the boundary conditions are expressed. In the general case, this transformation involves a rotation and a translation of the spheroidal coordinate system.

The multiple scattering coefficients of each spheroid being numerically determined, we can express by summation the total field produced by all the scatterers (denoted $\mathbf{E}^s{}_t$) at any point of observation.

If the scattered field of interest is the far field, it becomes transverse and it is possible to define a bistatic multiple scattering cross-section as

$$\sigma\,(\theta,\phi) = \lim_{r\to\infty} 4\pi\, r^2\, \left(|\mathbf{E}^s{}_t|^2/|\mathbf{E}^i|^2\right)$$

which can be normalized [7] dividing, for example, by $\sigma_G = \pi\,(b^4/a^2)$ if all the prolate spheroids are identical.

To conclude, the bistatic multiple scattering cross-section depends on :

1) the characteristics of the incident plane wave (angle β_i and wave number $K = 2\pi/\lambda$) and its direction of polarization,

2) the individual characteristics of each scattering prolate spheroid (ξ_0 or b/a and d by means of the parameter c = kd),

3) the geometrical characteristics of the set of scatterers (distances between the centers of the spheroids and angles between their axes of revolution).

To illustrate the above calculations, multiple scattering cross-sections have been computed in various particular planes for two identical prolate spheroids, characterized by c = 5. and ξ_0 = 1.1547 (namely a/b = 2). Their axes of revolution are either perpendicular

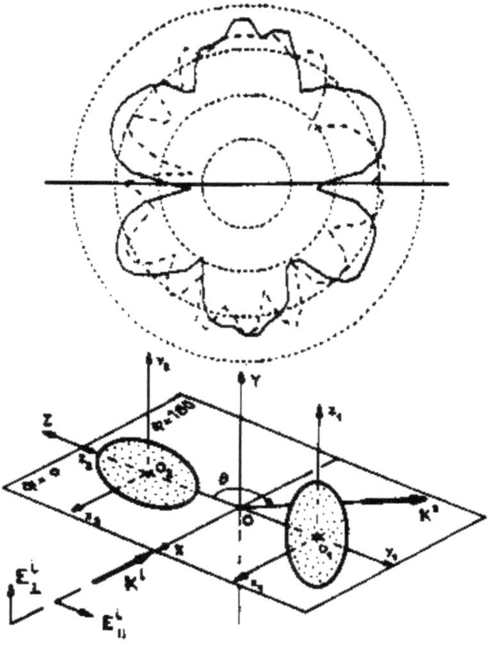

Fig. 2

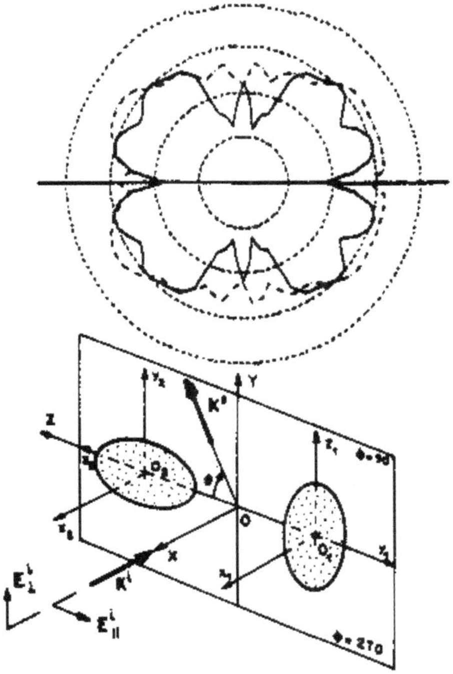

Fig. 3

(Figs. 2 and 3) or parallel (Figs. 4 and 5). On each pattern, the logarithmic normalized bistatic multiple scattering cross-section is plotted using polar coordinates. The corresponding geometry is represented and the distance between the centers of the spheroids is indicated. In addition, in Figs. 2 and 3 the solid line corresponds to E^i directed along the z-axis and the dashed line to E^i directed along the y-axis. For all patterns, the four circles correspond to 10, 20, 30 and 40 dB levels, respectively : the range of variation of $\sigma_\perp$ and $\sigma_{/\!/}$ is approximately the same.

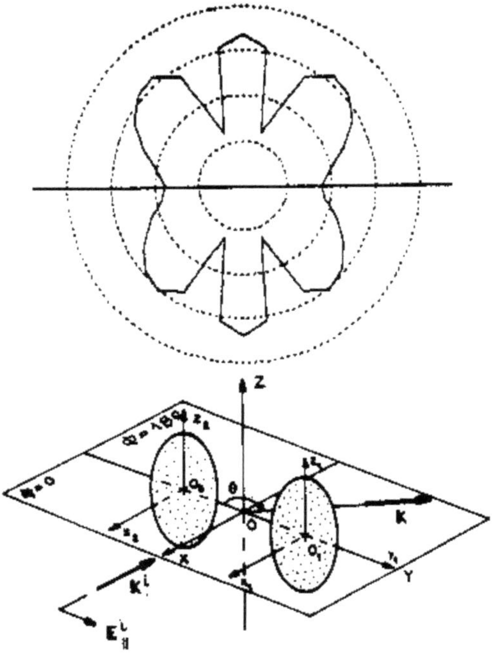

Fig. 4

The scattering is essentially dependent on the plane under consideration [(0,180)-plane or (90,270)-plane] and the direction of polarization of the incident field. In each case the scattering diagrams exhibit well shaped lobes (more thant 20 dB of amplitude variation). The main difference between the two geometries is that the bistatic multiple scattering cross-section is null for several directions which depend on the spacing of the centers of the spheroids when their axes of revolution are parallel [8] or in the same direction [9].

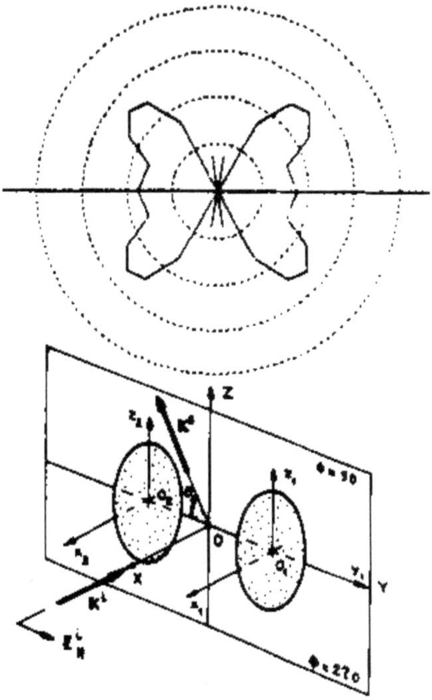

Fig. 5

Calculations on multiple scattering by prolate spheroids having arbitrary of their axes of revolution require the use of addition theorems of the spheroidal vector wave functions involving a rotation *and* a translation. Consequently, the computations are the most complicated among those which can be executed in multiple scattering theory. In addition, during the computation, infinite multiple summations and infinite systems of linear equations must be truncated.

For these reasons, computational studies can be carried out in the hope of favourable results if geometries and characteristic parameter values are chosen beforehand in order to ensure future efficiency in the computations; therefore, the present study is limited to two identical prolate spheroids with perpendicular axes of revolution. Nevertheless, this study should be extended to other spacings between the centers of the spheroids and other directions of propagation of the incident field, to take into account some specific Radar applications.

REFERENCES

[1] Statton, J.A. Electromagnetic Theory, MacGraw-Hill Book Company (1941).

[2] Flammer, C., Spheroidal wave functions, Stanford University Press (1957).

[3] Dalmas, J., Optica Acta, 28 n°7, 933-948 (1981).

[4] Dalmas, J., Optica Acta, 28 n°9, 1277-1287 (1981).

[5] Dalmas, J. and Deleuil, R., Quart. Appl. Math, 44 n°2, 213-222 (1986).

[6] MacPhie, R.H,. Dalmas, J. and Deleuil, R., Quart. Appl. Math, 44 n°4, 737-749 (1987).

[7] Dalmas, J. and Deleuil, R., Optica Acta, 27 n°5, 637-649 (1980).

[8] Dalmas, J. and Deleuil, R., Radio Science, 20 n°3, 575-581 (1985).

[9] Dalmas, J. and Deleuil, R., Optica Acta, 29 n°8, 117-1131 (1982).

SELECTION RULES INDUCED BY SYMMETRIES IN THE DIFFRACTION PATTERN OF A CYLINDER

Patrick VINCENT

Laboratoire d'Optique Electromagnétique, CNRS UA-843
Université Aix-Marseille III, Centre de St-Jérôme,service 262
F-13397 Marseille Cedex 13, FRANCE

1 ABSTRACT

This paper studies the symmetry properties of the resonances of the diffracted electromagnetic field linked to poles of the scattering matrix. It is shown on the square dielectric cylinder example that a simple irreducibility axiom allows to exhibit forbidden directions for the excitation of resonances as well as for the diffracted field. The method used is based on the orthogonality relations between the characters of the irreducible representations of the symmetry group of the object.

2 INTRODUCTION

The diffraction pattern of a symmetrical object must reflect in some way the symmetries of the object: we intend to show in this paper how the features of the finite group representation can be used to determine selection rules on the excitation of resonances or diffraction directions of resonance. For the purpose of simplicity, a simple diffraction problem is examined but the method can be easily extended to more complex cases, avoiding often cumbersome calculations.

3 PROBLEM

The typical diffraction problem used to demonsrrate the determination of selection rules is the computation of the harmonic diffracted field by a dielectric square cylinder in TE polarization, i.e. when the electric field vector is parallel to the cylinder axis.

3.1 Notations

We use (r,θ,z) cylindrical coordinates with Oz as the axis of the cylinder (fig. 1). The electromagnetic field is written taking into account an $\exp(-i\omega t)$ time dependence. The cylinder material is an isotropic dielectric with relative permittivity ϵ_r real or complex and permeability μ_0 . The object is surrounded by vaccuum (ϵ_0, μ_0).

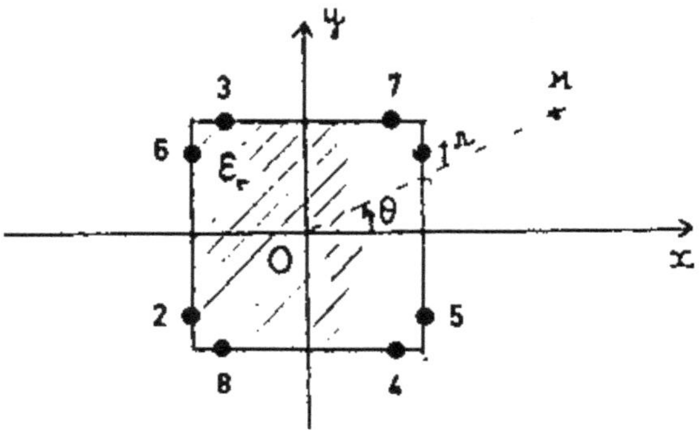

Figure 1
The cylinder cross section

3.2 Propagating equations

In TE polarization, the diffraction problem reduces to a scalar one and the electromagnetic field is entirely determined by its E_z component. The propagation equation, derived from Maxwell harmonic equations, and valid in the sense of distributions is:

$$\Delta E_z + k^2 E_z = 0 \tag{1}$$

where k^2 takes two different values: k_1^2 inside the cylinder and k_0^2 outside.

Projected onto the $\{\exp(in\theta)\}$ Fourier basis, this partial derivative equation yields:

$$E_n'' + \sum_n (k^2)_{n-m} E_m - n^2 E_n = 0 \tag{2}$$

with:

$$E_z(r,\theta) = \sum_n E_n(r) \exp(in\theta) \qquad n \in \mathbb{Z} \tag{3}$$

$$k^2(r,\theta) = \sum_n (k^2)_n \exp(in\theta) \qquad (4).$$

For r greater than the half-diagonal r_0 of the square cross section, the E_n Fourier coefficients are Hankel functions:

$$E_n(r) = e_n^+ H_n^+(k_0 r) + e_n^- H_n^-(k_0 r) \ . \qquad (5)$$

thus it is possible to separate the electromagnetic field in an incoming part described by the e_n^- coefficients, and an outgoing one given by the e_n^+ coefficients. The linearity of Maxwell equations implies that the outgoing part is related to the incoming one by a linear operator: the scattering operator. Onto the Fourier-Hankel basis used here, this operator is represented by the scattering matrix S giving the $\{e_n^+\}$ set of coefficients from the $\{e_n^-\}$ one.

4 S MATRIX SINGULARITIES

The scattering matrix S is a function of the circular frequency ω of the harmonic field, thus depends from $k_0 = \omega(\epsilon_0\mu_0)^{1/2}$. We assume that the relative permittivity ϵ_r of the square cylinder is frequency independent, thus that k_1 is proportionnal to k_0. Physical incident fields correspond to real values for k_0, but it is possible to look for the S matrix when this variable is made complex and get from its properties in the complex k_0 plane valuable information on its behaviour on the real axis.

4.1 S matrix analytic continuation

The definition of incoming and outgoing waves from relation (5) is obvious when k_0 is real. The extension of this definition to complex values of k_0 implies the use of appropriate branch cut in the complex plane. We shall not discuss here the choice of such elements and their physical signification because the only requirement for the following considerations is that the branch cuts do not lie between the singularities of the S matrix and the real axis segment we are interested in. In fact we suppose that the values of the matrix on the real axis can be deduced from the singularities in the vicinity of k_0 and that no branch cuts are found in this region.

Physically, the existence of singularities for the S matrix are suggested by resonance peaks in scattered cross section curves. In the case of metallic scatterers, the use of singularities to compute the diffracted field has been developped under the "Singularity Expansion Method" (SEM) denomination [1]. It can be done also for dielectric scatterers and, in this case, the resonance peak separation is better when the permittivity is high[2].

4.2 Pole location in the complex plane

The S matrix determination is done numerically, using a shooting method and a step-by-step integration of equations (2) from r=0 to $r=r_0$. The matching with the Hankel expansion (5) gives either the S matrix or its inverse. The pole determination is done in an iterative manner, searching for the zeros of the determinant of S^{-1}. The poles found for a square dielectric cylinder with dielectric real permittivity ϵ_r varying from 2 to 6 can be found in figure 2.

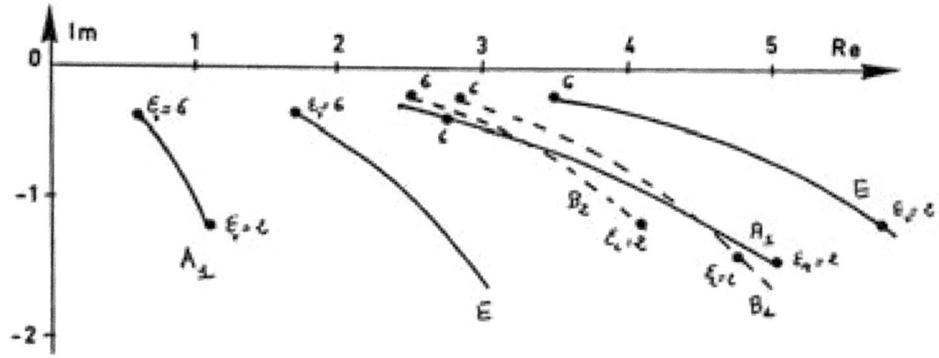

Figure 2
The complex poles of a square cylinder with permittivity varying from 2 to 6.

It is known that a perfectly conducting sphere has only simple poles [3], and it seems likely that the same applies for finite-size objects. Our cylinder diffraction problem is, from a mathematical point of view, a two dimensional one, thus the finite size cross-section is similar to finite size object requirement in 3-dimension. This leads us to assume that the poles of the S matrix are simple ones.

4.3 S matrix residues

Let us introduce the eigenvector $|pq\rangle$ of S^{-1} and the associated eigenvalue c_p defined by:

$$S^{-1} |pq\rangle = c_p |pq\rangle , \qquad (6)$$

where the letter q is used to span the eigensubspace when the eigenvalue has a multiplicity greater than one. Assuming that the S matrix is regular enough, the spectral decomposition theorem yields:

$$S = \sum_{p,q} \frac{|pq\rangle \langle pq|}{c_p} , \qquad (7)$$

where $\langle pq|$ is the eigencovector associated to $|pq\rangle$ and satisfying the normalization relation:

$$\langle pq|pq\rangle = \delta_{ip}\, \delta_{jq} \qquad (\delta \text{ is the Krônecker symbol }) .$$

The poles of S are the complex frequency values that reduces the eigenvalues to zero. Assuming that the transition operator T=I-S is compact, the Fredholm analytic theorem[4] apply thus S is a meromorphic function of k_0 and the residues at the poles are finite rank dyadic, i.e. the index q takes only a finite number of values for a given eigenvalue. Taking into account our hypothesis of simple poles, we write:

$$R_p = \lim_{k_0 \to k_p} (k_0 - k_p)\, S(k_0) \qquad (8).$$

Using expansion (7), we get:

$$R_p = \lim_{k_0 \to k_p} (k_0 - k_p) \sum_q \frac{|pq\rangle\langle pq|}{c_p} . \qquad (9)$$

The eigenvalue c_p tends towards zero, thus we get:

$$R_p = \frac{dk_0}{dc_p} \sum_q |pq\rangle\langle pq| . \qquad (10)$$

4.4 Singularities expansion

Using Mittag-Leffler expansion for meromorphic functions, we write in the vicinity of poles:

$$S(k_0) = \sum_p \frac{R_p}{k_0 - k_p} + S_0(k_0) \ , \qquad (11)$$

where S_0 is analytic in a neibourhood around the poles. For lossless dielectric, the permittivity ϵ_r is real and the S matrix is unitary, thus its determinant is null for values $\bar{k}_p$ conjugated of the poles values. Numerical experiments have shown that it is possible to represent with a good accuracy the S matrix by taking into account the zeros values using the improved version of expansion (11) :

$$S(k_0) = \sum_p \frac{k_0 - \bar{k}_p}{k_0 - k_p} R_p' \ . \qquad (12)$$

5 RESIDUES AND SYMMETRY GROUP

Let us write G_i the matrix representation of a geometrical transformation which left the cylinder cross section invariant, thus commutes with the S matrix. Consequently, G_i commutes also with the residue R_p' and we get:

$$R_p \ G_i \ |pq\rangle = G_i \ R_p \ |pq\rangle = c_p \ G_i \ |pq\rangle \ , \qquad (13)$$

thus, $G_i |pq\rangle$ is also an eigenvector of R_p for the same eigenvalue c_p. This implies that the (sub)space associated to eigenvalue c_p is left invariant by the symmetry group of the object and, in the following, we assume that this space defines an irreducible representation of the group [5]. Thus it is possible to link each pole to an irreducible representation of the group and to deduce selection rules for the resonances corresponding to the poles from the orthogonality properties of the characters of the representation. Let us demonstrate how on the square cylinder example.

5.1 The symmetry group of the square

Let us recall briefly the main properties of the square symmetry: it is left invariant by the following geometrical transformations:

1) identity: I,
2) rotation: $R(\pi)$,
3) rotation: $R(\pi/2)$,
4) rotation: $R(-\pi/2)$,
5) symmetry with respect to Ox: S_x,
6) symmetry with respect to Oy: S_y,
7) symmetry with respect to the first diagonal: S_u,
8) symmetry with respect to the second diagonal: S_v,

which are the elements of the symmetry group. The diagonal elements sums (i.e. the trace) of the matrices representing the elements are the characters of the transformations. Two equivalent representations of the group have the same set of characters and two elements of the group having the same characters belong to the same class. Here we get five different classes: $E=\{I\}$, $C_2=\{ R(\pi) \}$, $2C_4=\{ R(\pi/2), R(-\pi/2) \}$, $2\sigma=\{ S_x, S_y \}$, $2\sigma'=\{ S_u, S_v \}$. The number of inequivalent irreducible representations is equal to the class number and the following table of characters can be computed[5]:

C_{4v}	E	C_2	$2C_4$	2σ	$2\sigma'$
A_1	1	1	1	1	1
A_2	1	1	1	-1	-1
B_1	1	1	-1	1	-1
B_2	1	1	-1	-1	1
E	2	-2	0	0	0

Practically, the irreducible representation associated to a given pole is deduced from the symmetry properties of the corresponding eigenvectors $|pq\rangle$. If the eigenvalue is not simple, the representation is the E one. For simple eigenvalues, we must decide which of the four remaining irreducible representations is the good one. This is done by looking for the action of the different geometrical transformations on the eigenvector: if the vector is unchanged, the character is 1, if it is changed in the opposite, the character is -1. Thus we get the set of characters of the irreducible representation and we can identify it with one of the row of the group table. Let us now deduce the selection rules.

5.2 Selection rule theorem

It can be shown the following theorem:
let S be an operator invariant with respect of all symmetry transformations and $|i\rangle$ and $|j\rangle$ two vectors belonging to two different representations. The matrix elements $\langle j|S|i\rangle$ are zero if the two representations are irreducible inequivalent representations of the same group.

5.3 Selection rules for the resonance excitation

It is possible to know if a given incident field excites a given resonance using the preceding theorem when the symmetry group of the incident field is a subgroup of the group of the diffracting object. For instance, if the incident field is a plane wave TE polarized, propagating in the Ox direction, it is invariant with respect to the S_x symmetry, and the incident symmetry group is a two elements one with the following character table:

C_s	E	σ
A	1	1
B	1	-1

The irreducible representation of the square symmetry group C_{4v} attached to the resonance is also a possibly reducible representation of the incident symmetry group C_s . The number of occurence of given irreducible representations of C_s in the reduction of a given representation of C_{4v} is given by the formula:

$$n = \frac{1}{g} \sum_{i=1}^{g} \chi(G_i)\ \chi'(G_i) \qquad (14)$$

where g is the number of elements G_i of the group C_s , χ and χ' the characters of the elements in the two representations. For instance a B_2 resonance illuminated by Ox propagating plane wave correspond to the characters $\chi(E)=1$, $\chi(2\sigma)=-1$, $\chi'(E)=1$ and $\chi'(\sigma)=1$ thus n=0 and this resonance is not excited by the incident field. Similar considerations and the use of the character table of the square show that resonances A_1 , B_1 and E are excited by this field but the A_2 is not (figure 3).

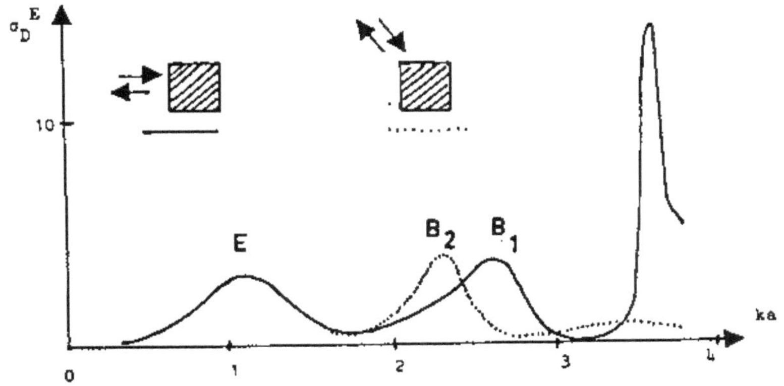

Figure 3
Backscattering of the square cylinder for two different incident plane waves. The E resonance is excited but a selection rule implies that the plane wave propagating along Ox does not excites the B_2 resonance, which is excited when the plane wave propagate towards the diagonal.

5.4 Selection rules for the diffracted directions

The computation of the scattering differential cross-section[6] shows that in TE polarization the field diffracted in a given direction exhibit the plane wave symmetry and the method described in the preceding section also holds. The direction founds are forbidden directions for the diffracted field. This result is consistent with the reciprocity theorem and a resonance cannot be excited by a plane wave propagating in a direction where it does not radiate.

5.5 Degenerate resonances

When the irreducible representation attached to a resonance pole is multi-dimensional, the resonance is degenerate. In this case, the selection rules determined by methods described in section 5.3 and 5.4 holds, but additionnal rules due to a cross effect between the incident and diffracted directions can also be found. Let us write $|i\rangle$ and $|d\rangle$ the incident and diffracted vectors. For a resonance with a degeneracy order of two like the one associated to the E irreducible representation of the square symmetry group, the cross section is proportional to the quantity $| \langle d|p1\rangle\langle p1|i\rangle + \langle d|p2\rangle\langle p2|i\rangle |^2$. For some incident and diffracted directions, it is possible to only nullify the scalar products <p1|i> and <d|p2> simultaneously. This selection rule is different from the preceding ones because the resonance is excited and can, if it is excited with another incident field, scatters energy in the observation direction. For the square cylinder example, it is found that the E resonances are excited by a plane wave propagating in the Ox direction in such a way that they do not radiate in the Oy direction (figure 4).

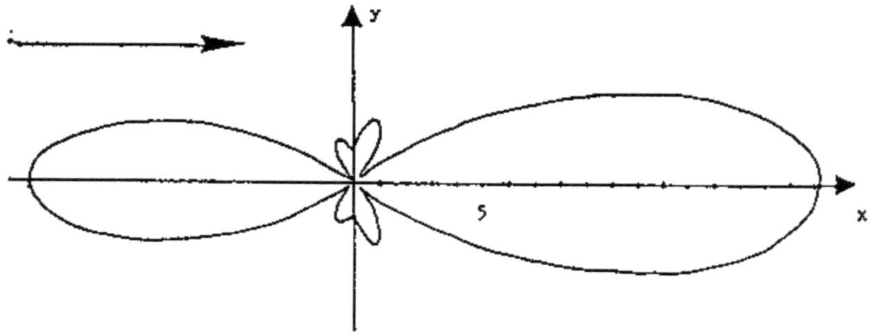

Figure 4
Diffraction diagramm of the square cylinder near a E resonance. The incident plane wave propagate toward the Ox axis and it can be seen that in this case the E resonance does not radiate in the Oy direction. The residual value of the field on the Oy axis is caused by another pole farther than the E one.

6 CONCLUSION

We described briefly the methods used to exhibit selection rules for the field diffracted by a cylinder in TE polarization. The use of finite group representation properties can be easily extended to the TM polarization. In this case the symmetry group of rhe diffracting object remains of course the same but the symmetry properties of the incident field are changed, due to the pseudo vector nature of the magnetic field. Thus the irreducible group representation of C_s for the incident plane is B instead of A and, consequently, the selection rules differs but the principle of the method remains the same. Another generalization to finite-size 3-dimensional object such cubes, etc... is also possible and, the use of a systematic procedure avoid tedious computation in this more complicated cases.

REFERENCES

1] Baum, C.E., in *Transient Electromagnetic fields*, ed. L. Felsen, Topics in Applied Physics, vol. 10, Springer (1976).

2] Vincent, P., Applied Physics, 17, 239-248 (1978).

3] Stratton, J.A., *Electromagnetic Theory*, McGraw Hill Book, New York (1941).

4] Reed, M., Simon, B., *Methods of Modern Mathematical Physics*, vol 1, Academic Press, New York (1972).

5] Landau, L., Lifchitz, E., *Mecanique quantique*, Editions Mir, Moscou (1967).

ACOUSTIC SCATTERING BY THIN SPHERICAL SHELLS

BRIZE-TRAMONI laurence, SESSAREGO jean-pierre
CNRS LMA Equipe ultrasons 31 ch. joseph AIGUIER13402 MARSEILLEcedex 9
FRANCE

INTRODUCTION

Lots of theoretical and experimental studies have been written concerning the acoustical scattering by thin elastic spherical shells [1,2,3,4]. In this paper, we have studied the influence of thickness and nature of the material on the response of the target.

First a theoretical study based on the modal theory permits us to characterize the acoustic response of the target by means of the farfield form function. At the surface of the spherical shell, we show the existence of surface waves which can be splitted in two types depending on the nature of their support.

Then a numerical calculation of this function gives us a representation of the response of the spherical shell. These representations show rough variations of amplitude linked with the presence of resonances because of phenomenons of interferences of surface wave. The resonances are important because they permit to characterize the oscillatory behaviour of the target. Many methods were worked out to locate these resonances. We developped two of them: the modal theory and the method derived from the application of the S-matrix theory. After having locatied the resonances, we have to classify them and to analyze the results we obtained. From the Regge's trajectories or from the dispersion curves of celerity, we observe variations of the drawing in relation with the thickness and the nature of the materials.These methods give results in good accordance with the experimental results.

The purpose of the study is to characterize the target with a view to make a non destructive test of materials with acoustical spectroscopy techniques.

I THEORETICAL PRESENTATION

a) Solution of the scattering problem

We consider a spherical shell with an outer radius "a" and an inner radius "b". This defines three media delimited by the geometry of the scatterer and characterized by the wave functions which are Helmholtz equation solutions. The shell is isonified by a plane wave of infinite extension in Ox direction (fig 1)

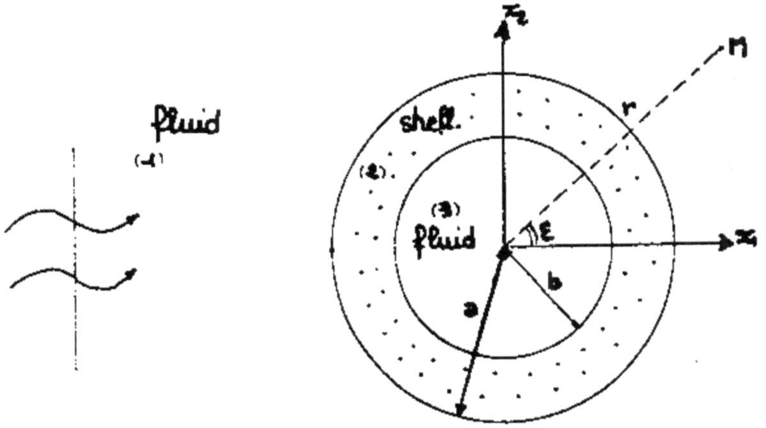

fig 1: Scattering geometry

The medium 1 is defined by a potential which represents the superposition of an incident wave and a scattered wave:

$$(1)\ \Phi_1 = \Phi_0 e^{-i\omega t} \times \sum_{\infty}^{n=0} [i^n(2n+1)\{j_n(k_1 r)+b_n \times h_n^{(1)}(k_1 r)\} \times P_n(\cos\epsilon)]$$

$$k_1 = \omega / c_1$$

the $j(.)$ functions are spherical Bessel functions, $h(.)$ spherical Henkel function of first kind and $P(.)$ Legendre polynomials.

At the boundaries, $r=a$ and $r=b$, the conditions are the followings:

*displacement and normal stress must be continuous

*tangential stress must be zero.

These conditions will permit us to determine the bn unkown quantities [5]. We write the farfield form function:

(2) $F_\infty(k_1a,\epsilon) = (2r/a)|(p_{1diff}/p_{1inc})|$

$$= (2/k_1a)\ |\Sigma_{\infty}^{n=0}(-i)(2n+1)b_n.P_n(\cos\epsilon)|$$

From a programme which calculates this function, we obtain curves which represent the acoustic response of the spherical shell (fig 2). These curves are not precise enough to observe the finest structure due to the resonances. Many methods to extract them already exists. We chose two of them: the first one coming from the S-Matrix theory and the other one from the modal theory.

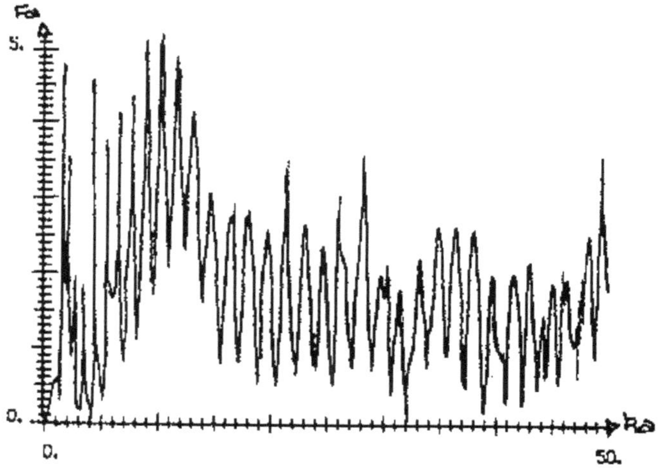

fig 2: Farfield form function (brass spherical shell b/a=0.65)

b) S-Matrix

The S-Matrix has its origin in Particle Physics. It is a scattering operator which characterizes the relation between the whole of initial status and the whole of final status [6,7,8].

This S-Matrix for a potential scattering is linked with the b_n scattering coefficients as follows:

$$(3)\ S_n = 1 + 2b_n = 1 + 2\times(\det{}^*M(n)/\det M(n))$$

detM(n) is the 6×6 Matrix determinant coming from the equation system given by the limit conditions. And det*M(n) is the previous determinant in which the b_n unknown quantities are replaced by the coefficients of the second item.

We take an interest to the factorisation property of the scattering matrix:

$$(4)\ S_n = S_n^{(0)} \times S_n^{(*)}$$

$S_n^{(0)}$ describes the potential scattering which characterizes the non resonant base and $S_n^{(*)}$ contains the information concerning the resonances. This function permits to identify the vibrations. Therefore, with a programm we are able to represent the peaks of the resonances. It consists of a difference, mode per mode, between the farfield form function $F_\infty(k_1a)$ for an elastic spherical shell and the farfield form function $F_\infty^{(0)}(k_1a)$ for a non-resonant base (rigid or soft)[5].

We realize the difference (5) $\left| F_\infty(x_1) - F_\infty^{(0)}(x_1) \right|$

*In case of a spherical shell extremely thin, the target has a soft behaviour (6) $S_n^{(0)}=S_n^{(s)}= -[h_n^{(2)}(x_1)/h_n^{(1)}(x_1)]$

*In case of spherical shell thick enough (fig 3), the target has a rigid behaviour (7) $S_n^{(0)}=S_n^{(r)}= -[h_n^{(2)'}(x_1)/h_n^{(1)'}(x_1)]$

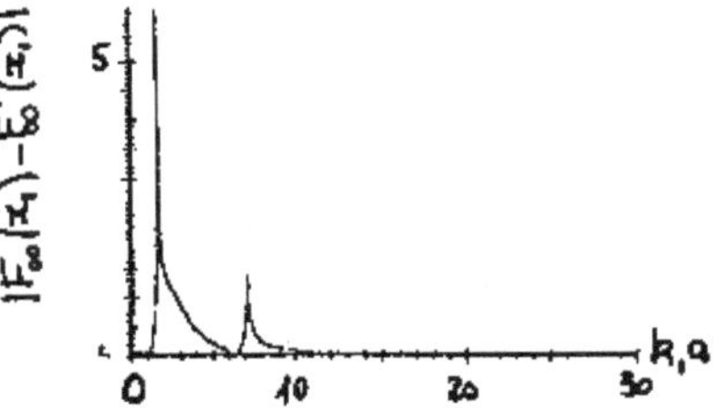

mode 2

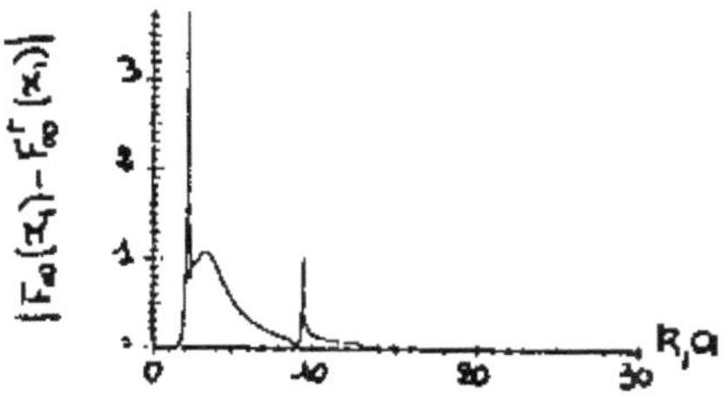

mode 3

fig 3: Method of S-matrix response (rigid background b/a=0.86)

This method gives good results in agreement with those obtained in [9] (search of pole in complex plane) in case of very thin or very thick shells. Some difficulties appear in case of intermediate thickness for which the behaviour "rigid" or " soft" are wrong approximations.

c) Mode per mode method

It consists of drawing each item of the modal series (2). These are a function of the no-dimensional parameter $x_1=k_1a$. We analyze each curve (fig 4) to extract for each mode n, the position x_1 of each resonance.

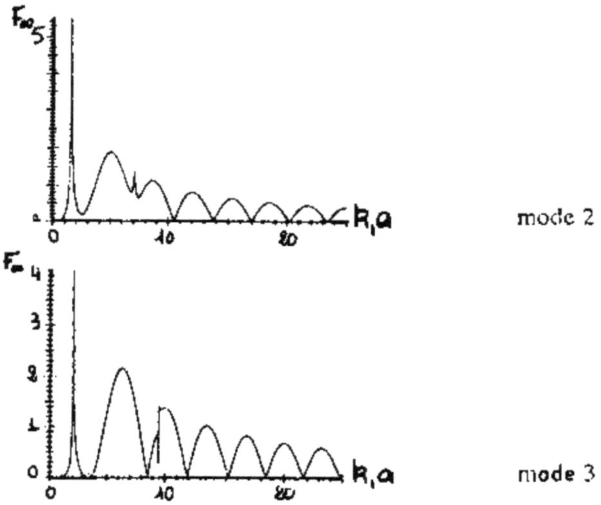

fig 4: Modal response (spherical shell of brass b/a=0.86)

These results permit us to draw the Regge's trajectories and the dispersion curve of the celerity. In a graph, we draw the mode n in function of the reduced frequency k_1a for each resonance ℓ (ℓ represents the appearance class for the function drawn mode per mode). The points of each class ℓ gives the Regge's trajectories. These curves permit to classify the resonances and differents kinds of trajectories appear which correspond to the different behaviours. As a matter of fact, the resonance depends on the formation, at the circumference of the sphere, of a stationary wave which appears during the propagation around the sphere.

II SIMULATED RESULTS ANALYSIS

a) Resonances study in relation to the thickness of the wall

We drew the Regge's trajectories in case of brass and of aluminium for differents values of b/a in a range of 0.8 to 0.99. We notice the variations of each resonance ℓ=1,2,3 for different vibration modes (n from 1 to 11)(fig 5).

Table

Numerical values for different used materials:			
materials	mass	longitudinal celerity	transversal celerity
brass	8.7	4350	2050
aluminium	2.7	6419.86	3039.86

In both case, we notice for ℓ=1 (resonance linked with Rayleigh wave) that thiner is the shell, more the trajectories draws up. The peaks slide to low frequencies. These remarks can be observed in the following table (mode n=4 and ℓ=1)

Table

	b/a	0.76	0.86	0.9	0.95	0.99
aluminium	k_1a	5	3	2.5	2	1.2
brass	k_1a	3.25	3	2.5	2	1.2

For the brass, ℓ=3, the resonance linked with second Whispering Gallery wave, remove to the high frequencies when the shell become very thin.

In both case, (brass and aluminium), the resonance ℓ=2 linked with the first Whispering Gallery wave are unchanged. We take in light an invariant in comparison with the thickness of the shell.

Table

l=2 n=6								
aluminium	k_1a	/	25	24.9	24.5	24.2	24	23.8
laiton	k_1a	16.5	16.5	/	16.5	16.5	16	16

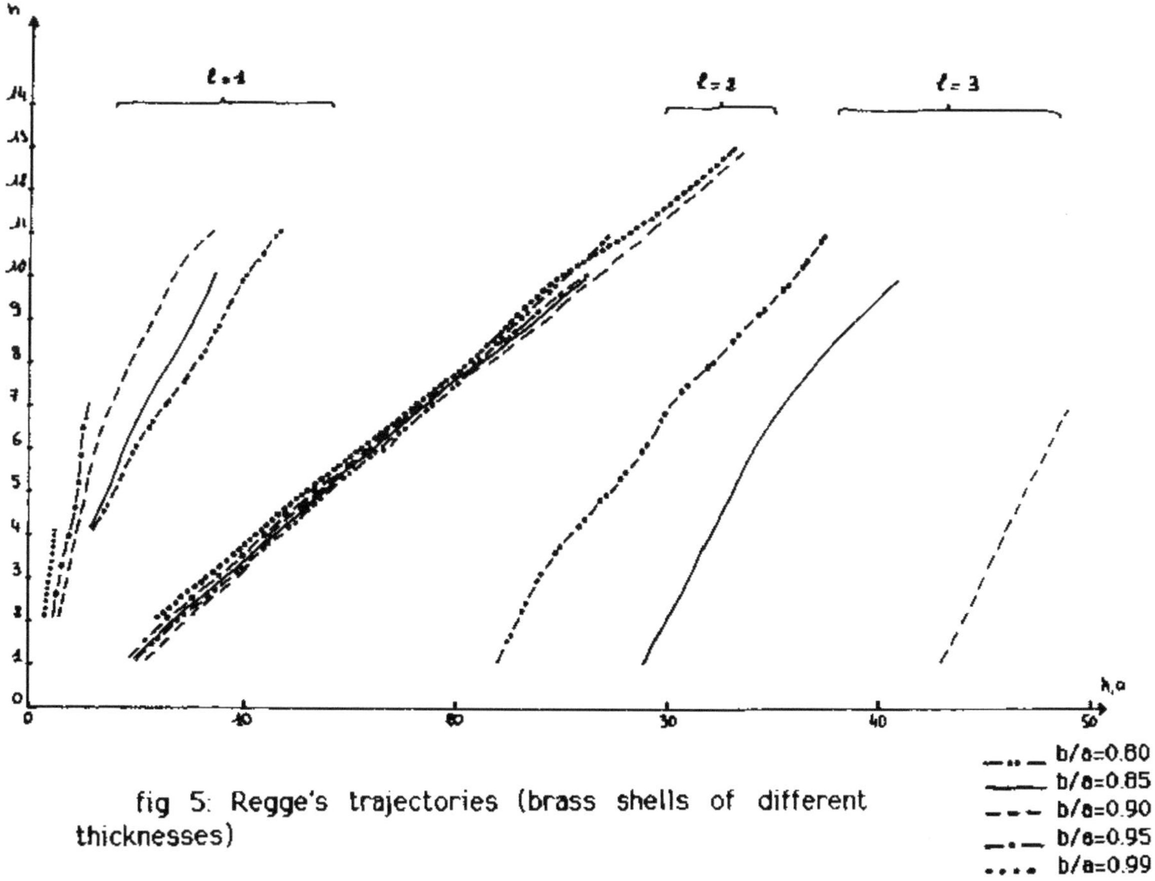

fig 5: Regge's trajectories (brass shells of different thicknesses)

We calculate the velocity of the surface wave from knowledge of resonant mode by the relation $c_\ell/c_1=x_1/(n+1/2)$. We reach the same conclusions as we draw the dispersion curves (fig 6) of the celerity for the wave in the liquid medium. We put in clearness the stability for the resonance ℓ=2, the parallel remoteness for the resonance ℓ=3. For the resonance ℓ=1, we find a trend to a decrease of c_ℓ/c_1 as b/a grows less

b) Nature variation of materials

In this part, we examine the farfield form function mode per mode of material series, in order to see the variation of the surface wave resonance positions. These materials had been chosen following the parameters which characterize them: longitudinal wave celerity, shear wave celerity and density. We carried out three testing types

1)Testing about neighbouring longitudinal celerity element

$$c_\ell(\text{iron}) < c_\ell(\text{aluminium}) < c_\ell(\text{beryllium})$$

$$c_\ell(\text{silver}) < c_\ell(\text{brass}) < c_\ell(\text{platinium})$$

2)Testing about neighbouring shear celerity element

$$c_t\ (\text{magnesium}) < c_t\ (\text{aluminium}) < c_t\ (\text{iron})$$

$$c_t\ (\text{uranium}) < c_t\ (\text{brass}) < c_t\ (\text{cast iron})$$

3)Testing about neighbouring density element

$$\rho\ (\text{quartz glass}) < \rho\ (\text{aluminium}) < \rho\ (\text{flint glass})$$

$$\rho\ (\text{iron}) < \rho\ (\text{brass}) < \rho\ (\text{nickel silver})$$

We fixe deliberately the parameter b/a which characterizes the thickness of the shell to the value b/a=0.9. The range of the reduced frequency is between 0 and 25 which corresponds to a frequency band varying between 0-500 kHz. This study is numerical one and the purpose is to understand the evolution of the resonances according to the celerity c_ℓ,celerity c_t and the density ρ.

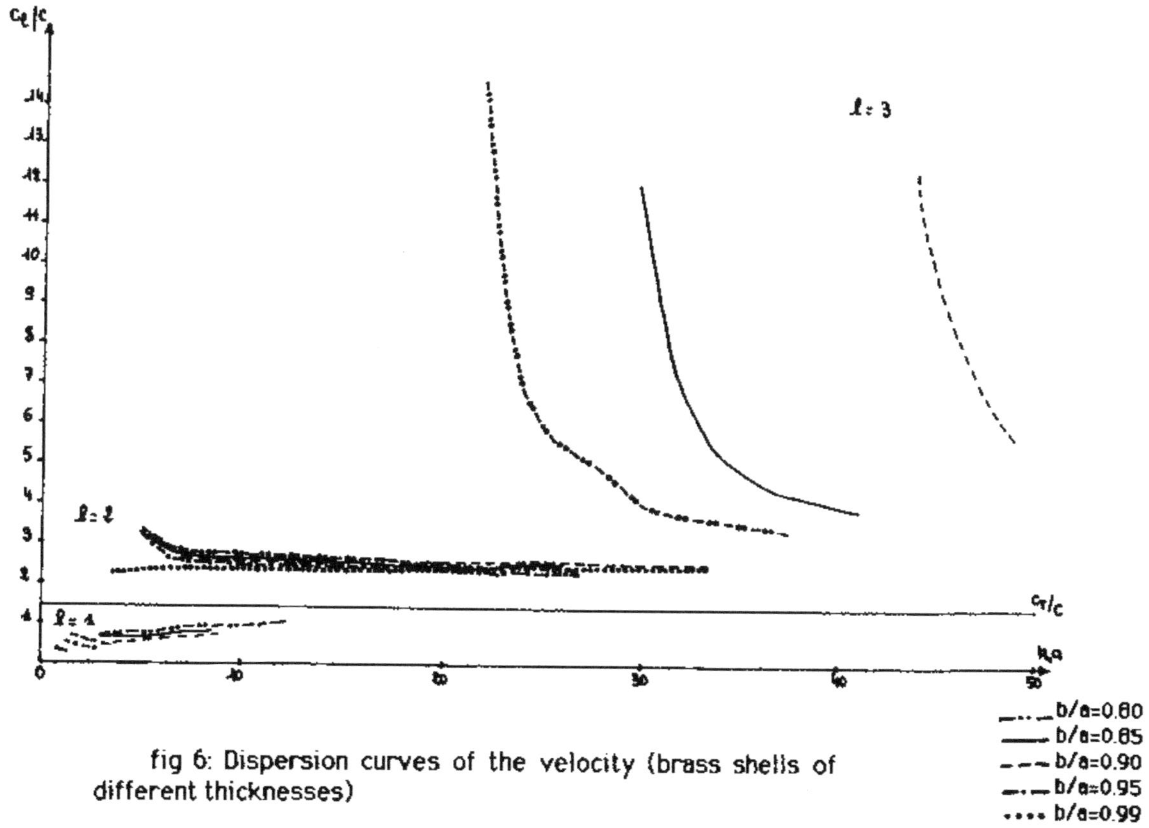

fig 6: Dispersion curves of the velocity (brass shells of different thicknesses)

Table:

materials nature	longitudinal celerity	transversal celerity	density
beryllium	12900	6800	1.8
aluminium	6300	3080	2.7
iron	5850	3230	7.8
nickel silver	4760	2160	8.4
quartz glass	5660	3515	2.6
magnesium	5770	3050	1.7
cast iron	4600	2160	7.2
flint glass	4260	2560	3.6
platinium	3960	1670	21.4
brass	3830	2050	8.1
silver	3600	1590	10.5
uranium	3370	2020	18.7

We draw the Regge's trajectories for these six different cases. The main observation concern the location of one curve from another. From the dispersion curves (fig 7), we set a table (ℓ=2) following the transversal celerities. We remark that c_t decreases as the rate c_ℓ/c moves. As for ℓ=1, the previous curves give us no information.

Table:(given by straight reading on the figure 7) (c=1480 m/s) ℓ=2

nature	c_t	c_ℓ	ρ
quartz glass	3515	5660	2.6
iron	3230	5850	7.8
aluminium	3080	6300	2.7
flint glass	2560	4260	3.6
cast iron	2160	4600	7.2
brass	2050	3830	8.1
uranium	2020	3370	18.7
platinium	1670	3960	21.4
silver	1590	3600	10.5

CONCLUSION

In this study, we are induced to calculate the form function which characterizes the response in frequency of thin spherical shell. This function permitted to put in clearness the peaks of the resonances. These

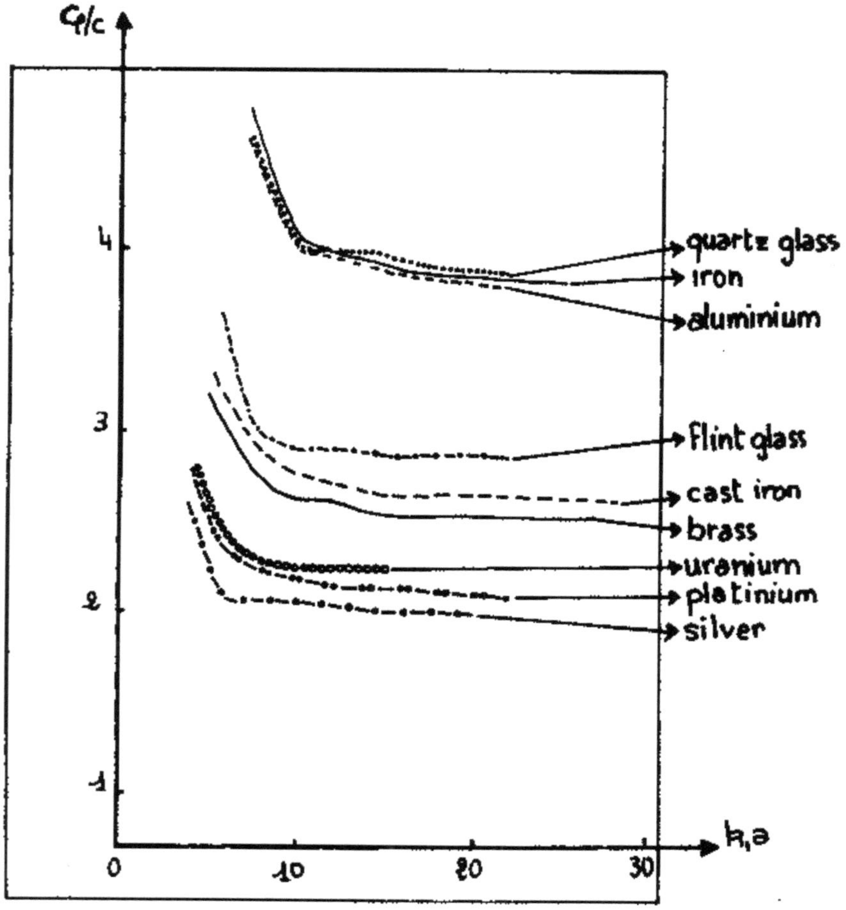

fig 7: Dispersion curves of the velocity (shells of differents materials, ℓ=2, b/a=0.90)

were calculated in accordance to the Regge's trajectories. Both theorical study and experimental one show the presence of surface waves. (Rayleigh and Whispering Gallery waves) due to the formation of stationary phenomenons on the circumference of the sphere.

The resonances are very important because they permit to understand the mechanisms of the scattering. Some researchs have been taken on the physical methods which put them in light. From these methods, we have developped two of them: the modal method and the one derived from the applying of the S-Matrix theory. As the conclusion, the experimental testing [9,10] meet the theorical results.

In case of thickness variation, from the Regge's trajectories, we determine the evolution of the resonances in a wide frequency band. We observe a stability of the curve ℓ=2, a straightening to the low frequencies for l=1 and a removing to the high frequencies for ℓ=3. The celerity curves give the same results.

The dependance between, the variation of resonances position and the materials nature is more complex. The dispersion curves give us a good classification for transversal celerities for ℓ=2, the first Whispering Gallery wave.

BIBLIOGRAPHY

1 R.HICKLING,An analysis of echoes from a solid elastic sphere in water,JASA n°34/2,p1582-1592 (1962)

2 W.G.NEUBAUER,R.M.VOGHT and L.R.DRAGONETTE,Acoustic reflection from elastic spheres,JASA n°55/6,p1123-1137 (1974)

3 C.GAZANHES,J.P.SESSAREGO,J.P.HERAULT and J.LEANDRE,Etudes des premieres résonances d'une sphére élastique, Revue du Cethedec,70,1-15,(1982)

4 A.GERARD and J.L.ROUSSELOT,Diffusion à basse fréquence d'une coque sphérique immergée,Revue du Cethedec,78,122-133,(1984)

5 L BRIZE-TRAMONI,J P SESSAREGO,J SAGELOLI,Etude de la diffusion acoustique par des coques spheriques à parois minces, J. d'Acoustique,1,177-184,(1988)

6 L FLAX -L R DRAGONETTE-H ÜBERALL Theory of elastic resonance excitation by sound scattering JASA n°63,p 723 -731 (1978)

7 J D MURPHY - J GEORGE - A NAGL - H ÜBERALL Isolation of the resonant component in acoustic scattering from fluid -loaded elastic spherical shells JASA n°65,p 368-373 (1979)

9 J L ROUSSELOT,A GERARD, J P SESSAREGO, J SAGELOLI, Reponse basse frequence d'une coque sphérique mince immergée,Rapport final,Contrat DRET n°86.501 (1987)

8 A.DEREM, Diffusion acoustique par des corps cylindriques immergés:aspects généraux, Revue du Cethedec,n°80,p221-245 (1980)

10 C.GAZANHES, J.P.SESSAREGO, J.SAGELOLI, Etudes des premieres résonances d'une sphére élastique creuse, Rapport final de synthese,LMA,Contrat DRET n°84.041 (1986)

ACKNOWLEDGMENT

We would like to thank the DRET and more particullary GROUPE VI B.POIREE for the financial support.

SCATTERING OF AN ACOUSTIC WAVE FROM FINITE ELASTIC SCREENS

Christian AUDOLY, Groupe d'Etudes et de Recherches de Détection Sous-Marine, DCAN TOULON, D.C.N., 83140 - SIX-FOURS.

ABSTRACT

The calculation of scattering of an acoustic wave by rigid screens can be done classicaly using integral equation methods. With a simplified modeling of the structure, the calculation is extended here in the case of screens made with one thin layer of elastic or viscoelastic material. Theoretical and experimental results on steel and soft material plane panels placed in water show the importance of edge effects and diffraction phenomena. These effects must be taken into account when the purpose is to characterize materials using standard panel measurements in near field.

INTRODUCTION

The performances of materials used as reflectors or screens in underwater acoustics are in most cases evaluated by measuring the steady-state nearfield pressure in front of and behind a test panel submitted to an acoustical wave. In fact, the aim is to obtain the intrinsic characteristics of the tested material, i.e. as if the sample were infinite. Unfortunately, the measurements are pertubed by the scattering effects due to the finite size of the panel. A theoretical model, based on the Helmholtz integral equation and its derivative form, is used to investigate the scattering effects from soft, rigid and elastic screens in two dimensions (infinite strip), and three dimensions (rectangular screen).

1 - THEORY

1.1 - LIST OF SYMBOLS AND ASSUMPTIONS

a,b	: dimensions of the screen (see figure 1).
h	: thickness of the screen.
ρ_S	: volumic mass of the material.
c_L	: longitudinal celerity in the material.
c_T	: tranverse celerity in the material.
ρ, c	: volumic mass and celerity in the fluid medium.
λ, k	: wave length and wave number in the fluid.
f, ω	: frequency and circular frequency.

The panel is submitted to an incident wave at steady-state. In all the following, complex variables with convention exp ($i\omega t$) are used. The total surface S' of the planar screen is divided in two sides : side 1 is facing the incident wave. It is assumed that the thickness of the panel is small i.e. $h \ll \lambda$. Let Q1 be a point of side 1 and Q2 the symmetric on side 2. With this model, Q1 and Q2 have the same geometrical coordinates : the point Q on surface S. The frequency domain of interest is the low frequency one ($ka < 10$). In this case, integral equation methods are well adapted.

1.2 - INTEGRAL EQUATION IN THE FLUID

For a body of arbitrary geometry, the Helmholtz integral is :

$$p(M) = p_o(M) + \int_{S'} \left(p(Q) \frac{\partial G}{\partial n_Q}(M, Q) - \frac{\partial p}{\partial n}(Q)\, G(M, Q) \right) dS'(Q)$$

$$M \in S', Q \in S'$$

with : $G(M, Q) = \exp(-ikr)/4\pi r$ (3D).

$G(M, Q) = -i/4\, H_o^{(2)}(kr)$

$r = |MQ|$

$H_o^{(2)}$ Hankel function of order O.

P_o Incident field.

When M is now a point of the surface S and the thickness of the body is neglected, we have, following Terai [1] :

$$\frac{1}{2}(p(M_1) + p(M_2)) = p_o(M) + \int_S \left(\sigma(Q) \frac{\partial G}{\partial n_Q}(M, Q) - \mu(Q)\, G(M, Q) \right) dS(Q)$$

Where: $\sigma(Q) = P(Q_2) - P(Q_1)$ is the pressure discontinuity.

$\mu(Q) = \frac{\partial p}{\partial n}(Q_2) - \frac{\partial p}{\partial n}(Q_1)$ is the normal derivative discontinuity

n is the outgoing normal on side 2.

In the case of a planar screen, $\frac{\partial p}{\partial n}(M,Q) = 0$ for $M \in S$ and $Q \in S$, and the integral equation becomes :

$$\frac{1}{2}(p(M_1) + p(M_2)) = p_o(M) - \int_S \mu(Q)\, G(M, Q)\, dS(Q) \tag{1}$$

To make appear the density σ, an other integral equation can be derived.

$$\frac{1}{2}\left(\frac{\partial p}{\partial n}(M_1)+\frac{\partial p}{\partial n}(M_2)\right)=\frac{\partial p_o}{\partial n}(M)+\int_S \sigma(Q)\,\frac{\partial^2 G}{\partial n_M \partial n_Q}(M,Q)\,dS(Q) \qquad (2)$$

1.3 - PARTICULAR CASES OF INFINITELY SOFT AND RIGID PANELS

For an infinitely soft panel, we have the following condition : $P(M_1) = P(M_2) = 0$.

Then, we have :

$$\begin{cases} P_o(M) = \int_S \mu(Q)\, G(M,Q)\, dS(Q) \\ \sigma(M) = 0 \end{cases}$$

For an infinitely rigid panel, $\frac{\partial p}{\partial n}(M_1) = \frac{\partial p}{\partial n}(M_2) = 0$.

Then :

$$\begin{cases} \frac{\partial p_o}{\partial n}(M) = -\int_S \sigma(Q)\,\frac{\partial^2 G}{\partial n_M \partial n_Q}(M,Q)\,dS(Q) \\ \mu(M) = 0 \end{cases}$$

The integral equations found for these special cases have yet been studied in the literature, for example by Filippi [2].

1.4 - MODELING OF THE ELASTIC PANEL

To solve the fluid-structure coupled problem, relationships between p and $\frac{\partial p}{\partial n}$ on the surface, must be obtained.

A classical method is to use finite elements techniques. However, in our case, the dynamical behavior of the screen is yet in the high frequency domain (i.e. the number of resonance modes is large) and finite element techniques are uneasy and costly to use. So an approximate method is proposed here alternatively.

The variables are first decomposed in plane wave components using a spatial Fourier transform (for example $\tilde{p}_1 = \mathcal{F}(p_1)$). Then, the infinite plane model [3] is used to obtain a transfer matrix between the two sides for each Fourier component :

$$\begin{bmatrix} \tilde{u}_y^1 \\ \tilde{u}_x^1 \\ \tilde{\sigma}_{xx}^1 \\ \tilde{\sigma}_{xy}^1 \end{bmatrix} = |A| \begin{bmatrix} \tilde{u}_y^2 \\ \tilde{u}_x^2 \\ \tilde{\sigma}_{xx}^2 \\ \tilde{\sigma}_{xy}^2 \end{bmatrix} \qquad (3)$$

The continuity relationships at the fluid-structure interface are :

$$\tilde{\sigma}_{xx}^1 = -\tilde{p}_1 \qquad \tilde{\sigma}_{xx}^2 = -\tilde{p}_2 \qquad \tilde{\sigma}_{xy}^2 = \tilde{\sigma}_{xy}^1 = 0$$

$$\tilde{\mu}_x^1 = \frac{1}{\rho\omega^2}\frac{\partial \tilde{p}_1}{\partial n} \qquad \tilde{\mu}_x^2 = \frac{1}{\rho\omega^2}\frac{\partial \tilde{p}_2}{\partial n}$$

Then, transforming (3) introducing the variables μ and σ :

$$\begin{aligned} \tilde{\mu}^1 &= k_1\tilde{\sigma} + k_3\tilde{\sigma}' \\ \tilde{\sigma}^1 &= k_2\tilde{\mu} + k_4\tilde{\mu}' \end{aligned} \qquad (4)$$

With :

$$\widetilde{\mu'} = \frac{1}{2}\left(\frac{\partial \widetilde{p}_1}{\partial n} + \frac{\partial \widetilde{p}_2}{\partial n}\right)$$

$$\widetilde{\sigma'} = \frac{1}{2}(\widetilde{p}_1 + \widetilde{p}_2)$$

In fact, $K_3 << K_1$ and $K_4 << K_2$

Then, using an inverse Fourrier transform and combining (1), (2) and (4), two decoupled integral equations for unknowns σ and μ are obtained.

$$\mathcal{F}^{-1}(K_2\,\mathcal{F}(\mu)) + \int_S \mu\,(Q)\,G\,(M,Q)\,dS(Q) = p_o(M)$$

$$\mathcal{F}^{-1}(K_1\,\mathcal{F}(\sigma)) - \int_S \sigma\,(Q)\,\frac{\partial^2 G}{\partial n_M \partial n_Q}(M,Q)\,dS(Q) = \frac{\partial p_o}{\partial n}(M) \qquad (5)$$

1.5 - SIMPLIFIED MODEL

In some cases K_1 and K_2 do not vary much with the spatial wavenumber. Assuming K_1 and K_2 constant avoids the calculation of Fourrier transforms. In the case of screens submitted to waves at normal incidence, the equations become :

$$\frac{\rho\omega}{\rho_S\,c_L}\,\frac{\sin k_L h}{2\,(1-\cos k_L h)}\,\sigma\,(M) - \int_S \sigma\,(Q)\,\frac{\sigma^2 G}{\partial n_M\,\partial n_Q}(M,Q)\,dS(Q) = \frac{\partial p_o}{\partial n}(Q)$$

$$-\frac{c_L\,\rho_S}{\omega\rho}\,\frac{\sin k_L h}{2\,(1-\cos k_L h)}\,\mu\,(M) + \int_S \mu\,(Q)\,G\,(M,Q)\,dS(Q) = p_o\,(Q)$$

with $k_L = \frac{\omega}{c_L}$.

This set of equations is solved numerically using a piecewise constant approximation. When $\rho_s\,c_L \to 0$ or $\rho_s\,c_L \to \infty$, the cases of perfectly soft end regid screens are recovered as limiting cases.

1.6 - TOTAL PRESSURE AROUND PANEL

The Helmholtz is used, once σ and μ known :

$$p(M) = p_o(M) + \int_S \left(\sigma(Q) \frac{\partial G}{\partial n_Q}(M,Q) - \mu(Q)\, G(M,Q)\right) dS(Q)$$

2 - RESULTS AND EXPERIMENTS

2.1 - THEORETICAL RESULTS IN 2D SPACE

In all the following, the surrounding fluid is water ($c = 1\ 500$ m/s, $\rho = 1\ 000$ kg/m^3). Figures 2 and 3 give contour maps of the modules of the total pressure field (incident + scattered) normalized to incident field for perfectly soft and rigid screens submitted to a plane wave at normal incidence. The reduced frequency is $ka = 5$. In the case of the soft screen, the shadow zone appears clearly, and the stationary wave are parallel to the screen. For the rigid screen, on the contrary, the scattering effects are very important : the pressure behind the panel is large in some places, and the field in front of the screen is distorted.
However, perfectly soft and rigid screens are extreme case that do not appear in practice in underwater acoustics. Figures 4 and 5 give the contour maps for the following materials :

"Microvon"	: $\rho_S = 700$ kg/m^3	$c_L = 250$ m/s
Steel	: $\rho_S = 7\ 800$ kg/m^3	$c_L = 5\ 870$ m/s

The dimension of the screens were $a = 0.45$ m , $h = 3$ cm and the frequency was $ka = 5$.
The same conclusions than above hold but in a fainter way.

2.2 - MEASUREMENTS ON A SQUARE STEEL PANEL

The panel has dimensions 0,9 m x 0,9 m, 3 cm thickness and was made with steel. The experimental configuration is shown on figure 1. The transmitter generates a spherical wave and a small hydrophone measures the pressure on a line perpendicular to the panel for fixed coordinates y and z and x varying between - 0,3 m end 0,3 m. The distance between the transducers is constant : 3 m. In front of the panel, stationnary wave phenomena are observed and behind the panel, we have the shadow region.

Figure 6 shows the results for three different positions for the frequency f = 2 000 Hz. (Position 1 : facing the center of the panel ; position 2 : shifted 0,2 m ; position 3 : near the edges). In each case, the model is compared to the experiments. In most cases, the scattering effects are important, except for position 2 where the results are very close to the infinite plane model.

Figure 7, for f = 3 500 Hz leads to similar remarks.

CONCLUSION

A theoretical model, based on the Helmholtz integral equation and its derivative form, is used to calculate the scattering of acoustic waves from thin elastic panels in water. The resuts, in good agreement with experiments, show that the scattering effects due to the finite size of the panels are important, mainly for "rigid" materials. This model can be used to study and improve acoustical measurement techniques on test panels.

BIBLIOGRAPHY

[1] T.TERAI, On calculations of sound fields around three dimensional objects by integral equation methods.
Journal of Sound and Vibration, 69 (1) 71 - 100 (1980).

[2] P.FILIPPI, G.DUMERY, Etude théorique et numérique de la diffraction par un écran mince.
Acustica 21, 343-350 (1969).

[3] D.L.FOLDS, C.D.LOGGINS, Transmission and reflexion of ultrasonic waves in layered media.
J. Acoust. Soc. Amer. 62 (5) 1102-1109 (1977).

FIGURE 1 a. INFINITE STRIP

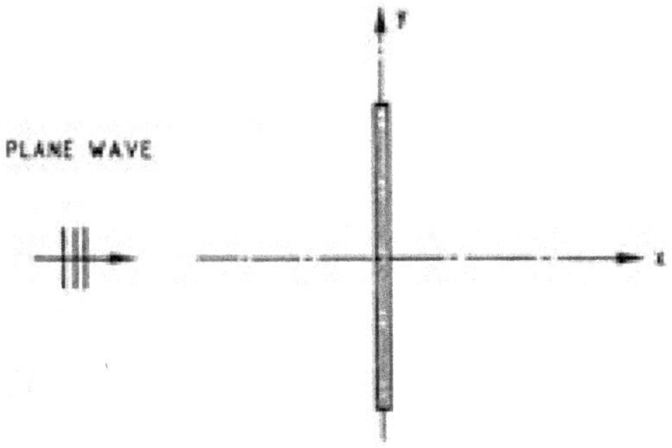

FIGURE 1b RECTANGULAR SCREEN

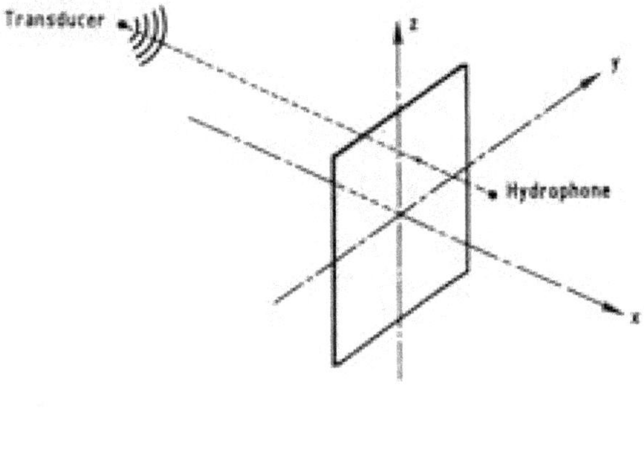

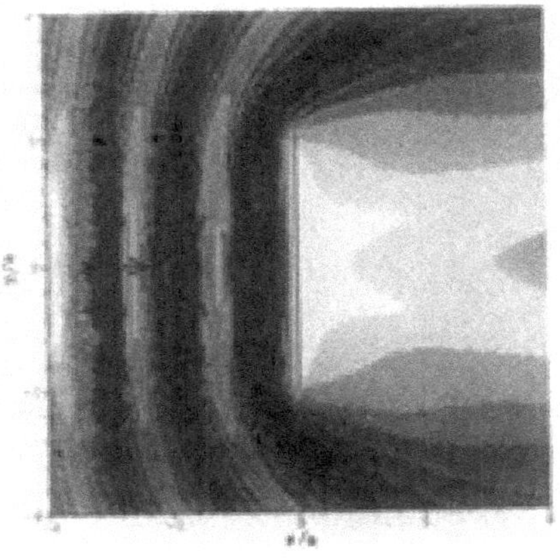

FIGURE 2 :
SOFT SCREEN ka=5

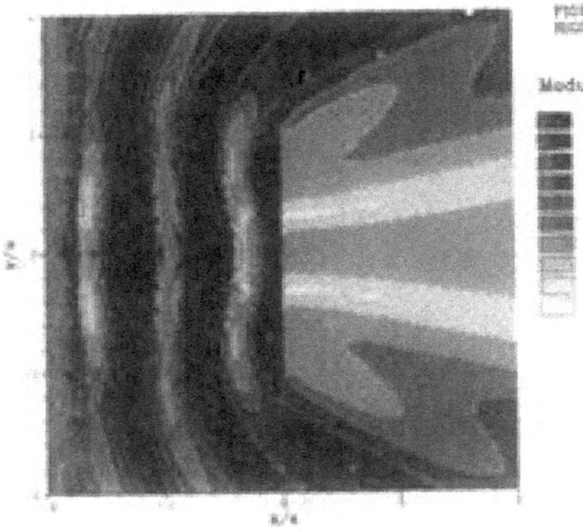

FIGURE 3 :
RIGID SCREEN ka=5

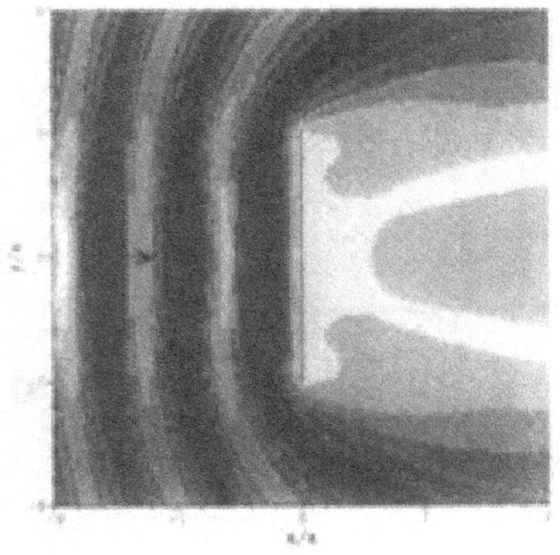

FIGURE 4

Modulus of total pressure

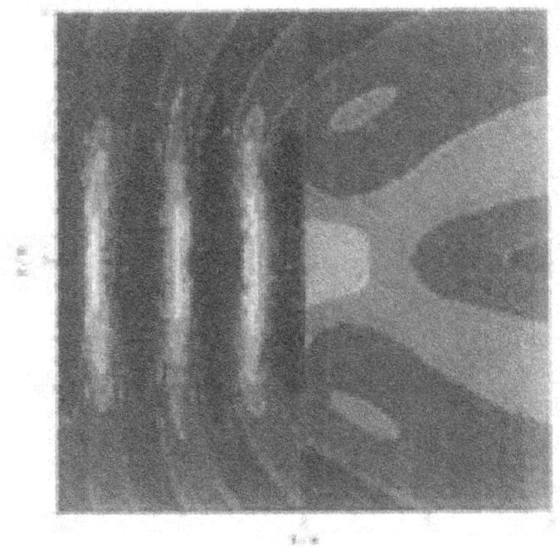

FIGURE 5
STEEL

Modulus of total pressure

FIGURE 8

■ Experiment
- - - - - Infinite plane
——— Present method

(A) Position 1

(B) Position 2

(C) Position 3

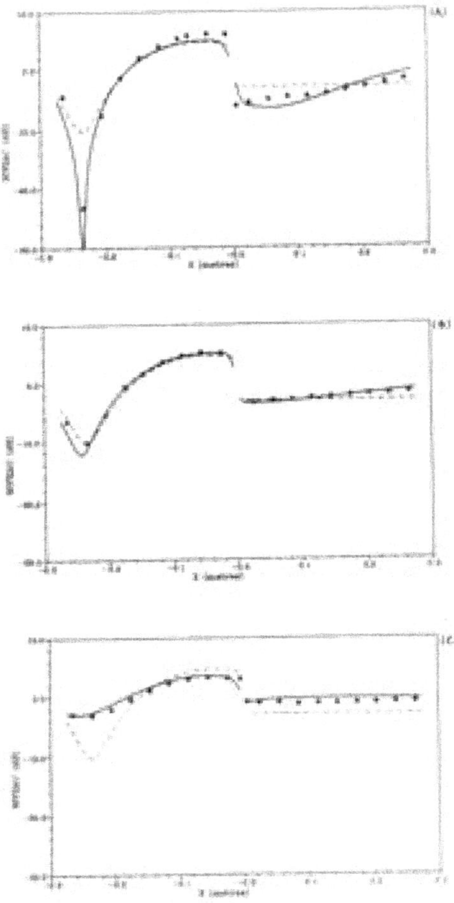

FIGURE 7 :
Steel screen 0.9 m x 0.9 m
Thickness 3 cm
Frequency 3000 Hz
◆ Experiment
------ Infinite plane
——— Present method

(A) : Position 1
(y=0. z=0.)

(B) : Position 2
(y=0.2 m z=0.)

(C) : Position 3
(y=0.2 m z=0.4 m)

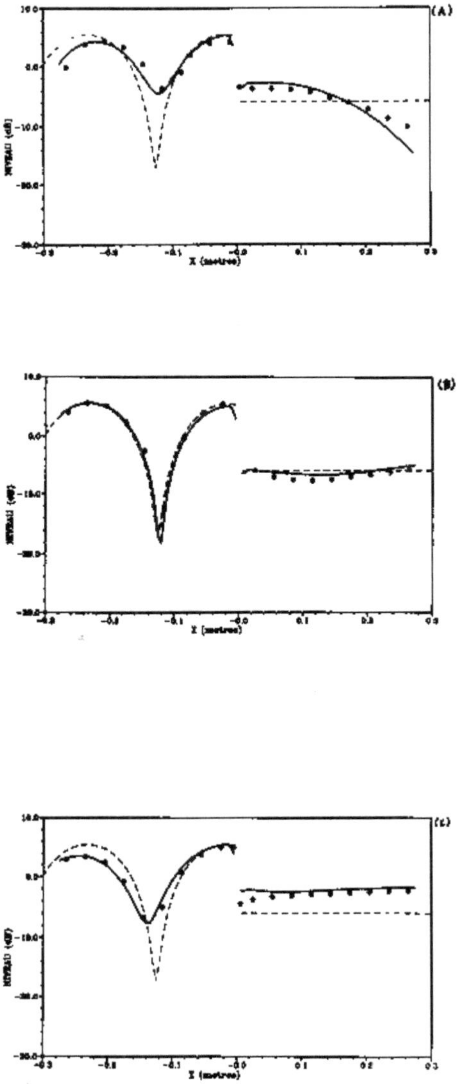

EFFET ACOUSTIQUE D'ELEMENT EN ELASTOMERE

Auteur : D. VAUCHER DE LA CROIX
Société METRAVIB R.D.S.
64, Chemin des Mouilles
69132 - ECULLY CEDEX - FRANCE

ABSTRACT

A simple theoretical model has been developed in order to estimate the perturbation generated by an elastomer target on the propagation of an arbitrarily incident plane acoustic wave.

We assume that shear waves in the material can be neglected (hypothesis that seems justified as the ratio shear waves to the longitudinal waves speeds is much lower than unity), which means a quasi-acoustic material behavior, with known density and celerity. Doing so, the problem is reduced to the influence of a fluid target on the propagation of an incident plane wave.

The solution is then obtained by applying a well-known BORN approximation method to the integral formulation of the problem. In fact, Neumann series expansion is not needed in the case of weak contrast between ambiant fluid and imbedded inhomogeneity.

Theoretical results agree quite well with experimental data (insertion loss measurements) for material properties according to the weak contrast hypothesis.

1 - PROBLEME CONSIDERE

Il s'agit donc d'évaluer la perturbation engendrée par la présence d'un prisme en élastomère sur la propagation d'une onde acoustique plane en milieu fluide.

En supposant un prisme de très grande dimension h (hauteur) devant les côtés définissant sa section triangulaire, on se limite à la description d'un problème bidimensionnel.

D'autre part, en se plaçant dans une hypothèse de faible contraste fluide/hétérogénéité, on identifie le triangle considéré à une cible fluide de caractéristiques ρ_2, c_2. Pour le type de matériau dont il est question, une telle simplification trouve sa justification dans le rapport des célérités longitudinales et transversales c_L/c_T élevé. La présence d'ondes de cisaillement dans l'élastomère est donc occultée dans cette étude.

La figure 1 illustre le problème bidimensionnel posé, ainsi que les notations utilisées.

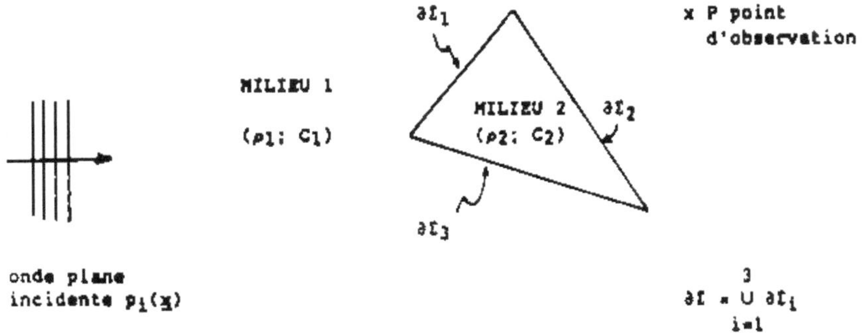

Figure 1 : Position du problème.

2 - FORMULATION MATHEMATIQUE

La propagation de l'onde incidente $p_{inc}(x)$ dans le milieu fluide ambiant perturbé par la présence de la cible fluide est régie par les équations de mouvement suivantes :

(1.1) $\Delta\Phi_1 - 1/(c_1^2)\ \partial^2\Phi_1/\partial t^2 = S\ (\underline{x},t)$ Fluide 1

(1.2) $\Delta\Phi_2 - 1/(c_2^2)\ \partial^2\Phi_2/\partial t^2 = 0$ Fluide 2 (= Perturbation)

où Φ_i est le potentiel de vitesse associé au milieu fluide i (i = 1,2). S ($\underline{x}$,t) représente la source acoustique considérée (onde incidente). Les conditions aux limites associées s'écrivent :

(1.3) $\bar{n}.\nabla\ \Phi_1 = \bar{n}.\nabla\ \Phi_2$

ou

(1.3') $\partial_n\Phi_1 = \partial_n\Phi_2$ ($\partial_n = \partial/\partial_n$, $\bar{n}$ normale extérieure)

(1.4) $\rho_1\partial_t\Phi_1 = \rho_2\partial_t\Phi_2$

(1.5) Condition de rayonnement de Sommerfeld (ondes sortantes).

Ainsi, la relation (1.3) traduit l'égalité des composantes normales des vitesses sur $\partial\Sigma$, et (1.4) celle des pressions.

En passant dans le domaine fréquentiel (dépendance temporelle en exp(+jωt)), on obtient les équations d'onde suivantes, exprimées désormais en terme de pressions ($p_i = \rho\partial\Phi_i/\partial_t$) :

(1.6) $\Delta\hat{p}_1 + k_1^2\ \hat{p}_1 = S_p\ (\underline{x},\omega)$

(1.7) $\Delta\hat{p}_2 + k_2^2\,\hat{p}_2 = 0$

où $k_i = \omega/c_i$ désignent les nombres d'onde caractéristiques des milieux en présence ($i = 1,2$). $\hat{p}_i$ dénote la transformée de Fourier des pressions correspondantes. Par la suite, la description se restreignant au seul domaine fréquentiel, on omettra cette précision.

Exprimées en termes de pression, les conditions imposées à l'interface $\partial\Sigma$ deviennent :

(1.8) $(\rho_1)^{-1}\,\partial_n\,p_1 = (\rho_2)^{-1}\,\partial_n\,p_2$

(1.9) $p_1 = p_2$

Il convient alors de résoudre le système couplé, constitué des équations (1.6)-(1.9) assorties de la condition de rayonnement de Sommerfeld. On en déduit l'expression suivante de la pression dans le milieu fluide ambiant, compte tenu de la perturbation engendrée par l'hétérogénéité de type cible fluide :

(1.10) $$p_1\,(\underline{x}_0) = \int_{\partial\Sigma} [p_1(\underline{x})\partial_{n_x}(G_2-G_1) - ((\rho_2/\rho_1)G_2-G_1)\partial_{n_x}\,p_1(\underline{x})]\,d\sigma_x + \text{Source}$$

Cette dernière relation vaut également à l'interface $\partial\Sigma$.

Les fonctions de Green bidimensionnelles associées aux deux milieux en présence sont des fonctions de Hankel de seconde espèce, conformément au choix de la dépendance temporelle évoqué plus haut.

Formellement, l'expression (1.10) permet d'évaluer la pression p_1 en tout point du domaine fluide 1. Toutefois, la solution exacte du problème présuppose la connaissance de p_1 et de sa dérivée normale à l'interface $\partial\Sigma$. En d'autres termes, il conviendrait de résoudre pour $x_0 \in \partial\Sigma$ le système intégro-différentiel constitué de l'équation (1.10) et de son analogue exprimant la dérivée normale $\partial_n p_1(\underline{x}_0)$ sous forme intégrale. Survient alors la question de l'évaluation de termes en dérivées secondes des fonctions de Green, à comportement singulier lorsque le point d'intégration $\underline{x}$ tend vers le point de calcul $\underline{x}_0$.

L'ambition limitée de cette étude ainsi que la situation expérimentale particulière qu'elle est appelée à décrire (faible contraste fluide ambiant/hétérogénéité), nous ont incité à recourir à une approximation de type BORN pour déterminer les grandeurs d'intérêt.

Le principe d'une telle approximation repose sur une décomposition bien connue de la pression $p_1(\underline{x}_0)$ (cf. [1] p. ex), que nous limitons au premier ordre en écrivant :

(1.11) $$p_1(\underline{x}_0) = p_{inc}(\underline{x}_0) + p_1^{(1)}(\underline{x}_0)$$

avec

(1.12) $$p_1^{(1)}(\underline{x}_0) = \int_{\partial\Sigma} [p_{inc}(\underline{x})\partial_{n_x}(G_2-G_1) - ((\rho_2/\rho_1)G_2-G_1)\partial_{n_x} p_{inc}(\underline{x})]\, d\sigma_x$$

(le terme source, soit la pression incidente, est noté $p_{inc}(\underline{x}_0)$.)

Le domaine de validité d'une telle approximation (restreinte à l'ordre le plus faible) peut se définir par les conditions suivantes que doivent satisfaire les deux paramètres du modèle (ρ_2 : masse volumique de la cible fluide, c_2 sa célérité) relativement au milieu fluide ambiant.

(1.13) $\quad \Delta_\rho/\rho_1 \ll 1, \; \Delta_c/c_1 \ll 1.$

Il s'agit donc de l'hypothèse de faible contraste. A densité ρ_2 donnée, seul subsiste le paramètre de célérité c_2. La simulation numérique dont quelques résultats sont présentés ci-après consiste alors à faire varier c_2 autour de c_1 pour approcher au mieux la mesure.

Cette manière de procéder se justifie si l'on prend en considération la remarque suivante. La célérité longitudinale caractéristique des prismes étudiés (élastomère d'utilisation courante en applications sous-marines) est en général mesurée à une fréquence supérieure à celle où est effectuée l'observation, et affectée d'une incertitude qui peut atteindre 10 %. De plus, c_2 varie sensiblement avec la température.

Or, ainsi que le montre le modèle, les résultats numériques s'avèrent très sensibles à de faibles variations de c_2 (dans le cadre de l'approximation de BORN d'ordre 1). La célérité c_2 ne pouvant être considérée comme une donnée certaine du problème, elle est donc <u>naturellement</u> considérée en tant que paramètre du modèle.

3 - COMPARAISON CALCUL/MESURE

Deux géométries ont été retenues pour l'expérimentation. Le premier prisme est à base triangulaire isocèle, alors que le second ne présente aucune symétrie particulière. Leurs dimensions restent cependant voisines.

Le schéma de la figure 2 illustre le dispositif expérimental et précise quelques dimensions relatives. L'écart entre hydrophones de mesure peut être réduit par translation de la rampe (± δ). L'ensemble des mesures exploitées collecte des informations obtenues en 3 positions différentes de cette rampe (→ 33 hydrophones), ce qui assure un échantillonnage spatial suffisant sur la quasi totalité de la bande fréquentielle étudiée (jusqu'à ka = 40)

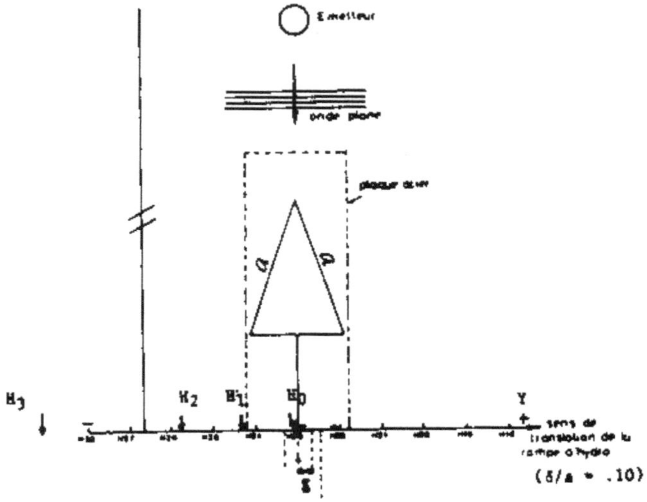

Figure 2 : Dispositif expérimental (cas symétrique)

La comparaison calcul/mesure porte sur la grandeur $Q(\omega)$

(1.14) $Q(\omega) = 20 \log_{10} |p/p_0|$ (niveau en dB)

où p_0 est la pression mesurée (ou calculée) en l'absence d'hétérogénéité et p son homologue en présence de cible fluide.
$Q(\omega)$ est en fait l'opposé de la perte par insertion $I_{loss}(\omega)$.

On envisage deux types de représentation, à savoir :

a) évolution fréquentielle $Q(\omega,x_k)$ pour une position x_k d'hydrophone donnée
b) évolution spatiale $Q(\omega_1,x)$ à fréquence ω_1 fixe

Les figures 3 à 5 illustrent diverses comparaisons calcul/mesure. En variant le paramètre c_2, il apparaît que le résultat numérique s'approche au mieux des valeurs expérimentales pour une célérité de cible fluide légèrement supérieure à la valeur homogène c_1 (1500 ms^{-1}). En effet, les écarts n'excèdent que rarement 1 dB, tant par l'évolution en fréquence que sur la dépendance spatiale de $Q(\omega,x)$ à fréquence donnée.

L'observation des résultats obtenus sur le second prisme (base triangulaire quelconque) confirme cette appréciation. On peut souligner que la valeur optimale c_2 = 1530 ms^{-1} s'inscrit en accord avec les valeurs expérimentales disponibles, compte tenu des réserves émises plus haut quant à la mesure d'une telle grandeur.

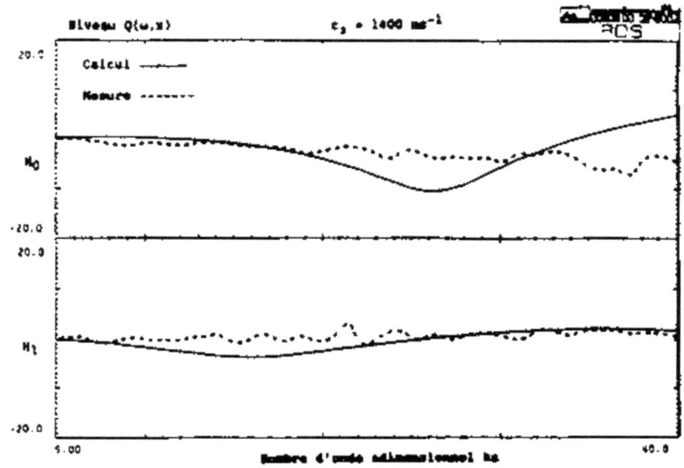

Figure 3.1 : Evolution fréquentielle $Q(\omega, x_0)$ pour deux positions d'hydrophone données. Cas symétrique - Célérité c_2 non optimale.

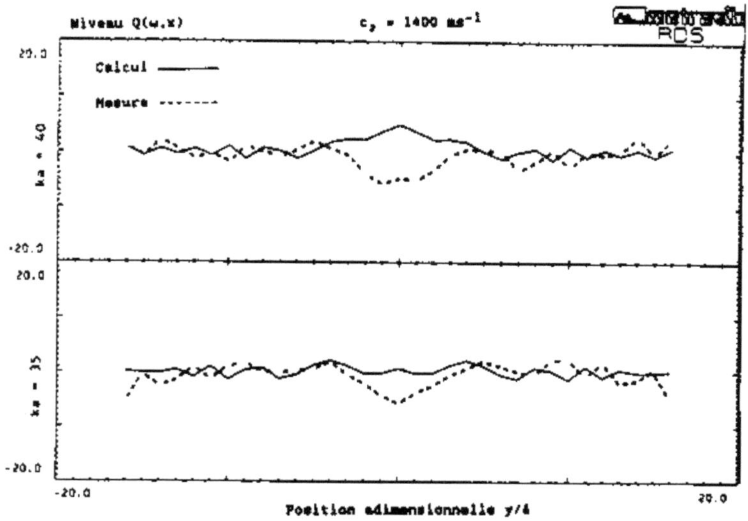

Figure 3.2 : Evolution spatiale $Q(\omega_1, x)$ à fréquence fixe. Cas symétrique - Célérité c_2 non optimale.

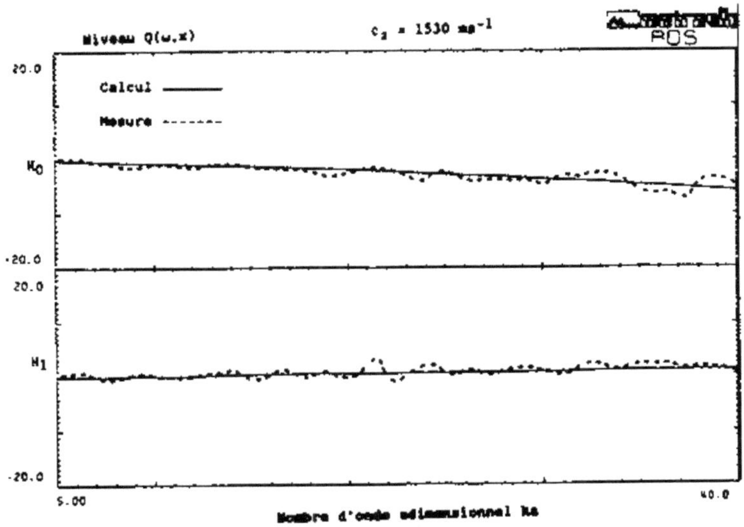

Figure 4.1 : Evolution fréquentielle $Q(\omega, x_k)$ pour deux positions d'hydrophone données. Cas symétrique - Célérité c_2 optimale.

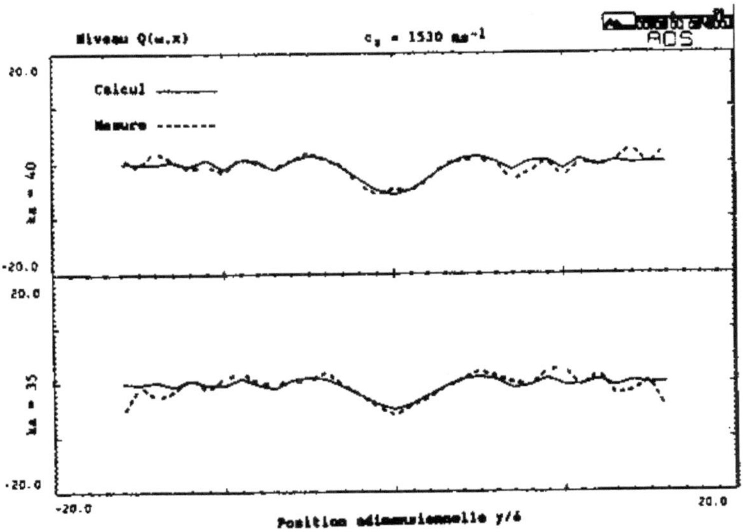

Figure 4.2 : Evolution spatiale $Q(\omega_1, x)$ à fréquence fixe. Cas symétrique - Célérité c_2 optimale.

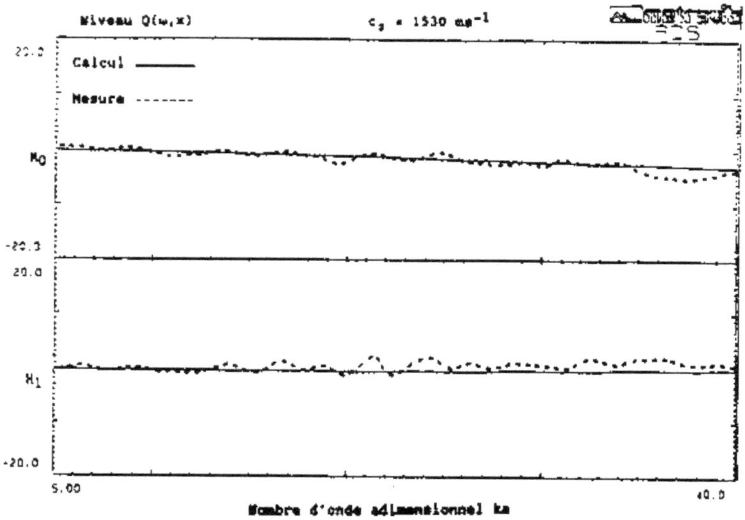

Figure 5.1 : Evolution fréquentielle $Q(\omega, x_k)$ pour deux positions d'hydrophone données. Cas asymétrique - Célérité c_2 optimale.

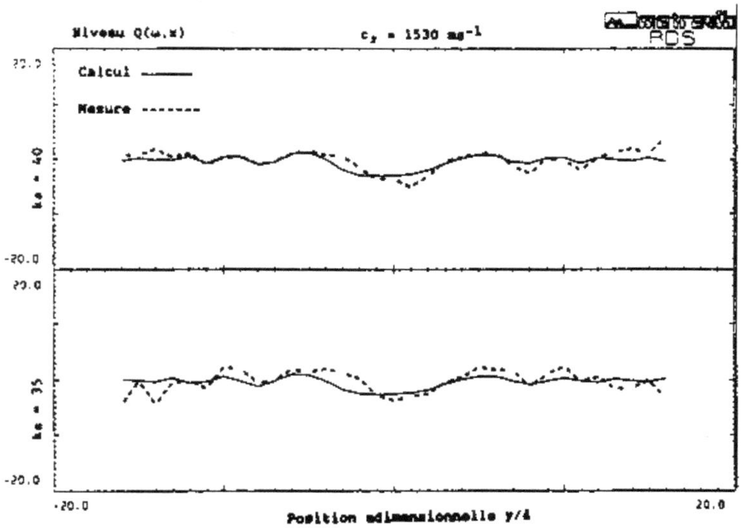

Figure 5.2 : Evolution spatiale $Q(\omega_i, x)$ à fréquence fixe. Cas asymétrique - Célérité c_2 optimale.

4 - CONCLUSION

Un modèle théorique bidimensionnel, reposant sur une approximation de BORN d'ordre 1, a été élaboré pour évaluer la perturbation inhérente à la présence d'un élément en élastomère sur la propagation d'une onde acoustique en milieu fluide.

Une hypothèse de faible contraste permet de traiter l'hétérogénéité en tant que cible fluide, de densité ρ_2 connue. La célérité c_2 associée à cette cible fluide constitue alors le paramètre du problème. La comparaison à deux séries de mesures expérimentales permet d'en obtenir une bonne approximation pour une valeur "optimale" de c_2.

D'éventuels prolongements à cette étude (d'ambition limitée, rappelons-le) pourraient prendre en compte une dépendance fréquentielle $c_2(\omega)$, de même qu'une extension de l'approximation de BORN à des ordres supérieurs, voire une résolution exacte du problème posé.

5 - REMERCIEMENTS

Cette étude a bénéficié du soutien du GERDSM, qui s'est en outre acquitté de la partie expérimentale. A ces divers titres, nous tenons à le remercier.

6 - BIBLIOGRAPHIE

[1] P.M. MORSE - K.U. INGARD - "Theoretical Acoustics", Mc Graw Hill, 1968.

CARACTERISATION OF ACOUSTIC SCATTERING IN THE TIME-FREQUENCY PLANE

Jean-Pierre SESSAREGO
Laboratoire de Mécanique et d'Acoustique
C.N.R.S.
31 Ch. Joseph-Aiguier
F-13402 MARSEILLE CEDEX 9

No written contribution received.

ACOUSTICAL SCATTERING BY A LIMITED CYLINDRICAL SHELL

Gérard MAZE
L.E.A.H.-U.S.
Université du Havre
Place Robert Schuman
76610 LE HAVRE

No written contribution received.

RESONANCE SCATTERING THEORY FOR OBLIQUE INCIDENCE ; SCATTERING BY AN ELASTIC INFINITE CYLINDER.

J.M CONOIR

Le Centre Thomson d'Applications Radar (L.C.T.A.R)
6 rue Nieuport - BP. 16
78143 VELIZY VILLACOUBLAY CEDEX FRANCE

ABSTRACT

A resonance scattering theory has been developed [1-9] for the problem of acoustic-wave scattering from elastic cylinders immersed in a fluid, when the incident plane wave direction of propagation is perpendicular to the cylinder axis. The present study is concerned with the scattering of an incident plane wave of oblique direction of propagation, and with the evolution of the "resonance spectra" with the oblique incident angle α (figure 1). In this paper, the eigenfrequencies of the cylindrical obstacle are interpreted as resonances due to phase matching of circumnavigating helical surface waves. A refraction effect is found to take place between the incident and the helical wave direction, which explains the resonance shift towards high frequencies when angle α increases. Moreover, we analytically prove that supplementary resonances are caused by oblique insonification. These resonances have been already detected experimentally [10], and are related to helical propagations guided in a direction parallel to the cylinder axis.

THEORY OF THE SCATTERING PROCESS

The total acoustic field in the presence of an infinite elastic cylinder is given in cylindrical coordinates (r , ϕ, z) by [11]

$$(1)\quad p = \frac{1}{2}\sum_{n=0}^{+\infty}(2-\delta_{n0})i^n\left[H_n^{(2)}(K_\perp r)+S_n H_n^{(1)}(K_\perp r)\right]\cos n\phi \times e^{i(K_z z-\omega t)}$$

where

$$(2)\quad K_\perp = K_1\cos\alpha \qquad K_z = K_1\sin\alpha$$

and S_n (n = 0,1,2,...) are the eigenvalues of the S-matrix ($K_1 = \frac{\omega}{c_1}$ is the wave number of the incident plane wave and c_1, the phase velocity in the external fluid). For a rigid cylinder, $S_n = S_n^{(R)}$ with :

$$(3)\quad S_n^{(R)} = -H_n^{(2)'}(x_\perp) / H_n^{(1)'}(x_\perp)$$

and $x_\perp = K_\perp a$ (a is the cylinder radius). The diffracted field amplitude for large values of r is given by [4]

$$(4)\quad |F_\infty| = \frac{2}{\sqrt{\pi x_\perp}}\left|\sum_{n=0}^{+\infty}(2-\delta_{n0})\left[\frac{S_n-1}{2}\right]\cos n\phi\right|$$

and its evolution for many oblique incident angles α is plotted on figures(2-7). Typically four different behaviours can be observed

- At normal incidence (α =o), we find the classical back scattering again ($\phi = \pi$) and resonances are detected when the curve varies rapidly (figure 2).

- When α is small (figure 3), we can observe some new resonances indicated by arrows.

- When α reaches 31° we observe a destructive interference between the Rayleigh surface wave and the specular reflexion (figure 6).

- Finally, beyond $\alpha \simeq 31°$, all resonances vanish and we find again the scattering by a rigid cylinder with a few modulations caused by Franz surface waves (figure 7).

It may be noticed that resonances are also shifted towards high frequencies when α increases (figures2-5). To explain all these effects, we must introduce the Sommerfeld-Watson transformation [12-15] . It consists in rewriting the normal-mode sum (1) as a contour integral in the complex ν -plane

$$(5) \quad p = \frac{i}{2} \mathcal{P} \int_{\mathcal{C}} \frac{d\nu}{\sin \pi\nu} e^{-i\nu\frac{\pi}{2}} \left[H_\nu^{(2)}(K_\perp r) + S_\nu H_\nu^{(1)}(K_\perp r) \right] \times \cos \nu\phi \; e^{i(K_z z - \omega t)}$$

where $\mathcal{C}$ tighly surrounds the positive real axis in the complex ν -plane, passing through ν =o ($\mathcal{P}$: principal value). The contour $\mathcal{C}$ is then deformed in the complex ν -plane to surround the poles of the S-function S_ν ("Regge poles"), these poles are denoted by ν_ℓ (ℓ =1,2,3... labeling their multiplicity). Here from, one finds

$$(6) \qquad p = p_{inc} + p_{geom} + p_{cw} + p_{ew}$$

where :

- p_{inc} is related to the incident pression
- p_{geom} is related to the geometrical waves
- p_{cw} is related to the creeping waves (Franz waves)
- p_{ew} is related to the elastic waves (Rayleigh wave and whispering gallery waves).

The phase factor $\Phi_\ell = \phi \,\mathrm{Re}\,\nu_\ell + K_z z$ of the elastic waves ℓ shows that these surface waves are helical ones [16] , and that their wave fronts have normals making an angle γ_ℓ with the x-axis (figure 8). This defines the refraction law

$$(7) \qquad K_\ell \sin \gamma_\ell = K_\perp \sin \alpha$$

where K_ℓ is the wave number of the helical wave ℓ :

$$(8) \qquad K_\ell = \left[\left(\frac{\mathrm{Re}\,\nu_\ell}{a} \right)^2 + K_z^2 \right]^{1/2} .$$

It may be shown that [11] a resonance related to the normal mode n occurs when

$$\mathcal{R}e\, \nu_{\ell} = n \tag{9}$$

which is the same relation as in the case of normal incidence. We then denote $x_n^{\ell}(\alpha)$ the resonance frequency. Relation (9) is also written [11] as

$$n \lambda_e = d \tag{10}$$

where λ_e is the wave length of the elastic surface wave and d is the geodesic distance between two points M and N (figure 9). Under these assumptions, resonance is due to phase matching of elastic circumnavigating helical surface waves. This is a very important point.

If we consider an elastic surface wave ℓ with phase velocity $C_{\ell}(0)$ at normal incidence, the resonances $x_n^{\ell}(\alpha)$ due to this wave at oblique incidence and $x_n^{\ell}(0)$ at normal incidence, are related by [11]

$$x_n^{\ell}(\alpha) = \frac{c_1}{c_{\ell}(0)} \; \frac{\mathrm{tg}\,\gamma_{\ell}}{\sin\alpha} \; x_n^{\ell}(0) \; . \tag{11}$$

Relation (11) shows that resonance shifting is proportional to $\mathrm{tg}\,\gamma_{\ell}$ and that resonances vanish (to infinity) when the elastic surface wave direction of propagation is parallel to the cylinder axis ($\gamma_{\ell} = \pi / 2$). In that case, there is no phase matching.

This explains the fact that beyond $\alpha \simeq 31°$ the elastic cylinder behaves like a rigid one. Indeed, all phase velocities of elastic surface waves are greater than the phase velocity in the external fluid. The Franz surface wave phase velocity is close to $C_{\perp}$, so that no refraction effect takes place between the incident and the helical wave direction ($\gamma_e \simeq \alpha$). Consequently, we still find the Franz waves beyond $\alpha \simeq 31°$ (see the modulations on figure 7).

Since we have proved that resonances are detected when relation (9) is true, we can use the S-matrix theory [9, 11] to calculate the resonance frequencies and plot the Regge trajectories related to elastic surface waves (figures10-12). When we compare figures 10 and 11, we observe the emergence of supplementary elastic surface waves (dotted line). Those waves related to helical propagations guided in a direction parallel to the cylinder axis [10] are responsable of the new resonances observed in oblique incidence. At $\alpha = 30°$ ($\simeq 31°$), we detect only one elastic surface wave (figure 12), the Rayleigh wave. All the whispering gallery waves having a more rapid phase velocity have already vanished. This is the reason why there is a destructive interference, as well as in the case of a plane interface between a perfect fluid and an elastic solid [17]. When α is greater than the critical excitation angle of Rayleigh wave ($\alpha \simeq 31°$), the last elastic surface wave vanishes with resonances.

BIBLIOGRAPHY

[1] J.J FARAN,
Sound scattering by solid cylinders and spheres,
J.A.S.A, 23 , 1951, 405-418.

[2] L. FLAX, L.R DRAGONETTE, H. ÜBERALL,
Theory of elastic resonance excitation by sound scattering,
J.A.S.A, 63 , 1978, 723-731.

[3] G.V FRISK, J.W DICKEY and H. ÜBERALL,
Surface wave modes on elastic cylinders,
J.A.S.A, 58, 1975, 996-1008.

[4] A. DEREM, JL. ROUSSELOT, G. MAZE, J. RIPOCHE, A. FAURE,
Diffusion d'une onde acoustique plane par des cylindres solides immergés : étude expérimentale et théorie des résonances,
ACUSTICA, 50, 1982, 39-50.

[5] H. ÜBERALL, L.R DRADONETTE, L. FLAX,
Relation between creeping waves and normal modes of vibration of a curved body,
J.A.S.A, 61, 1977 , 711-715.

[6] A. DEREM, JL. ROUSSELOT,
Résonances acoustiques du cylindre élastique infini et leur relation avec les ondes de surface,
Revue du CETHEDEC, 67, 1981, 1-34.

[7] J.W DICKEY, H. ÜBERALL,
Acoustic high frequency scattering by elastic cylinders,
J.A.S.A, 66, 1979, 275-283.

[8] G. MAZE, J. RIPOCHE,
Méthode d'Isolement et d'Identification des Résonances (M.I.I.R) de cylindres et de tubes soumis à une onde acoustique plane dans l'eau ,
Rev. Phys. Appl, 18, 1983, 319-326.

[9] N. GESPA,
La diffusion acoustique par des cibles de forme géométrique simple, Théories et expériences,
CEDOCAR, 1987.

[10] G. MAZE, J.L IZBICKI, J. RIPOCHE,
Diffusion ultrasonore par un cylindre insoné obliquement,
Rev. du CETHEDEC, 78-1, 1984, 95-105.

[11] J.M CONOIR,
Diffusion acoustique par un cylindre, théorie modale et approximation de l'acoustique géométrique,
Thèse de l'Université PARIS VI, 1987.

[12] G.N WATSON,
The diffraction of electric waves by the earth,
Proc. Roy. Soc. (London), A95, 1918, 83-99.

[13] W. FRANZ,
Über die Greenschen funcktionen des zylinders und der Kugel,
Z. Naturforsch. A, 9, 1954, 705-716.

[14] H. POINCARE,
Sur la diffraction des ondes hertziennes,
Rend. circ. Matem. Palermo, 29, 1910, 169-259.

[15] R.D DOOLITTLE, H. ÜBERALL, P. UGINČIUS,
Sound scattering by elastic cylinders,
J.A.S.A, 43, 1968, 1-14.

[16] A. NAGL, H. ÜBERALL, P.P DELSANTO, J.D ALEMAR and E. ROSARIO,
Refraction effects in the generation of helical surface waves on a cylindrical obstacle,
wave Motion, 5 , 1983, 235-247.

[17] H.L BERTONI, T. TAMIR,
Unified theory of Rayleigh - angle phenomena for acoustic beams at liquid-solid interfaces,
Appl. Phys, 2, 1973, 153-172.

FIGURE 1

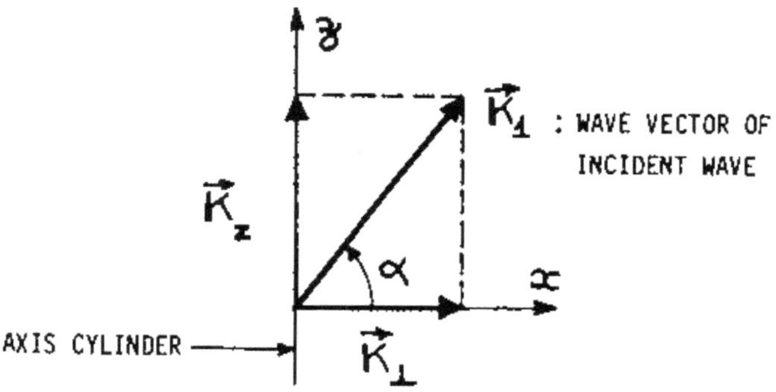

FIGURE 2

BACK SCATTERING

ALPHA = 0°

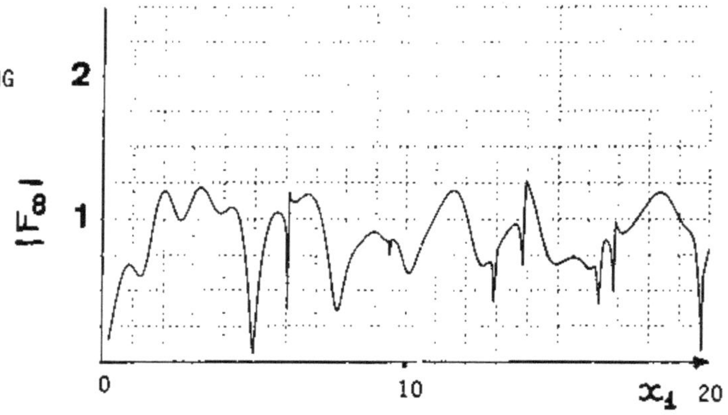

FIGURE 3

BACK SCATTERING

ALPHA = 3°

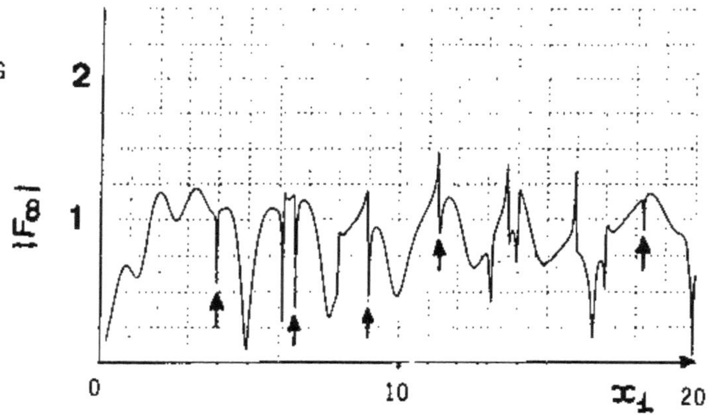

FIGURE 4

BACK SCATTERING
ALPHA = 9°

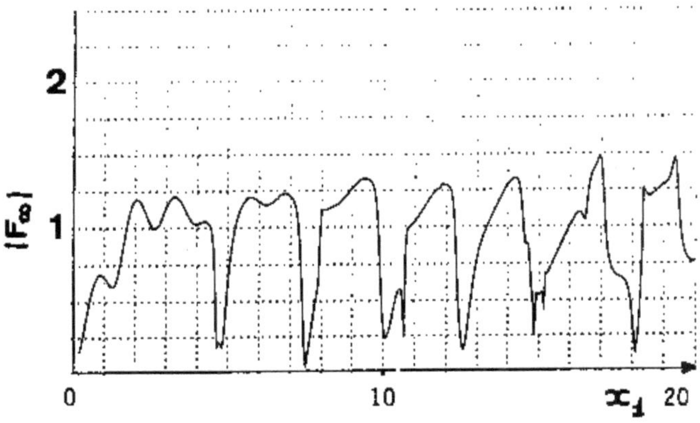

FIGURE 5

BACK SCATTERING
ALPHA = 24°

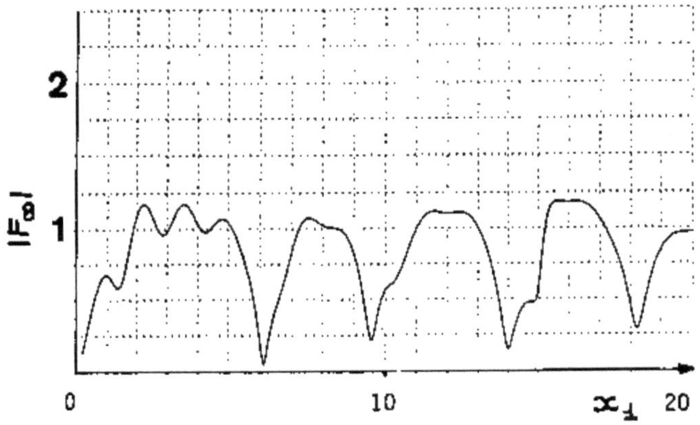

FIGURE 6

BACK SCATTERING
ALPHA = 31°

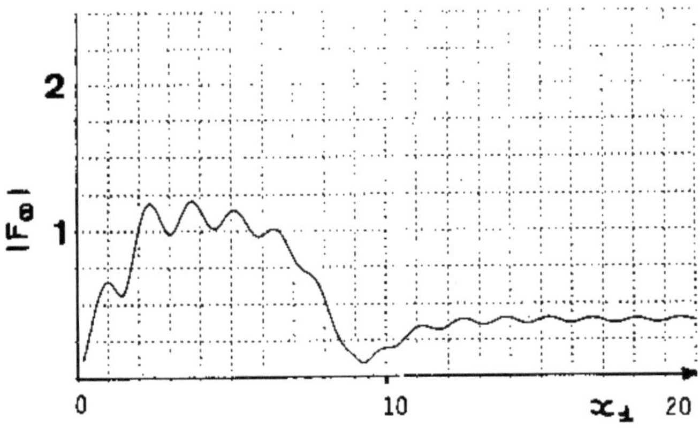

FIGURE 7

BACK SCATTERING
ALPHA = 45°

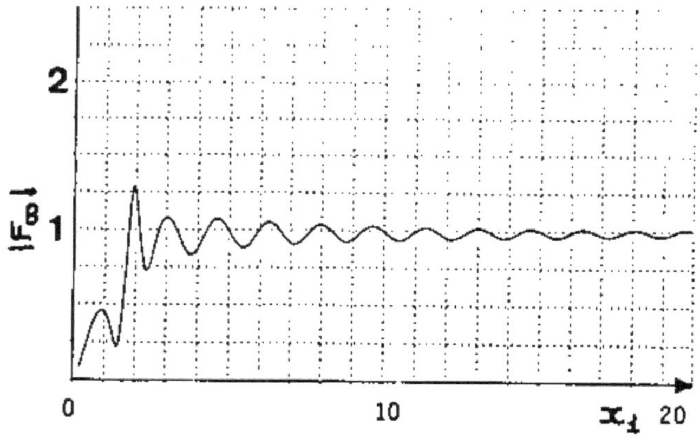

FIGURE 8

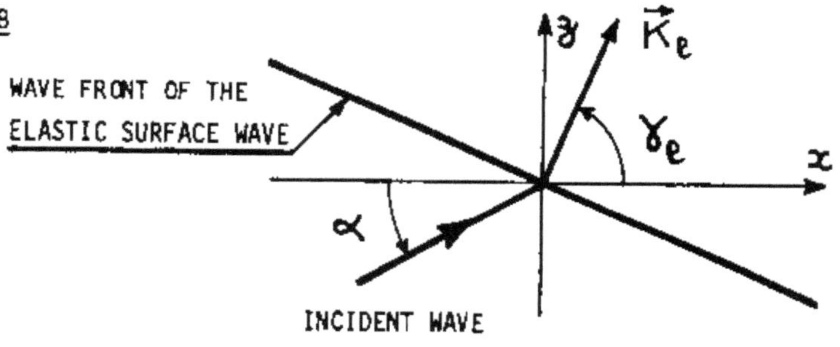

FIGURE 9

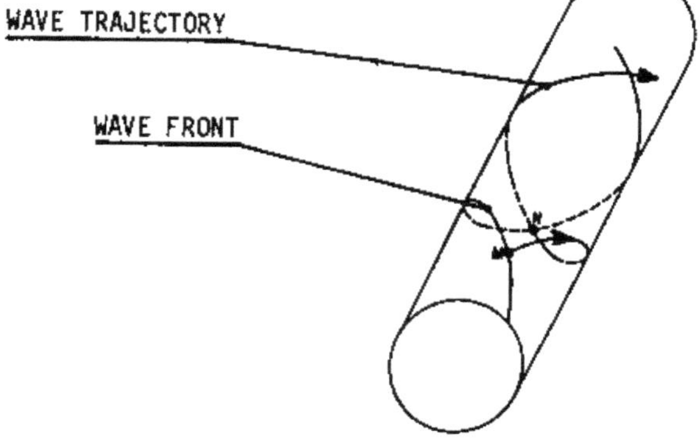

FIGURE 10

REGGE TRAJECTORY ALPHA = 0° (ALUMINIUM CYLINDER)

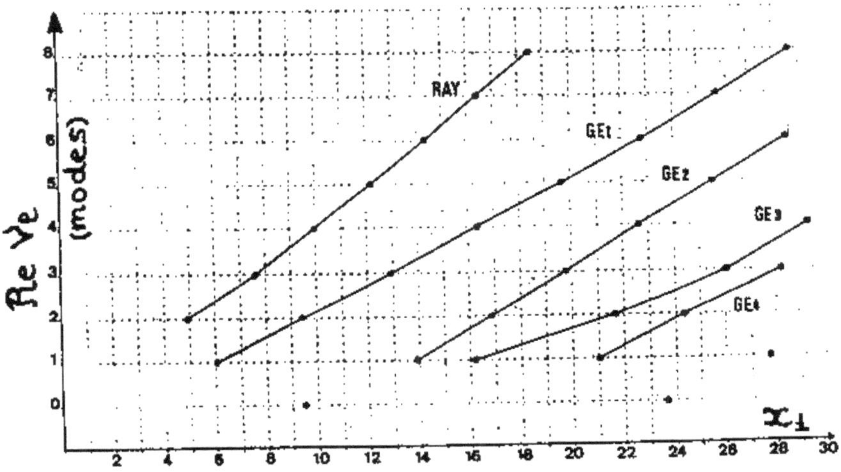

FIGURE 11

REGGE TRAJECTORY ALPHA = 3° (ALUMINIUM CYLINOER)

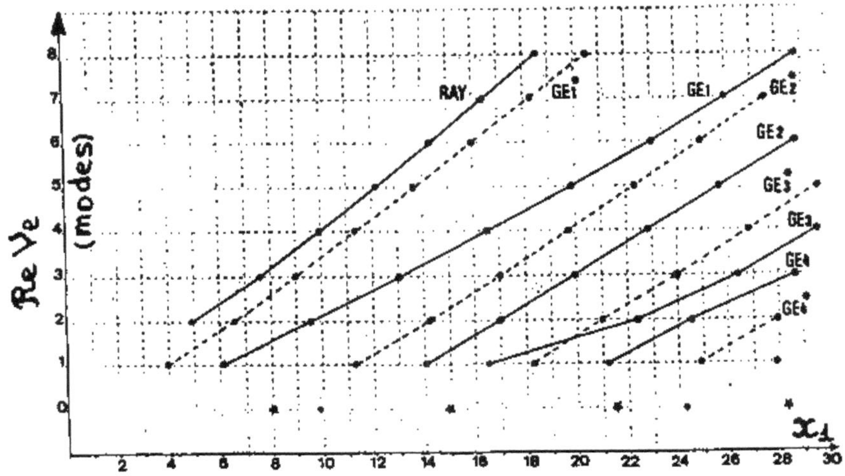

FIGURE 12

REGGE TRAJECTORY ALPHA = 30° (ALUMINIUM CYLINDER)

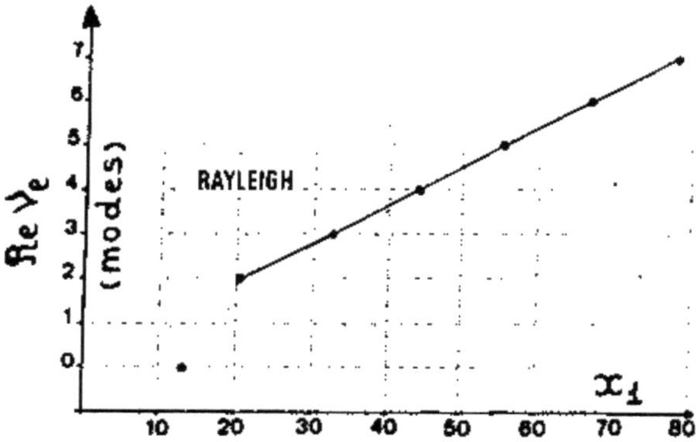

IMAGING

ACOUSTICAL IMAGING OF 2D FLUID TARGETS BURIED IN A HALF-SPACE: A DIFFRACTION TOMOGRAPHY APPROACH USING LINE-SOURCES INSONIFICATION

B. DUCHENE*, D. LESSELIER*, W. TABBARA*,**

*Laboratoire des Signaux et Systèmes
C.N.R.S. - E.S.E.
** Laboratoire d'Electronique Générale
Université Pierre et Marie Curie

E.S.E., Plateau de Moulon
91192 Gif-Sur-Yvette Cedex, France.

Abstract: The imaging of two-dimensional fluid targets buried in a homogeneous fluid half-space and insonified by a finite set of time-harmonic line-sources placed above and parallel to the half-space interface is studied. The data are the anomalous pressures observed for each source at discrete points along a probing line above and parallel to the interface. A reflection-mode diffraction tomography procedure which relates these data to the sources induced within the target by spectral processing is derived and discussed. Its performances are illustrated by the results of simulations led from exact synthetic data for various geometries and sound velocities of targets in different embeddings.

INTRODUCTION:

Imaging cylindrical targets concealed in a homogeneous half-space when they are illuminated by line-sources located above the half-space in the target near-field can be achieved by means of reflection-mode diffraction tomography techniques. To do so, we generalize an approach we previously developed in the case of a planar insonification [1)]. We consider a (more realistic) situation where a finite number of time-harmonic emitters radiating cylindrical waves at same frequency are located on a line above and parallel to the interface, the anomalous pressure being collected at a finite number of receivers on the same line or an other parallel one. We assume we are dealing with fluid media where only P-waves are propagated and we neglect the variations of density. Simulations are conducted using exact synthetic data. The imaging procedure provides us with qualitative results, but, theoretically, we should retrieve quantitative ones (the object function) when Born's approximation holds for every illumination. The formulation is an extension of the one used in the planar case and the algorithms utilized to compute the images are also based on similar principles. Some of the steps in the derivation of the fundamentals equations become more delicate, whereas, if the computational burden of the simulations steeply increases, this is essentially due to the large number of data we have to calculate beforehand by solving the direct scattering problem for each line-source, and not due to the algorithms complexity. Therefore, we write down the formulation in detail but give only a general outline of the numerical algorithms. Then, we illustrate the behavior of the imaging procedure by images of several targets with different shapes and various contrasts with respect to their environment. We observe that the images of point targets are fairly good and that, for more complicated targets, high-pass filtered versions are retrieved, as is usual in the reflection-mode. We conclude that this approach allows imaging of small isolated scatterers and to some extent discrimination between larger ones especially with sharp corners, but that results are less satisfying than in the planar case.

As already indicated, this work is a continuation of the one we presented earlier[1]. The literature on diffraction tomography and related non-destructive testing procedures is now abundant, with applications proposed in civil and biomedical engineering, and in geophysics. A unified description of their mathematical fundations and a number of characteristic examples in acoustics can be found in ref. 2. Within the framework of diffraction tomography, not so many references consider non-planar wavefields; noteworthy, among others, are: (i) ref. 3, which presents one of the first extensions of the theory to general geometries of the sources and receivers, in particular to cylindrical wavefields, but does not show any numerical results, (ii) ref. 4, where reconstruction algorithms equivalent to those of ref. 3 but more compact are applied to the imaging of a cylindrical target lying between two parallel arrays of line-sources and receivers, (iii) ref. 5, where the feasibility of geophysical imaging is discussed from various simulations led using different configurations of the source and detector arrays, with strong emphasis on the classical offset vertical seismic profiling geometry. Also, but in the framework of computerized tomography, fan-beam insonification is considered in refs. 6 and 7 where reflection-mode tomography techniques are developed in order to image defects in metallic and biological structures, respectively, and illustrated by numerous experimental results. (The first reference presents both compressional- and shear-wave scannings.) The case of unknown targets embedded in a known half-space and probed using sources and receivers located above the embedding has been studied recently by some authors interested into subsurface radars. For example, holographic formulations are employed in refs. 8 and 9 to obtain simulated and experimental images of dielectric or metallic pipes in soil from wide-band transient fields, and microwave imaging using diffraction tomography to process the data (multi-incidence multi-frequency anomalous fields) is investigated in ref. 10. A set-up operating in the X-band has also been constructed and shown to provide reliable identification of metallic pipes buried at shallow depths in cement using the approach of ref. 10, some experimental results being given in ref. 11. (This reference reviews the various studies led in our laboratory on the diffraction tomography imaging of two- and three-dimensional, buried or non-buried objects in electromagnetics and acoustics.)

I. THEORETICAL FORMULATION:

Let us refer to Fig. 1. We consider two homogeneous fluid half-spaces D_1 and D_2 with respective sound velocities C_1 and C_2 and attenuations α_1 and α_2 separated by a horizontal planar interface γ_{12} located at x=0. The fluid cylindrical target, of arbitrary cross-section D_Ω in the x-y plane and with sound velocity C_Ω and attenuation α_Ω varying with position (x,y), is embedded in the lower half-space and is illuminated, at the operating frequency f ($f=\omega/2\pi$), by a line-source located in the upper half-space. The density is constant throughout space and a time dependence $e^{-i\omega t}$ is assumed.

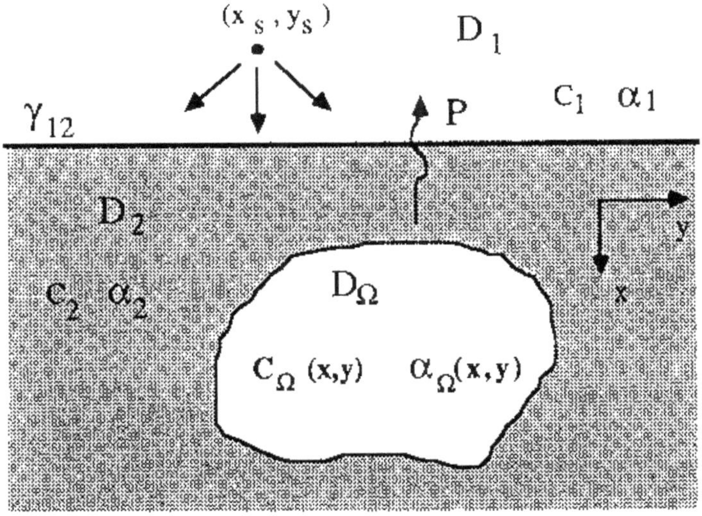

Fig.1: The geometry of the problem.

As is well-known, the scalar pressure field p(x, y) observed at point (x, y) when the target is illuminated by a line-source located at (x_S, y_S) satisfies, in the distribution sense, the wave equation:

$$\begin{aligned} &\Delta p(x, y) + k^2(x, y)\, p(x, y) = -\,\delta(x-x_S, y-y_S)\, \delta_{l,1} \\ &(x, y) \in D_l \quad , \quad k \equiv k_l \quad , \quad l = 1, 2, \Omega \end{aligned} \tag{1}$$

where k_l is the complex-valued propagation constant of medium l ($k_l=\omega/C_l+i\alpha_l$), δ is the Dirac distribution and $\delta_{l,1}$ is the Kronecker symbol ($\delta_{l,1}=1$ if $l=1$, $\delta_{l,1}=0$ if $l\neq 1$). The boundary conditions (continuity of the pressure and of its normal derivative) are implied, and the field satisfies a Sommerfeld radiation condition at infinity.

Diffraction tomography is based upon an exact integral formulation of the field p, which is obtained by applying Green's theorem to functions p and $G_{l,m}$, p satisfying equation (1) and $G_{l,m}$ being an auxiliary function, i.e., the Green function in layered media. This function satisfies the following wave equation in the distribution sense:

$$\begin{aligned} &\Delta G_{l,m}(x, y, x', y') + k_l^2\, G_{l,m}(x, y, x', y') = -\,\delta(x-x', y-y')\, \delta_{l,m} \\ &(x, y) \in D_l \quad , \quad (x', y') \in D_m \quad , \quad l = 1, 2 \quad , \quad m = 1, 2 \end{aligned} \tag{2}$$

$G_{l,m}$ satisfies the same boundary and radiation conditions as p. Let us note that $G_{l,m}(x,y,x',y')$ represents the field observed at (x,y) when a line source is located at (x',y'), the target being absent. $G_{l,m}$ is expanded into plane waves:

$$G_{l,m}(x, y, x', y') = (2\pi)^{-1} \int_{-\infty}^{+\infty} g_{l,m}(x, \alpha, x', y')\, e^{i\alpha y}\, d\alpha \tag{3}$$

and, in the case where $l=1$ and $m=2$ ($(x,y) \in D_1$, $(x',y') \in D_2$), $g_{l,m}$ is:

$$\begin{aligned} &g_{1,2}(x, \alpha, x', y') = i\ (\beta_1 + \beta_2)^{-1}\ e^{i\gamma}\ e^{-i\alpha y'} \\ &\gamma = \beta_2 x' - \beta_1 x \ , \ \beta_i = (k_i^2 - \alpha^2)^{1/2} \quad i = 1, 2 \end{aligned} \tag{4}$$

Applying Green's theorem to functions p and $G_{l,m}$ over a domain which covers the whole space, and using the fact that the target is entirely contained in the medium D_2 ($D_2 \supset D_\Omega$), we derive the exact integral formulation of the field

$$p(x, y) = p_0(x, y) + \int_{D_\Omega} J(x', y')\, G_{l,2}(x, y, x', y')\, dx'\, dy' \tag{5}$$

where $p_0(x,y)$ is the so-called incident field, i.e. the field that exists in the absence of target:

$$p_0(x, y) = G_{l,1}(x, y, x_s, y_s) \tag{6}$$

$J(x', y')$ represents the secondary sources induced inside the target by the incident wave:

$$J(x', y') = [k_\Omega^2(x', y') - k_2^2]\, p(x', y') \tag{7}$$

Let us assume now that we are able to measure the fields p and p_0 at different points (x_r,y_r) along a probing line $x=x_r$ parallel to the interface γ_{12} and located in the upper half-space ($(x_r,y_r) \in D_1$) for different locations of the source (x_s,y_s) ($(x_s,y_s) \in D_1$) along a line $x=x_s$ parallel to the interface. We denote $\psi(y_r,y_s)$ as the scattered field on the probing line for a given location of the source:

$$\psi(y_r, y_s) = p(x_r, y_r) - p_0(x_r, y_r) \tag{8}$$

and $\hat{\psi}(\alpha_r,\alpha_s)$ as its two-dimensional Fourier transform:

$$\hat{\psi}(\alpha_r, \alpha_s) = \iint_{-\infty}^{+\infty} \psi(y_r, y_s)\, e^{-i(\alpha_r y_r + \alpha_s y_s)}\, dy_r\, dy_s \tag{9}$$

Let us define an other quantity $\phi(x,y)$ which represents the normalized induced sources:

$$\phi(x, y) = J(x, y) / p_0(x, y) \tag{10}$$

and $\hat{\phi}(\mu,\nu)$ its two-dimensional Fourier transform:

$$\hat{\phi}(\mu, \nu) = \iint_{-\infty}^{+\infty} \phi(x, y)\ e^{-i(\mu x + \nu y)}\, dx\, dy \tag{11}$$

Using (8) and (10), we can rewrite (5) as:

$$\psi(y_r, y_s) = \int_{D_\Omega} \phi(x', y')\ G_{1,2}(x_r,y_r,x',y')\ G_{2,1}(x',y',x_s,y_s)\ dx'dy' \tag{12}$$

As $\phi(x',y') = 0$ when $(x',y') \notin \Omega$, the integration over D_Ω can be extended over the whole space. We introduce now the plane wave expansion (3) of the Green function and after some transformations we obtain a relationship between the Fourier transforms of ϕ and ψ:

$$\hat{\phi}(\mu, \nu) = -(\beta_{1s} + \beta_{2s})\ (\beta_{1r} + \beta_{2r})\ \ \hat{\psi}(\alpha_r, \alpha_s)\, e^{i\gamma}$$

$$\mu = -(\beta_{2r} + \beta_{2s}) \qquad \nu = \alpha_r + \alpha_s \tag{13}$$

$$\gamma = \beta_{1r}\, x_r - \beta_{1s}\, x_s\ ,\quad \beta_{ij} = (k_i^2 - \alpha_j^2)^{1/2}\ ,\quad i = 1, 2\ ,\quad j = r, s$$

This relationship is exact only when D_1 and D_2 are lossless (when k_1 and k_2 are real-valued) and does not hold for the evanescent spectrum ($|\alpha_j| > k_2$, $k_2 < k_1$). Let us note that the points (μ,ν) of the spectral plane where $\hat{\phi}$ can be estimated belong to a family of circular arcs with radius k_2 centered on a circle with radius k_2. These arcs cover the dotted domain depicted in Fig. 2. Once $\hat{\phi}(\mu,\nu)$ is known, an image of the target can be obtained by inverse Fourier transforming $\hat{\phi}$, as $\phi(x,y)$ depends upon the acoustical parameters of the target through $k_\Omega(x,y)$. However, as usual in the reflection mode, it can be seen on Fig. 2 that we don't get any information about the low spatial frequencies and about the frequencies relative to directions parallel to the interface. Consequently, the image will be a distorted high-pass filtered representation of the target. This is shown in Fig. 3 which depicts the image of a ring (internal radius = λ_2, external radius = $2\lambda_2$, λ_2 = wavelength in medium D_2) obtained by inverse Fourier transforming the spectrum of

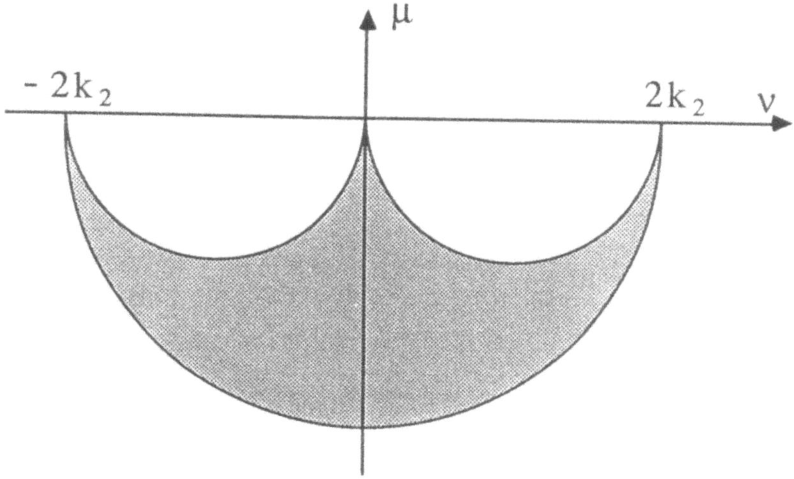

Fig. 2: Coverage of the (μ,ν) spectral plane.

the object function ($k_\Omega^2(x,y) - k_2^2$) filtered over the spectral domain depicted in Fig. 2. As is expected, only the front and back faces of the object, which are related to the high spatial frequencies along the μ axis, appear.

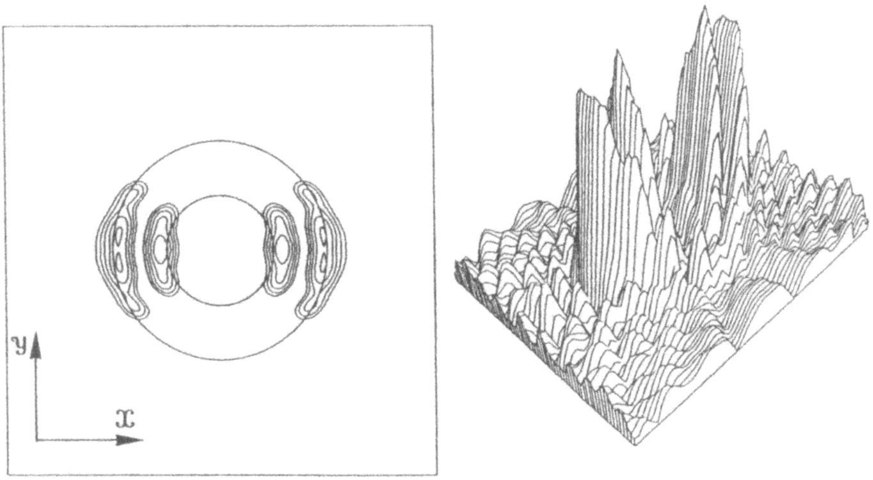

Fig. 3 : Theoretical image of a ring (internal radius = λ_2, external radius = $2\lambda_2$) obtained by filtering the object function in the spectral plane.

More generally, the imaging process is characterized by its theoretical point-spread function (the image of a point-like scatterer) depicted in Fig. 4. Due to the dissymmetry in filling up the spectral plane this point-spread function is not isotropic. Its longitudinal half-height width (along the x axis) is about $3\lambda_2/2$ whereas its transverse half-height width (along the y axis) is about $\lambda_2/2$.

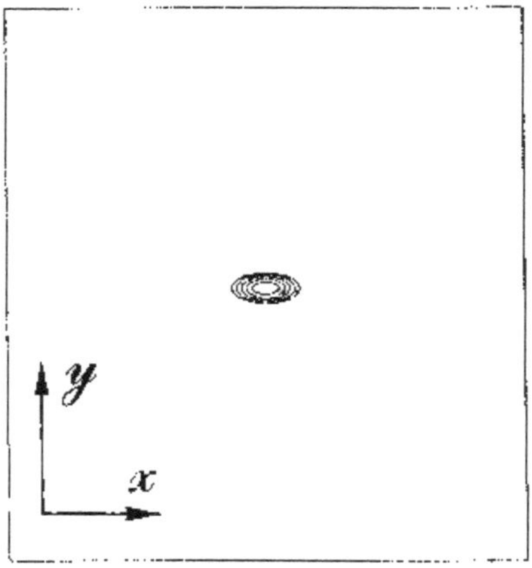

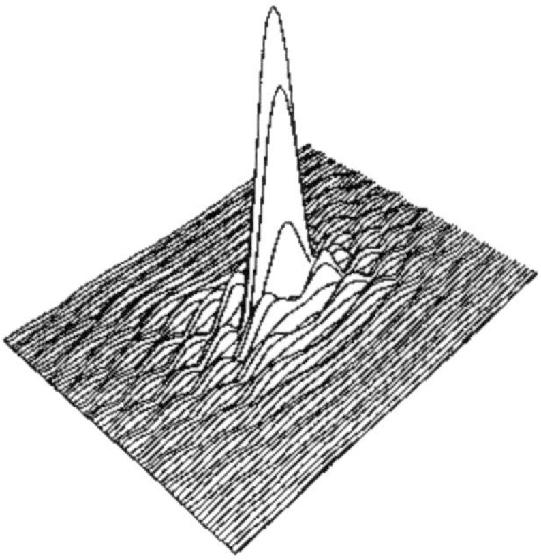

Fig. 4 : Theoretical point spread function.

Let us now point out some theoretical and numerical problems. The main theoretical problem comes from the illumination dependency of $\phi(x,y)$. Indeed, the secondary sources induced inside the object by a source (x_s,y_s) depend upon the location of this source; nevertheless, the image of the object is obtained by superposition, in the spectral plane, of the spectra $\hat{\phi}(\mu,\nu)$ corresponding to different locations of the source. Therefore, the reconstructed function has no physical meaning and is only a qualitative representation of the object. However, when Born's approximation holds, $\phi(x,y)$ does not depend anymore upon the field $p(x,y)$, and the reconstructed function is the object function $(k_\Omega^2(x,y) - k_2^2)$. The problem is then removed and a quantitative representation can be obtained. Some other limitations are linked to the acquisition and processing of the data. In particular, numerical problems arise when FFT algorithms are used to compute $\phi(x,y)$. Indeed, such algorithms can be used when the spectrum $\hat{\phi}(\mu,\nu)$ is known over a regular Cartesian mesh and the values of the spectrum over this mesh are interpolated from the sparse initial spectral data that are irregularly distributed in the spectral plane onto the circular arcs. Nearest neighbor or bilinear interpolation procedures are used to do so. The influence of the interpolation algorithm upon the quality of the image is important, as well as the influence of the data acquisition parameters, i.e. the number and spacing of sources and receivers, related to real-world measurements constraints. A previous study [12] has shown that a non-optimal choice of the sampling step and of the length of the probing line (the aperture) could damage the images in an important manner, whereas the number and spacing of the sources, which determine the number of initial spectral data and the coverage of the spectral plane, are, indeed, fundamental parameters of the images reconstruction.

NUMERICAL RESULTS :

Images of various buried targets, which are depicted herein, have been obtained from exact synthetic data. The fields are computed from the integral representation (5) by using either the method of moments and a conjugate gradient algorithm to solve the linear system we then obtain, or Neumann series expansion of the fields. The second method is used when the objects are "small enough" and/or have a low contrast with respect to the embedding medium but does not converge in other cases, whereas the first one, much more computationally expensive, can be used in every case.

The fields are computed on a line $x_r = -\lambda_1$ (λ_1 = wavelength in medium D_1) located above the interface, the sources being located on the same line. 65 sources are used and the fields are computed in 129 points with a $\lambda_2/2$ sampling rate (which is the optimal rate, since the evanescent spectrum is neglected). All the media are assumed to be lossless and D_1 and D_2 have respective sound velocities C_1= 1470 m/s and C_2= 1540 m/s. The images represent the modulus of induced sources $\phi(x,y)$ displayed using contour lines (the first line represents the half-height) over a $8\lambda_2$-sided square centered at the center of the object (located at a distance d under the interface). The interface is located at the left-hand side of the images. Fig. 5 depicts images of two targets buried at depth $d = 2\lambda_2$ under the interface with sound velocity C_Ω = 1540 m/s, which means a smooth transition compared to the host medium. Fig. 5 a) represents a $2\lambda_2$ sided square: as it has been previously mentioned, only the front and back faces (parallel to the interface), which correspond to high spatial frequencies on the μ axis, appear. Fig. 5 b) depicts the image of a thin inclined rectangle, $2\lambda_2$ in length and $\lambda_2/4$ in width, and shows that sharp corners are well reconstructed; however this image underlines the differences between cylindrical and planar insonifications. Indeed, in the second case, which has been previously studied [1],

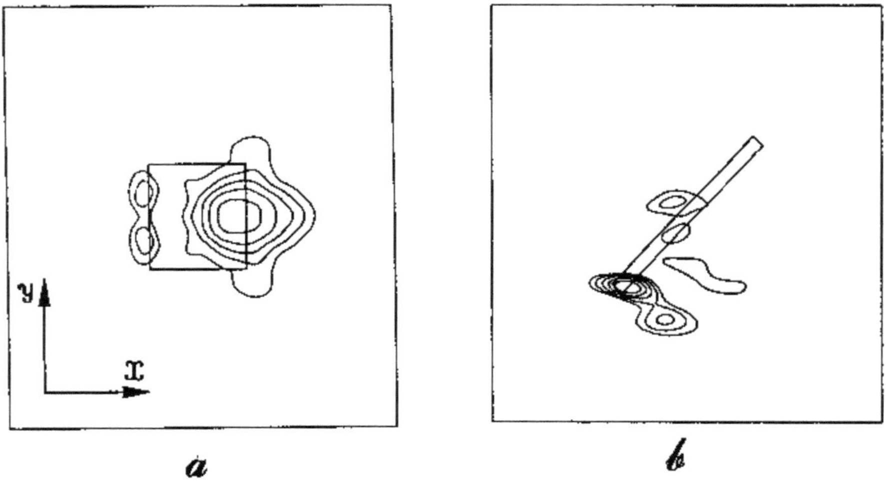

Fig. 5: Image of targets with sound velocity C_Ω = 1540 m/s buried at d = $2\lambda_2$.
a) a $2\lambda_2$ sided square and b) a thin inclined rectangle (length = $2\lambda_2$, width = $\lambda_2/4$)

all the corners appear whereas in the first case, due to energy dispersion, the back corners are invisible. Fig. 6 represents images of two thin rectangles, $2\lambda_2$ in length and $\lambda_2/6$ in width, $2\lambda_2$ distant from center to center. In Fig. 6 a) the rectangles are of same sound velocities C_Ω= 1800 m/s and are buried at depth d = $2\lambda_2$: the coupling between the two rectangles enhances the side lobes at the center of the figure and, due to the finite aperture of measurements, the back faces of the rectangles are shifted from their real positions. In Fig. 6 b) the upper and the lower rectangles are of different sound velocities, i.e., C_U = 1540 m/s and C_L = 1800 m/s respectively, and the upper rectangle, which is of very low contrast vis-à-vis the host medium, does not appear, whereas in Fig. 6 d), the two rectangles being of high contrast (C_U = 2500 m/s and C_L = 1800 m/s), both of them appear and the contrast between them is clearly depicted. Fig. 6 c) represents the same rectangles as in Fig. 6 d) buried at a depth d = $10\lambda_2$. Due to the reduction of the measurements aperture (compared to cases a), b) and d)), the delocalization of the front and back faces is increased and the lower rectangle disappears.

CONCLUSION :

The conclusions we can infer from these images are quite similar to those we wrote down in the plane wave insonification case [1]), i.e. :

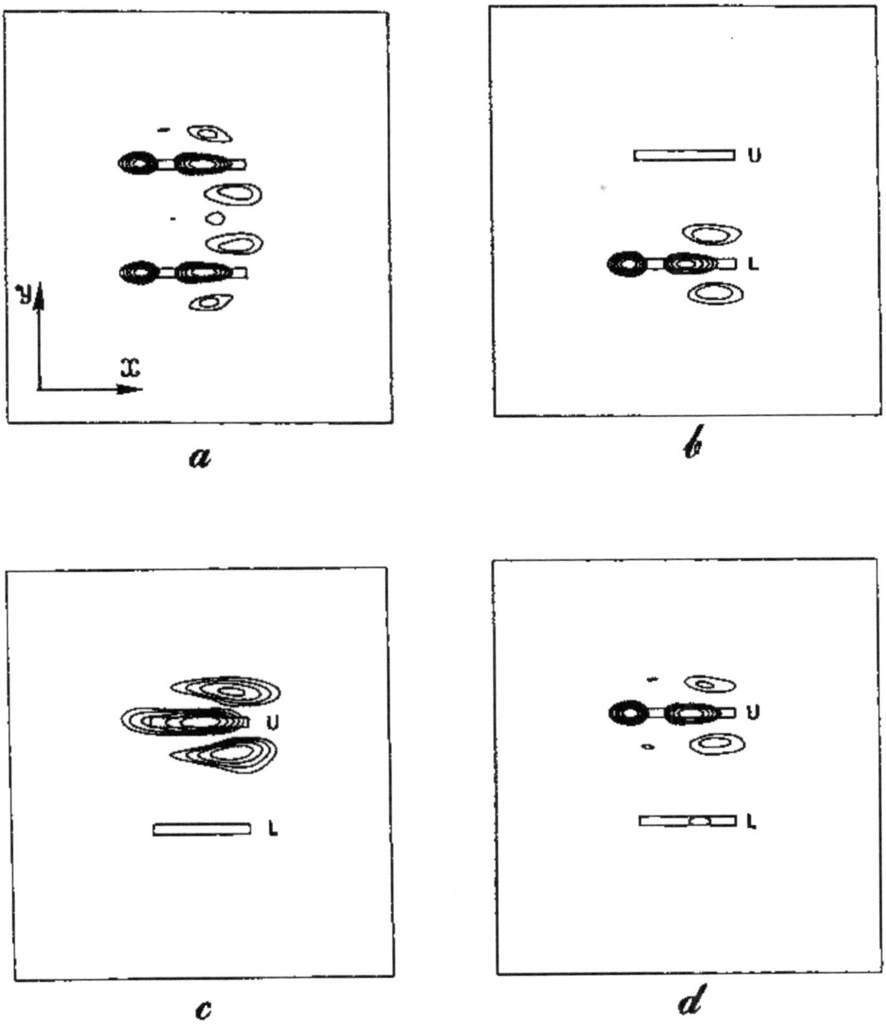

<u>Fig. 6:</u> *Images of two thin rectangles (length = $2\lambda_2$, width = $\lambda_2/6$) $2\lambda_2$ spaced from center to center : a) buried at $d = 2\lambda_2$ and with the same sound velocity $C_\Omega = 1800$ m/s b) buried at $d = 2\lambda_2$ with different sound velocities $C_U = 1540$ m/s and $C_L = 1800$ m/s c) buried at $d = 10\lambda_2$ with respective sound velocities $C_U = 2500$ m/s and $C_L = 1800$ m/s d) buried at $d = 2\lambda_2$ with respective sound velocities $C_U = 2500$ m/s and $C_L = 1800$ m/s*

- targets of different shapes are clearly discriminated.
- targets of same shape but different velocities can be discriminated from one another but may be not from the embedding medium.

However, the dispersion of energy due to point-source insonification leads to some results which can be very different from those obtained in the previous case, i.e., the deeper parts of the targets often become invisible.

Further conclusions would need other numerical investigations. In particular, filling-up the spectral plane over a larger domain, by varying the frequency of illumination, should improve the quality of the reconstruction.

REFERENCES:

1) Duchêne B. *et al.*, IEEE Trans. Ultrason. Ferroelect. Freq. Control UFFC-34, 540-549 (1987).
2) Langenberg K., in *Basic Methods of Tomography and Inverse Problems*, P. Sabatier Ed., Adam Hilger, Bristol, Part II, 127-467 (1987).
3) Devaney A. and Beylkin G., Ultrason. Imaging 6, 181-193 (1984).
4) Harris J., IEEE Trans. Geosci. Remote Sens. GE-25, 448-455 (1987).
5) Witten A. and Molyneux J., IEEE Trans. Geosci. Remote Sens. GE-26, 409-419 (1988).
6) Moshfeghi M., IEEE Trans. Ultrason. Ferroelect. Freq. Control UFFC-33, 299-314 (1986).
7) Dines K. and Goss S., IEEE Trans. Ultrason. Ferroelect. Freq. Control UFFC-34, 309-318 (1987).
8) Osumi N. and Ueno K., IEEE Trans. Antennas Propagat. AP-33, 152-159 (1985).
9) Michiguchi Y. *et al.*, IEEE Trans. Geosci. Remote Sens. GE-25, 11-15 (1987).
10) Chommeloux *et al.*, IEEE Trans. Microwave Theory Tech. MTT-34, 1064-1076 (1986).
11) Tabbara W. *et al.*, Inverse Problems 4, 305-331 (1988).
12) Duchêne B., Thèse de Doctorat-es-Sciences, Université Paris-Sud, Orsay (1986).

CONTRIBUTION OF RADAR POLARIMETRY IN RADAR TARGET DISCRIMINATION. THE "POINCARE PLANISPHERE": A NEW REPRESENTATION METHOD.

Eric POTTIER

Laboratoire de Communications Détection Systèmes

IRESTE
(Institut de Recherche et d'Enseignement Supérieur
aux Techniques de l'Electronique)

La Chantrerie-CP 3003
44087 NANTES Cédex 03
Tél. (33) 40.68.30.63

ABSTRACT

The study of the "Polarization Fork", defined by HUYNEN [1] is an important parameter in radar polarimetry.

The "Polarization Fork" is the optimal polarization vectors (XPOLL nulls and COPOLL nulls) representation on the POINCARE Sphere. These vectors are associated to the scattering matrix [S].

The representation of the "Polarization Fork" is a three dimensional display. Now, as is the case with all the three dimensional representations, the user is subject to many complex implementation problems and mainly interpretation problems.

The object of this paper is to develop a new method of the "Polarization Fork" representation, that will allow a better interpretation of results.

INTRODUCTION.

The contribution of radar polarimetry theory to radar target discrimination is presently, the subject of a study led by CELAR (Centre Electronique de l'Armement - Division ASRE/SRH - 35170 BRUZ - FRANCE) and by IRESTE (see the author's address). The object of this study is to emphasize the advantages of this theory for including radar polarimetry in high resolution electromagnetic image construction, to improve the discrimination of back-scattered elementary contributors.

In the fist part, we'll recall the basic principles in radar polarimetry, then we shall introduce classical representation methods, where we shall bring out the usual problems of implementation and interpretation.

In the second part we shall define a new display method, called the "POINCARE PLANISPHERE", and a new graphical method allowing an easy determination of the EULER angles values.

BASIC PRINCIPLES IN RADAR POLARIMETRY.[1]-[2]

In radar polarimetry, the complete scattering description of a target is obtained in terms of its scattering matrix [S], which relates the polarization of the scattered wave to that of the incident wave. If the wave incident on the target is expressed in terms of two orthogonal polarized components, say linear horizontal and vertical, then the wave scattered by the target is expressed in terms of the incident wave as follows:

$$\begin{vmatrix} Eh^s \\ Ev^s \end{vmatrix} = \begin{vmatrix} Shh & Shv \\ Svh & Svv \end{vmatrix} \cdot \begin{vmatrix} Eh^i \\ Ev^i \end{vmatrix} \quad \text{or } \underline{E}^s = [S] \cdot \underline{E}^i$$

The scattering matrix [S] is a characteristic of the target for one aspect and at one frequency. In a monostatic radar, the scattering matrix is symetric, with Shv=Svh. The elements of the scattering matrix are in general complex. Another important concept in radar polarimetry is the optimal vectors concept. These optimal vectors are the orthogonal polarization vectors of the incident wave which maximize or minimize the power in the receiver channels. They are obtained from the scattering matrix [S] by resolving an eigenvalue problem. They are, in fact, the polarization vectors of the incident wave, and are necessary to obtain a diagonal scattering matrix [S] or a null-diagonal one. The former are called "XPOLL nulls" or "CROSS-POL nulls", because they minimize the power received in the cross polarized channels and maximize the power received in the co-polarized channels. The latter are called the "COPOLL nulls" because they minimize the power received in the co-polarized channels or maximize the power received in the cross-polarized channels. They correspond to a perfect depolarization phenomenon due to the target.

These optimal polarization vectors constitute KENNAUGH's target characteristic operator also known as "Polarization Fork".

Adequate knowledge of the vectors is necessary to completely specify the scattering matrix of the target for one aspect and at one frequency.

BASIC REPRESENTATIONS OF A POLARIZATION VECTOR.[1]-[2]

A polarization vector can be displayed in the plane (0,x,y,z) by means of the polarization ellipse which is described by the wave resulting from its propagation along the z-axis (see figure n°1). This representation is a quite easily implemented, but however, it is relatively difficult to discriminate depolarization phenomena. Note two important parameters of the polarization ellipse which are the angle τ, known as "ellipticity", and the angle ϕ which is known as "ellipse orientation".

Another polarization vector representation consists in using a classical three dimensional display by introducing the POINCARE Sphere. In this case, we build the STOKES vector $\underline{G}$ (G0,G1,G2,G3) [1] which is expressed in terms of the polarization vector $\underline{H}$ (Hx,Hy) as follows:

$$G0 = |Hx|^2+|Hy|^2 = r^2$$
$$G1 = |Hx|^2-|Hy|^2 = r^2\cdot\cos(2\phi)\cdot\cos(2\tau)$$
$$G2 = 2\cdot Re(Hx\cdot Hy^*) = r^2\cdot\sin(2\phi)\cdot\sin(2\tau)$$
$$G3 = -2\cdot Im(Hx\cdot Hy^*) = r^2\cdot\sin(2\tau)$$

We can immediatly see that these expressions are functions of the geometric parameters of the polarization ellipse, say the angles τ and ϕ. Moreover, these expressions are equivalent to those of a point P (r,τ,ϕ) expressed in its spherical coordinates, where ϕ represents the longitude and τ the latitude.

From these expressions we can derive the relation (for completely polarized waves): $G0^2 = G1^2 + G2^2 + G3^2$. This is an important relation because it shows that a STOKES vector is a sphere generator.

THE POINCARE SPHERE.[1]-[2]

A polarization vector $\underline{H}$ is represented on the POINCARE Sphere by a point P (see figure n°2) where the precedently described angles ϕ and τ appear. On figure n°2 we can also see the canonical polarization vectors, say linear horizontal and vertical which are antipodal on the x-axis, linear ±45° which are antipodal and the y-axis and the circular right (CD) and circular left (CG) which are antipodal on the z-axis.

If we represent the "Polarization Fork" on the POINCARE's Sphere (see figure n°3) we can immediatly deduce two important remarks:

- The XPOLL nulls are antipodal, which mean that the XPOLL nulls vectors are orthogonal.
- The four optimal polarization vectors (XPOLL and COPOLL nulls) are located on the same great circle on the POINCARE Sphere.

This "Polarization Fork" is an important characteristic in radar Polarimetry and the goal is to find a really simple implementation method that allows a better interpretation of the "Polarization Fork" location on the Sphere.

THE EULER ANGLES.[1]-[2]

To describe the "Polarization Fork", it is obvious that we can use the geometric parameters (angles τ and ϕ) associated to each point of the fork, so it will give us eight parameters. To simplify this description we are going to introduce the EULER angles which allow us to describe the global location of the "Polarization Fork" on the POINCARE Sphere (see figure n°2). The EULER angles are the angles ϕ, τm and μ which are each defined along a cartesian reference axis, and a fourth, the angle δ, a spatial one, which is situated between the COPOLL nulls and the XPOLL nulls axis.

APPLICATION.

After having recalled the basic principles in radar polarimetry, we are now going to show an example.

Considering the scattering matrix $[S] = \begin{vmatrix} 2j & 1/2 \\ 1/2 & 2j \end{vmatrix}$

The optimal polarization STOKES's vectors (XPOLL and COPOLL nulls) are as follows:

$$X1 = \begin{vmatrix} 0.707 \\ 0 \\ 0.707 \end{vmatrix} \quad X2 = \begin{vmatrix} -0.707 \\ 0 \\ -0.707 \end{vmatrix} \quad CO1 = \begin{vmatrix} -0.330 \\ -0.882 \\ -0.330 \end{vmatrix} \quad CO2 = \begin{vmatrix} -0.330 \\ 0.882 \\ -0.330 \end{vmatrix}$$

The associated EULER's angles are equal to $\phi=0$, $\tau m=\pi/4$, $\mu=\pi/2$ and $\delta=\pi/3$. If we represent this "Polarization Fork" on the POINCARE Sphere (see figure n°4) we can easily see that it is difficult enough to discriminate the location of points CO1 and CO2 with any degree of accuracy. Are they coming towards us or are they going inside the drawing? Moreover, it is impossible to determine the EULER angles values.

THE "POINCARE PLANISPHERE".[3]

To improve the exploitation of the "Polarization Fork" representation, we are going to transform this three dimensional representation into a two dimensional one, applying modern and advanced methods of cartographic projection theory. The result of this operation is what we called the "POINCARE Planisphere" (see figure n°5). On this planisphere are displayed canonical polarization vectors, say linear horizontal (H) in the center (H is the reference projection point), circular left and right polarization vectors (CG,CD) at the top and bottom, and linear vertical (V) at the two extremities of the planisphere. A polarization vector is always indicated by a point P, defined by its two geometric parameters τ and ϕ.

BACK TO THE APPLICATION.

The "Polarization Fork" of our example is now projected and displayed on the "POINCARE Planisphere" (see figure n°6). We can immediatly see that the COPOLL nulls (CO1 and CO2) longitudes are greater than $\pi/2$, which proves that they were going inside the drawing when they were displayed on the POINCARE Sphere representation.

From the "Polarization Fork" projection on the planisphere, we can now determine, without any calculus, the values of two of the EULER angles, say angles τm and ϕ, only by studing the location of the XPOLL null vector X1. So we can read that $\phi=0$ and $\tau m=\pi/4$.

THE EULER ANGLES ABACUS.[3]

Now we are going to introduce a new graphical method which will allow us to determinate graphically, as easily as possible, the unknown values of the other two EULER angles.

Without going into calculus details, we have materialized two graphes networks which are necessary to determine the angles μ and δ.

These two networks are represented in figures n°7 and n°8. The first network (see figure n°7) represents the function $Y= \sin(A)+\sin(B)$ and the second (see figure n°8) the function $Y=2\cdot\sin(A)\cdot\cos(B)$.

UTILISATION METHOD.

We are now going to see how to use these two networks to determine the last two unknown angles.

If we look at figure n°6 again where the "Polarization Fork" is projected on the "POINCARE Planisphere", we can determine the two latitudes $\tau 1$ and $\tau 2$ associated to the two points CO1 and CO2, and we read $\tau 1=-20°$ and $\tau 2=-20°$. We now use the first network to determine the Y1 value of the function $Y1= \sin(\tau 1)+\sin(\tau 2)$ (see figure n°9). Once the Y1 value is obtained ($Y1=-0.7$) we transfer this value on the second network to determine the δ angle value (see figure n°10) corresponding to the function $Y1= 2\cdot\sin(-\tau m)\cdot\cos(\delta)$. The angle τm was determined by studing the XPOLL null X1 location ($\tau m=\pi/4$). So we read that $\vartheta=\pi/3$. It must be noted, there exist two opposite values which are $\vartheta=\pi/3$ and $\vartheta=-\pi/3$, but as ϑ is a spacial angle without any sign reference, we take the positive value.

The last unknown angle is μ. To obtain its value, we determine with the first network the Y1 value corresponding to the function : $Y1 = \sin(\tau 1)+\sin(-\tau 2)$, and with the second network we determine the Y2 value corresponding to the function: $Y2 = 2\cdot\cos(\tau m)\cdot\sin(\vartheta)$. With these two values, we can determine the μ angle value as follows:

$$\mu = \mathrm{Arccos}(Y1/Y2)$$

In our example the angles $\tau 1$ and $\tau 2$ are equal, so $Y1=0$ and we can immediatly determine μ with $\mu=\pi/2$.

We can now conclude that this graphical method enables us to find exactly the EULER angles values.

CONCLUSION.

To improve the study of the "Polarization Fork", which is an important parameter in radar polarimetry, we have determined a new representation method by constructing the "POINCARE Planisphere". This new two dimensional representation method allow a better implementation and mainly a better location interpretation of the projected "Polarization Fork".

Moreover, we have determined a new graphical method ,"the EULER angles abacus" which allows us to find, easily, whithout calculus, the EULER angles values associated to the "Polarization Fork".

REFERENCES.

[1] HUYNEN,J.R.,"PHENOMENOLOGICAL THEORY OF RADAR TARGET",Ph.D. Dissertation, DRUKKERIJ Broder-Offset, N.V ROTTERDAM (1970).

[2] BOERNER,W-M.,HUYNEN,J.R.,MATHUR,N.C,and al,"POLARIZATION UTILIZATION IN RADAR TARGET RECONSTRUCTION", Final Report, UIC-EECS-CL-EMID-ARO/EL.83.12.14. Contract N° US-ARMY DAAG-29-80-12-0027 (14 Décembre 1983).

[3] POTTIER,E.,"CONTRIBUTION DE LA POLARIMETRIE DANS LA DISCRIMINATION DE CIBLES RADAR". 1er Rapport d'avancement, Université de RENNES I-CELAR (Octobre 1987-Juin 1988).

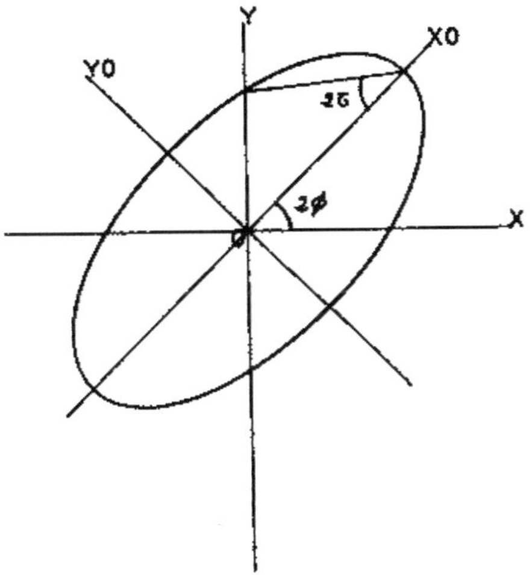

Figure n1

POLARIZATION ELLIPSE

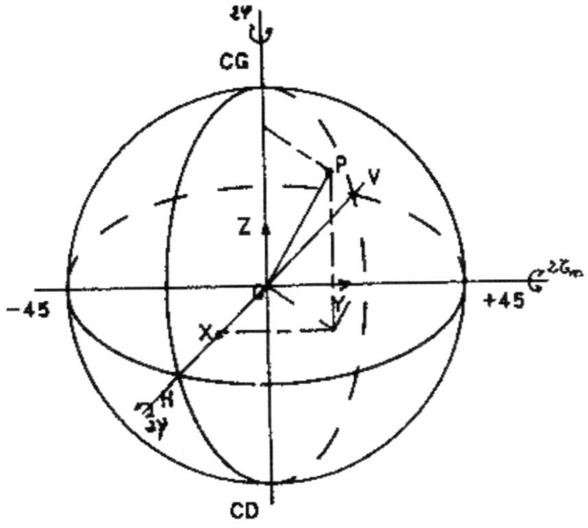

Figure n2

POINCARE SPHERE

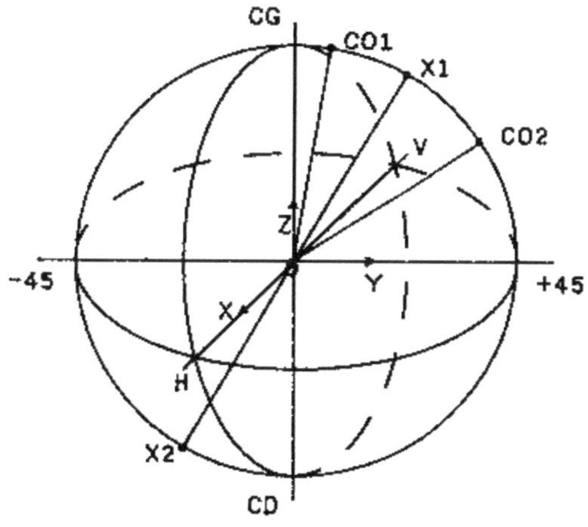

Figure n3

POLARIZATION FORK

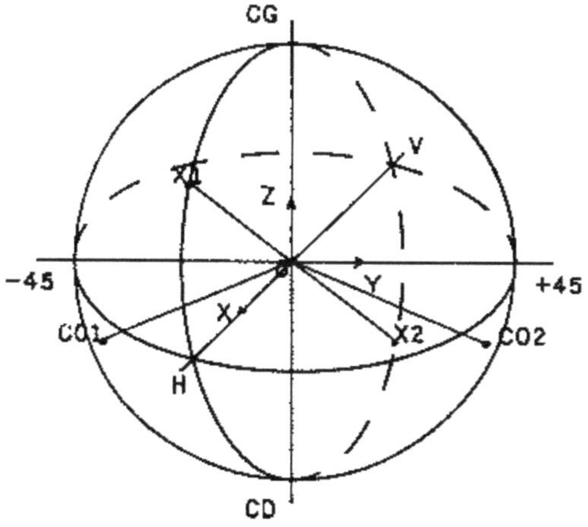

Figure n4

EXAMPLE

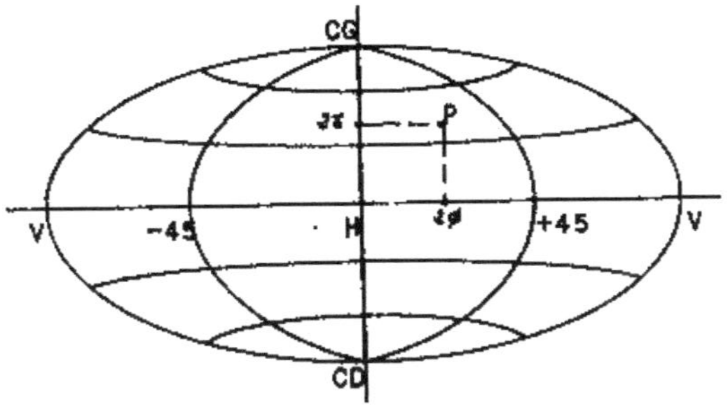

Figure n5

POINCARE's PLANISPHERE

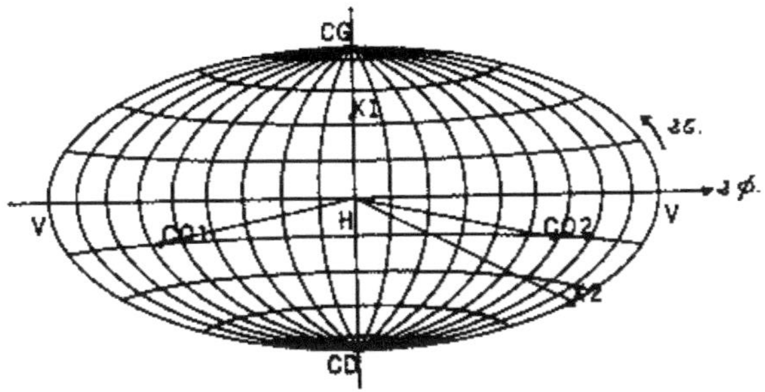

Figure n6

POLARIZATION FORK ON

POINCARE's PLANISPHERE

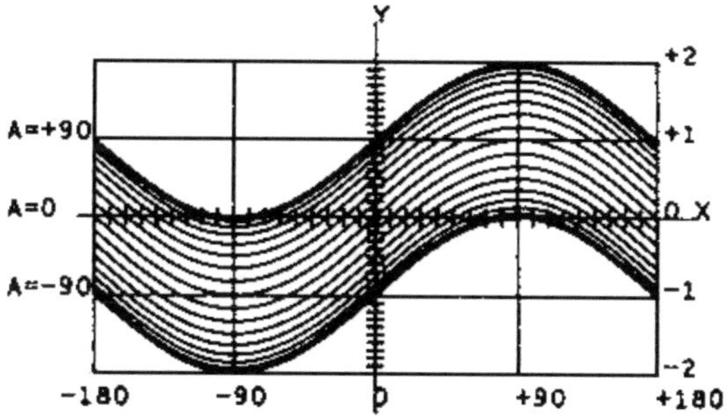

Figure n7

FUNCTION Y=SIN (A) +SIN (B)

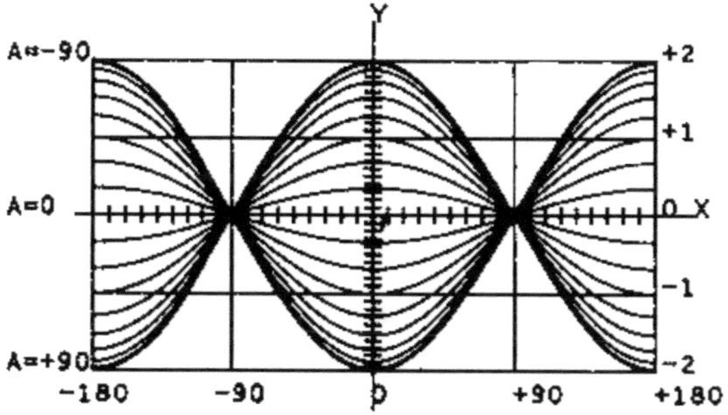

Figure n8

FUNCTION Y=2*SIN (A) *COS (B)

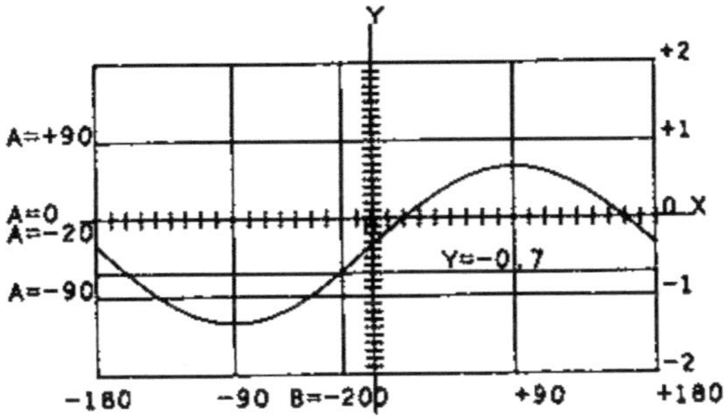

Figure n9

FUNCTION Y=SIN (A) +SIN (B)

(EXAMPLE)

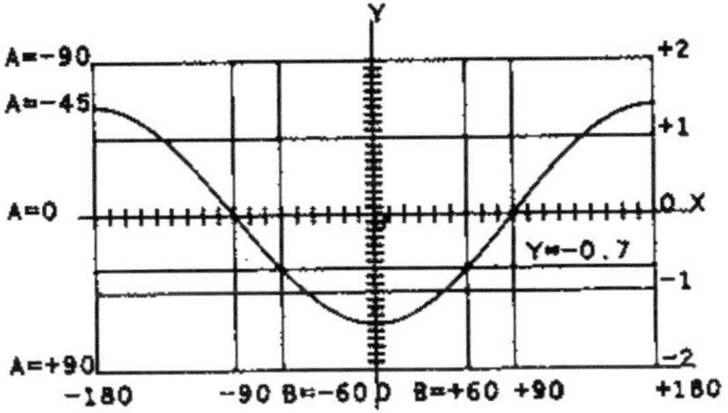

Figure n10

FUNCTION Y=2*SIN (A) *COS (B)

(EXAMPLE)

RCS HYPERRESOLUTION
in using spheroïdal functions

Joseph SAILLARD
Laboratoire de Communications Détection Systèmes
IRESTE
Institut de Recherche et d'Enseignement Supérieur
aux Techniques de l'Electronique
La Chantrerie - CP 3003
44087 NANTES Cédex 03
Tél. (33) 40.68.30.64

ABSTRACT

The objective of this paper shall be to show it is possible to increase discrimination from the same signal samples. This method has been inspired by that proposed by Thomson |1|. It is known that by Fourier transform the normalized resolution is 1/n where n stands for the number of samples. The new method gives us a resolution equal to 2/N, N being the size of fast Fourier transform.

1 - INTRODUCTION

For a set of fixed polarisation, one considers a target illuminated by a monochromatic wave of frequency f. The analysed band is situated between f_1 and f_2, the frequential increment is δf. One measures and records the ratio of diffracted and incident wave in amplitude and phase. It is assumed that the target illuminated by a plane wave, can be represented by N elementary contributors.

The diffracted field H(f) can be written as follows eqn 1 :

$$H(f) = \sum_{i=1}^{N} |a_i| \, e^{j\phi_i} \, e^{j4\pi\delta_i f/c}$$

One must be estimated the number N of contributors, the distance projected δ_i on the target-radar direction and the complex amplitude of every contributor i.e. $|a_i|$ and ϕ_i as accurately as possible.

It has been proved |2| that it is difficult to obtain completely accurate on all the parameters, if satisfactory precision is obtained on one of them this induces an error on the others. Further more, when the distance between the contributors is much greater than the wavelength, discrimination is possible |3|. But if this condition is not verified discrimination becomes impossible.

2 - DISCRETE PROLATE SPHEROIDAL WAVE FUNCTIONS

Many papers, in particular those by Thomson |1| and Slepian |4| show the interest in using eigenfunctions of Dirichlet kernel, better known as discrete prolate spheroidal wave functions and denoted as U_k (N,W,f).

k = 0, 1,...N-1. They are solutions of the equation (2) :

$$\int_{-w}^{+w} \frac{\sin N \pi (f - f')}{\sin \pi (f - f')} \cdot U_k(N,W;f')\, df' = \lambda_k(N,W) \cdot U_k(N,W;f)$$

where w is the bandwidth (0<w<1/2) and here w = 1/N.

Generally, these functions are ordered by their eigenvalues.

$$1 > \lambda_0(N,W) > \lambda_1(N,W) > \dots\dots\dots > \lambda_{N-1}(N,W) > 0 \quad (3)$$

The first 2 N W eigenvalues are extremely close to 1.

The eigenfunctions U_k(N,W,f) , k = 0,1,...N are doubly orthogonal over (-w,w) and over (-1/2,1/2).

Consequently, equations (4) and (4 bis) are written as follows :

$$\frac{1}{\lambda_k(N,W)} \int_{-w}^{+w} U_j(N,W;f) \cdot U_k(N,W;f)\, df = \delta_{j,k} \quad (4)$$

$$\int_{-1/2}^{1/2} U_j(N,W;f) \cdot U_k(N,W;f)\, df = \delta_{j,k} \quad (4\ \text{bis})$$

The Fourier transforms of the discrete prolate spheroidal wave functions are known as discrete prolate spheroidal sequences. |1|

3 - PROCESS

It is assumed that the signal complies with Expression number 5 :

$$y(f) = \mu \, \delta(f - f_o) \qquad (5)$$

One analyses this signal by using spheroidal windows. At f_0 frequency one obtains equation 6 for the filter of order k.

$$y_k(fo) = \mu \, U_k(N,W;0) \qquad (6)$$

One chooses K filters, o < k < K-1, by standard regression methods, the mean μ is given by equation 7 :

$$\mu = \frac{\sum_{k=0}^{K-1} U_k(N,W;0) \, y_k(f_o)}{\sum_{k=0}^{K-1} U_k^2(0)} \qquad (7)$$

One also assumes a spectral noise b(f) is added to the former signal. Every y_k (f) can be considered as an element of k orthogonal basis of all the others. In this new approximation K orthogonal functions space, one defines an approximation μ (f) of order K for μ (f) and equation 8 is obtained :

$$\mu(f) = \frac{\sum_{k=0}^{K-1} U_k(N,W;0) \, y_k(f)}{\sum_{k=0}^{K-1} U^2{}_k(0)} \qquad (8)$$

The approximation μ (f) is optimal for the least square mean the Fisher test |5| allows us to give a value similar to the existence probability of frequency line f. (9) :

$$F_1(f) = \frac{\text{Estimated Power Line Value}}{\text{Difference between real and estimated power line value}}$$

If the signal is slightly noised $F_1(f_o)$ is very important, but the smaller $F_1(f)$ is, the smaller its existence probability is.

A block diagram is represented in figure 1.

One notices that Uk (N,W:0) = 0 for k odd.

Now one assumes a signal complies with the following expression (10) :

$$y(f) = \mu\, \delta(f - f_1) + \nu\, \delta(f - f_2) + b(f) \qquad (10)$$

we put $d = f_2 - f_1$ $\quad d>0$

the length of the signal is N the expression of a samples is given by (11)

$$y_i = \mu\, s_i + \nu\, t_i + b_i \quad (11)$$

the system can be written as a matrix and Y - μs - νt is minimized in comparaison to s and t, hence :

$$\begin{bmatrix}(Y,t)\\(Y,t)\end{bmatrix} \quad \begin{bmatrix}||s||^2\\(s,t)\end{bmatrix} \quad \begin{bmatrix}(t,s)\\||t||^2\end{bmatrix} \quad \begin{bmatrix}\mu\\ \nu\end{bmatrix} \qquad (12)$$

$$\text{with } (a,b) = \sum_{i=1}^{K} a_i b_i^* \quad \text{and } ||a||^2 = \sum_{i=1}^{K} |a_i|^2$$

Knowing the values of y_i, s_i and t_i, we can calculate μ and ν expressions 13 and 14.

$$\mu = \frac{1}{\Delta}\left\{ ||t||^2 (y,s) - (t,s)(y,t) \right\} \quad (13)$$

$$\nu = \frac{1}{\Delta}\left\{ ||s||^2 (y,t) - (s,t)(y,s) \right\} \quad (14)$$

$$\text{with } \Delta = ||s||^2 - ||t||^2 - (t,s)(s,t)$$

Using the orthogonality properties of prolate functions, an orthogonal basis is established. The equivalents of s and t are the terms U_k (0), ... $U_k(\delta)$. As a result we obtain the expression 15 and 16 .

$$A = \Sigma(U_k(0)\, U^*_k(d))\, \Sigma(y_k(f)\, U^*_k(0))$$
$$B = \Sigma(U_k(d)\, U^*_k(0))\, \Sigma(y_k(f)\, U^*_k(d))$$

$$\mathop{\mu}_{(f,\delta)} = \frac{1}{\Delta}\left\{ A \quad - \quad B \right\} \tag{15}$$

$$C = \Sigma(U_k(0)\ U^*_k(0))\ \Sigma(yk_{(f)}\ U^*_k(d))$$
$$D = \Sigma(U_k(0)\ U^*_k(0))\ \Sigma(yk_{(f)}\ U^*_k(0))$$

$$\mathop{\nu}_{(f,d)} = \frac{1}{\Delta}\left\{ C \quad - \quad D \right\} \tag{16}$$

$$\text{with } \Delta = \Sigma\ |U_k(0)|^2\ \Sigma|U_k(d)|^2 - \Sigma|U_k(0)\ U_k(d)|^2$$

$$\Sigma = \Sigma_{k=1}^{K}$$

The values of $\mu(f_1,f_2)$ and $\nu(f_1,f_2)$ are the researched amplitudes for the data set (f1,f2).

Likewise, a value similar to the simultaneous existence of frequencies f_1 and f_2 is estimated by applying the Fisher test. (17)

$$\mathop{F_2(f_1,f_2)}_{(17)} = \frac{\sum_{k=0}^{K-1} |\mu(f_1,f_2)\ U_k(0) + \nu(f_1,f_2)\ U_k(d)|^2}{\sum_{k=0}^{K-1} |y_k(f) - \mu(f_1,f_2)\ U_k(0) - \nu(f1,f2)\ U_k(d)|^2}$$

The greater the value of F_2 (f_1,f_2), the greater possibility of existence of line Set (f_1,f_2).

The function $F_2(f_1,f_2) = F(f,\Delta f)$ is drawn with $\Delta f>0$.

In the figure 2 , the function $F_2(f,\Delta f)$ is represented using a signal having

$y = \mu\ \delta(f - 0.1953) + \nu\ \delta(f - 0.2031)$ as an expression.

In the case of a single spectrum line the smaller the size of FFT, the more rounded peak F_1 will be, but if the size of FFT is optimal peak F_1 will reach its nominal value and will not increase with the size of FFT.

To find near spectral lines, it is necessary to use the second method. Hence we calculate μ,ν, and $F_2(f_1;f_2)$. The more the FFT size increases, the better the peak separation. One assumes n length of signal and N the FFT size with N>n the new method just described allows us to separate, two frequencies distant of 2/N. A study must be done on the chosen frequency with a drift 5/N.

Having obtained F_1 and F_2, a decision must be taken. Indeed is there a single spectral line or two near spectral lines in the frequency area. To answer one observes the decrease of the peak. The narrower the peak is, the greater the value of existence probability.

In the area studied one calculates for F_1

$$D_1 = \frac{2A_im}{A_{(im-1)} + A_{(im+1)}}$$

and for F_2

$$D_2 = \frac{8\,A_{(im,jm)}}{\sum_{i=1}^{1} \sum_{j=1}^{1} A_{(im+i,\ jm+j)} - A_{(im,\ jm)}}$$

i_m and j_m are the coordinates of the Fisher test is maximum value and A_{im} or $A_{(im,\ jm)}$ the value of the Fisher test.

4 - RESULTS

The work presented above has been done in distance space and not in frequency space.

Two experimental results on a target having three contributors T_1 and another with two contributors T_2.

The analysis band is 8 to 13.2 GHZ by increment of 7MHZ, there are 740 samples for each recording.

On the figure 4 impulsional responses weighted by rectangular window and Hamming window have been drawn for the target T_1. The FFT size is 1024. One line appears about abcissa 0.008 in normalized distance. Furthermore two mixed lines appear about - 0.002 and 0.006. In the figure 5 we show the spectrum calculated with FFT a size of 4096 F_1 and F_2 functions by application of decision criteria. One notes that the target has only one contributor in area 1. The calculated values are $a_1 = 1.8189$, $\phi_1 = -175°$, $d_1 = 6.591 \times 10^{-3}$. For area 2 the spectrum and $F_1$and F_2 function are drawn.

One notes in figure 6 that the target has two contributors, the obtained values being $a_2 = 1.7109$ $\phi_2 = -3.6°$ $d_2 = -1.709 \times 10^{-3}$ $a_3 = 0.4346$ $\phi_3 = -30°$ $d_3 = -4.150 \times 10^{-3}$.

For target T_2 impulsional responses weighted by rectangular and Hamming windows are represented in figure 7. In looking at figures 4 and 7 a shape similar to the negative distances can be noted, the contributor situated about 0.008 is erased ; because it simply has been physically concealed. Further more an offset has been volountarily introduced in the data, the following values $a_2 = 0.2002$, $\phi_2 = -5.76°$, $d_2 = -1.709 \times 10^{-3}$, $a_3 = 0.1520$, $\phi_3 = -34.42$, $d_3 = -4.150 \times 10^{-3}$ have been obtained with the exception of amplitudes results similar to those of target T_1 have been acquired.

5 - CONCLUSION

From the process of real record it has been shown that it is possible to apply this procedure. The accuracy obtained in the results and the discrimination of near spectral lines are the principal merit of this method.

ACKNOWLEDGMENTS

The author is grateful to G. Bunel for carrying out the computations and ASRE department CELAR for help and financial support.

REFERENCES

1 - THOMSON D.
"Spectrum Estimation and harmonic analysis"Proc. IEEE, vol. 70, n°9, p. 1055-1096, (1982).

2 - MAX J., DIOT M., BIGRET R.
"Les analyseurs de spectre à FFT et les analyseurs de spectre à corrélation". Traitement du Signal, volume 3, n° 4-5, p. 241-256, (1986).

3 - SAILLARD J., COATANHAY J.L., GADENNE Ph.
"Extraction des caractéristiques d'une cible radar à partir d'une analyse fréquentielle. Extension de la méthode". 11ème Colloque GRESTI Nice,p. 345-348, (juin 1987)

4 - SLEPIAN D.
"Prolate spheroïdal wave functions, Fourier analysis and uncertainty. The discrete case". Bell System Technical Journal, vol. 57, p. 1371-1429, (1978).

5 - HAFNER
"Statistical methods for research workers". New-York, (1973).

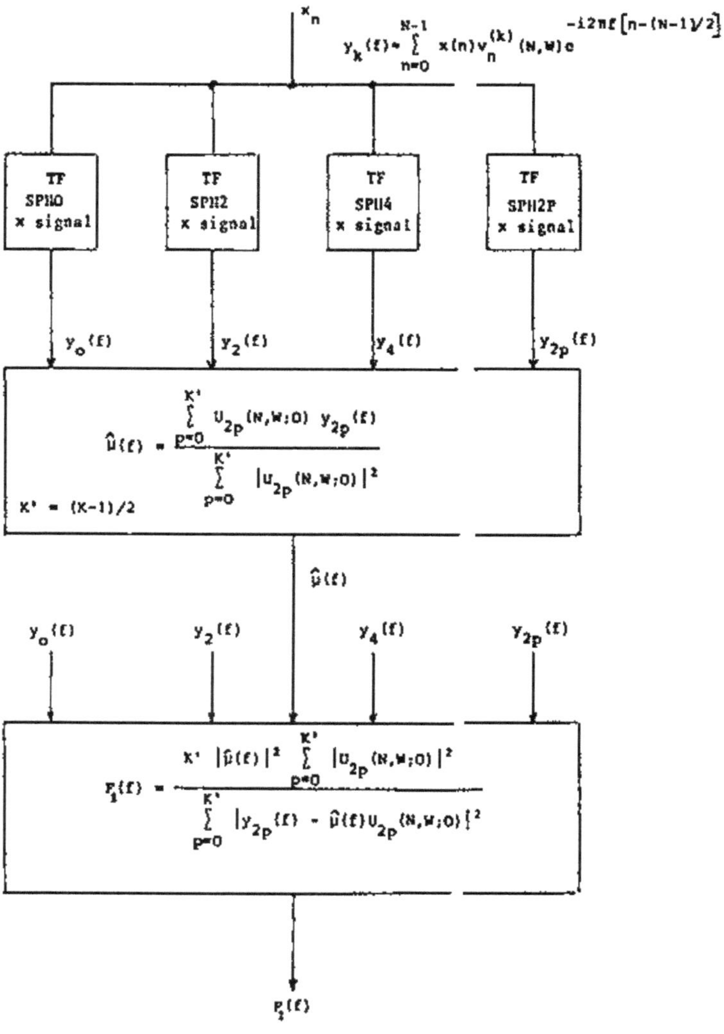

Figure 1 schematic diagram of process F_1

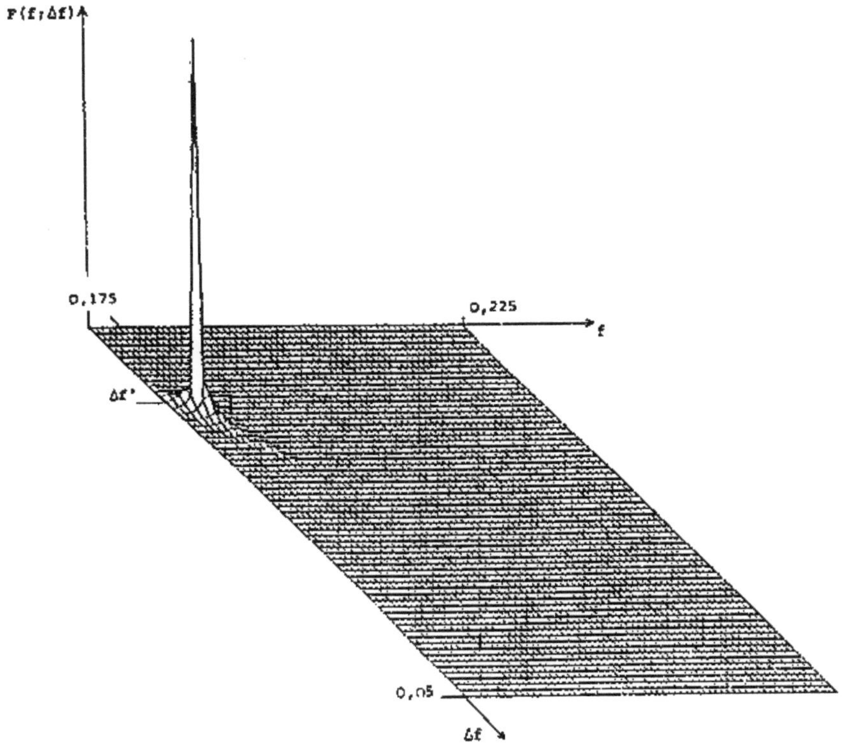

figure 2   Example of drawing F_2

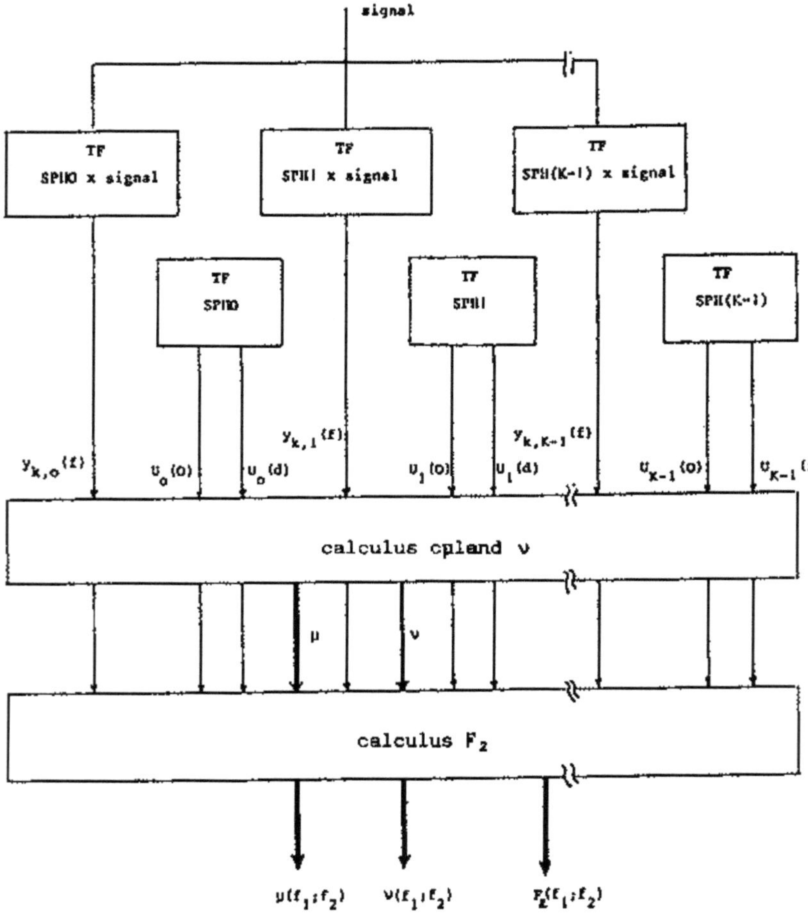

Figure 3 schematic diagram of process F_2

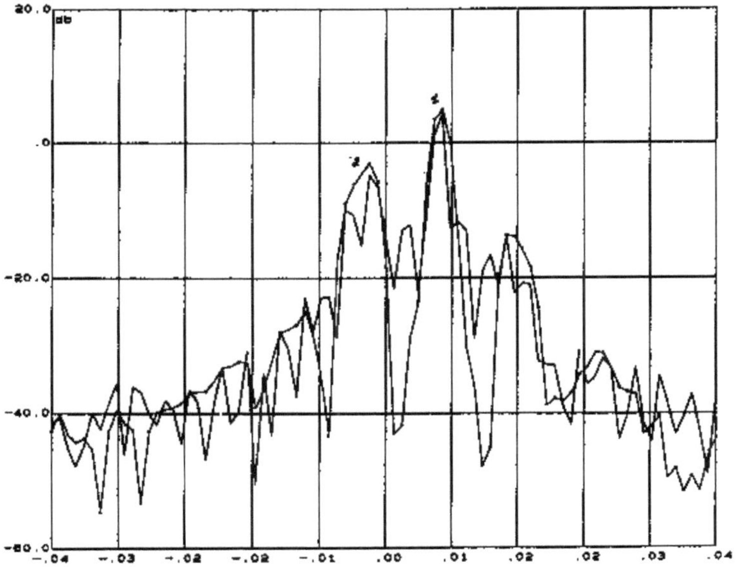

Figure 4 Target T_1 spectrum

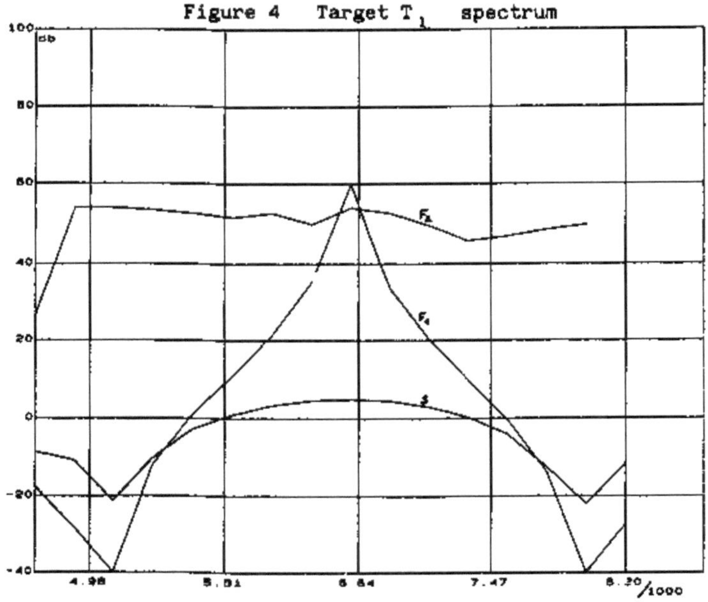

Figure 5 F_1, F_2, S as an function of normalized distance of area 1

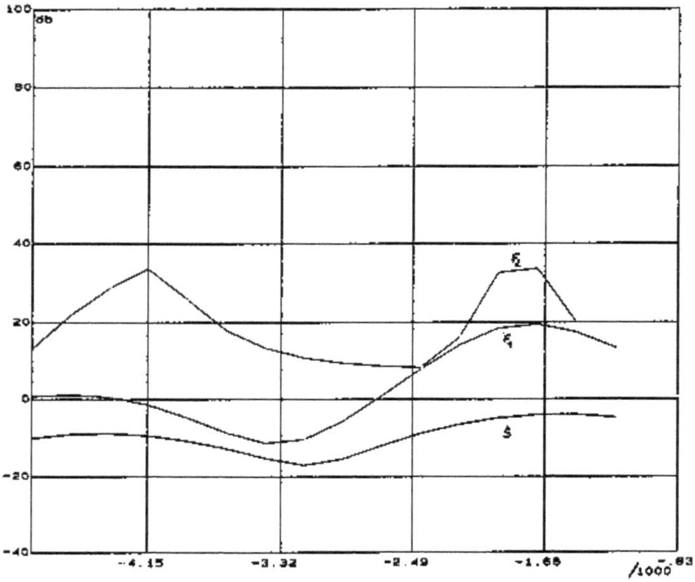

Figure 6 F_1, F_2, S as function of normalized distance for area 2

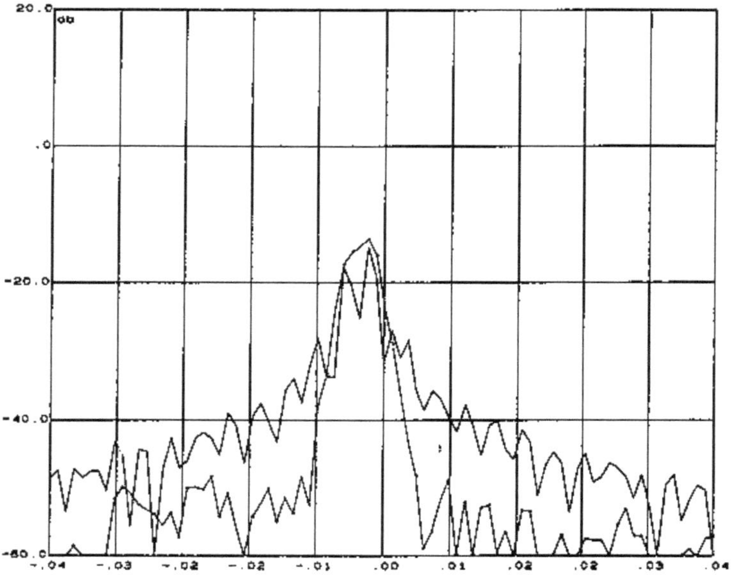

Figure 7 Target T_2 Spectrum

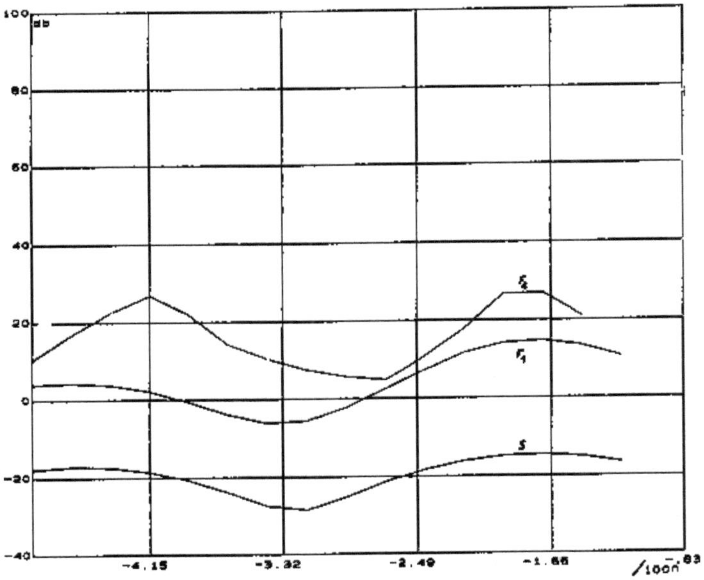

Figure 8 F_1 , F_2, S as function of normalized distance for maximal area

CONSTRUCTIVE GROUP ARGUMENTS IN MICROWAVE IMAGING

P.BERTRAND* and J.BERTRAND**

* ONERA-B.P. 72 - 92322 CHATILLON - France
** University Paris VII - LPTM - F 75251 PARIS

Reflectivity studies in radar imaging are usually dependent on the white isotropic bright point concept. By group analysis it is possible to obtain new imaging techniques by susbstitution of geometrical arguments.

1. INTRODUCTION.

Microwave imaging , also called radar imaging, is a technique of signal processing which was brought into existence by the development of coherent radars. Its object is to give a spatial description of target reflectivity in relation with the common experience of visual perception. The many technical and theoretical aspects of this topic are described in survey works[1-4] referring to a vast literature.

In a synthetic presentation, radar imaging may be related to an extension of holographic concepts beyond their optical basis. Following this way it must however be stressed that such an extension does not refer only to a broadening of the spectral domain but also to the irruption of specific operating performances of the microwave technique. This latter point is illustrated by recalling that radar observation allows direct resolution in the radial direction by using wide signal bandwidths. In fact, at the time being, microwave holography must be considered as a nontrivial generalization of the optical one. An effective version of this remark is given by considering the outcome of microwave imaging which consists mainly of generalized images without classical optical equivalents. The purpose of this paper is to stress the interest of transformation group studies for the construction and interpretation of such artificial images.

In Sec.2 the laboratory method of target imaging is recalled. As is well known the obtained reflectivity corresponds to a model of target built with white isotropic bright points. In Sec.3 a group charaterization of the method is given which permits to forget the basic model. At this point it becomes possible to obtain a new imaging process founded on a geometrical group only.

2. ROTATING PLATFORM OBSERVATION OF RADAR TARGETS AND ASSOCIATED IMAGING.

Laboratory studies of radar targets make generally use of a monostatic coherent radar operating as shown on Fig.1. In a simple typical application the target rotates around a fixed axis perpendicular to the radar line of sight and polarization dependence is disregarded. The acquired data are values of the reflection coefficient for different orientations of the target and different frequencies of irradiation. From a theoretical point of view the experiment is thus delivering samples of a complex function $H(f,\theta)$ defined on a plane where the frequency f and the orientation θ are polar coordinates of a point. This function is related to the scattering transformation of continuous waves by the relation

$$e^{2i\pi ft} \longrightarrow H(f,\theta)e^{2i\pi ft} \qquad (1)$$

Such a "one-axis" experiment permits only to characterize a projection of the target reflectivity on a plane perpendicular to the axis of rotation. As a result, the imaging problem is reduced to that of expressing the reflectivity of a planar object as a functional of the reflection coefficient $H(f,\theta)$.

The notion of reflectivity is associated to the modeling of targets by punctual reflectors which are all introducing the same phase factor at reflection. In the classical imaging these reflectors are moreover supposed to be white and isotropic. It is in fact difficult to imagine that a real opaque target can be well represented by a constellation of such non-directive and non-dispersive bright points! However, if the observations are limited to small intervals for both variables f and θ, the white and isotropic feature becomes acceptable and actually the model leads to relevant images.

The reflectivities met in radar applications correspond to experiments in which the distance of the target is much larger than the size of the target. In this case the range determination of the bright points may be founded on the simple fact that the reflexion coefficient (1) of a white mirror is:

$$H(f) = e^{i\phi}\, e^{-4i\pi f(x/c)} \qquad (2)$$

where ϕ is a phase and c the light velocity. In effect, for any θ_0 the reflection coefficient $H(f,\theta_0)$ can be represented as a sum of elementary contributions of type (2) by the formula:

$$H(f,\theta_0) = \int_{-\infty}^{+\infty} e^{i\phi}\, e^{-4i\pi f(x/c)}\, \rho(x,\theta_0)\, dx \qquad (3)$$

where $\rho(x,\theta_0)$ is given by:

$$\rho(x,\theta_0) = (2/c) \int_{-\infty}^{+\infty} e^{-i\phi}\, e^{4i\pi f(x/c)}\, H(f,\theta_0)\, df \qquad (4)$$

Taking the square of the modulus of $\rho(x,\theta_0)$ we obtain a positive function independent of the arbitrary phase ϕ. This function is interpreted as a one-dimensional reflectivity associated with a given orientation of the target. It depends on the frequency band of observation which acts as a window in the integral of equation (4). A high degree of range resolution is obtained by a broad-band processing and so requires a large use of the whiteness of the bright points. This remark exhibits the general trade-off between distance and color resolutions of the elementary reflectors. A similar trade-off exists between cross-range and angular resolutions. It is well illustrated by the so-called Doppler analysis which is based on the study of the signal reflected by a rotating target. If the irradiating field is a continuous wave of frequency f_0, this signal is obtained by an adaptation of (1) which can be written:

$$e^{2i\pi f_0 t} \longrightarrow H(f_0, \Omega t)\, e^{2i\pi f_0 t}$$

where Ω is an angular velocity. In this transformation, a monochromatic signal does not remain monochromatic and the spreading of the spectrum can be connected to the radial velocities of the bright points in the radar reference system. This leads to a cross-range resolution as illustrated on Fig.2. The operation can be achieved by use of a standard time-frequency representation of the reflected signal[5]. The two methods of range resolution and cross-range resolution can be combined in a general method called range-Doppler imaging. The main defect of this method is its flagrant limit of applicability due to the variation of the instantaneous frequencies when processing on large angles. This difficulty does not appear in the "tomographic" method which is a two-dimensional generalization of the above range determination. To introduce it we return to (4) and interpret the function $\rho(x,\theta)$ as the integral along isorange lines of a bidimensional complex amplitude $R(\vec{x})$ for punctual reflectors. If such integrals are known for all orientations θ, the function $R(\vec{x})$ can be obtained by inversion of a Radon transform. In the far-field approximation the isorange lines are straight lines perpendicular to the radar line of sight and the Radon transform is the usual one. The result is:

$$R(\vec{x}) = \int_{\mathbb{R}} e^{-i\phi} H(\vec{k}) e^{2i\pi\vec{k}.\vec{x}} d\vec{k} \quad (5)$$

where $\vec{k}$ defines the point with polar coordinates $(2f/c,\theta)$.

When isorange lines are circles centered on the radar the Radon inversion is also possible [6].

3. GROUP THEORETICAL APPROACH TO REFLECTIVITY.

In the search for an expression of the reflectivity we want to find a sort of density of bright points in a general space $(\vec{x},\vec{k})$ which can be interpreted as a position-color-directivity space. The problem of expressing this density as a functional of the reflection coefficient $H(\vec{k})$ is in fact vey similar to that of exhibiting phase space representations in quantum mechanics or in signal theory.

A possible $(\vec{x},\vec{k})$ description adapted to the white and isotropic bright points model is given by an application of the Wigner function to the imaging problem which reads[7]:

$$D(\vec{x},\vec{k}) = \int e^{-2i\pi\vec{u}.\vec{x}} H^*(\vec{k}+ \frac{\vec{u}}{2}) H(\vec{k}- \frac{\vec{u}}{2}) d\vec{u} \quad (6)$$

This function possesses the properties:

i) $\int D(\vec{x},\vec{k})\, d\vec{k} = |R(\vec{x})|^2$

ii) $\int D(\vec{x},\vec{k})\, d\vec{x} = |H(\vec{k})|^2$

iii) $H(\vec{k}) = e^{-2i\pi\vec{k}\cdot\vec{x}_0} \longrightarrow D(\vec{x},\vec{k}) = \delta(\vec{x}-\vec{x}_0)$

Relation i) expresses that the density of bright points, when disregarding their color and directivity is given by the squared modulus of $R(\vec{x})$ (i.e. equation (5)). In the same way relation ii) gives the radar cross-section as a density of bright points when disregarding their position. Relation iii) expresses the perfect localization of the white isotropic bright points.

The general form (6) is consistent with the interpretation of the transformation

$$H(\vec{k}) \longrightarrow H'(\vec{k}) = e^{-2i\pi\vec{k}\cdot\delta\vec{x}}\, H'(\vec{k}-\delta\vec{k}) \qquad (7)$$

as a translation $(\delta\vec{x},\delta\vec{k})$ in the $(\vec{x},\vec{k})$ space. The space translations $\delta\vec{x}$ of the points have a clear physical interpretation which, in the far field approximation, does not require any color or directive property for the reflectors. At the opposite the consistency with translations in the $\vec{k}$ subspace has no physical interpretation and results only from the white and isotropic character of the bright points. Among consequences it implies a bad localization for a perfect (non bireflecting) mirror as illustrated by Fig. 3. This remark leads to the conclusion that no reflectivity expression can be acceptable for description of opaque targets and at the same time be covariant by the entire group(7).

To solve the above difficulty we can try and substitute a more physical basic group to (7). In a first step we limit ourselves to one-dimensional situations where the reflection coefficient is a function H(f)defined for f>0. This function can be transformed by dilation of the form:

$$H(f) \longrightarrow H'(f) = H(af) \qquad (8)$$

with a>0. By simple dimensional analysis we can find that the transformed reflection coefficient H'(f) corresponds to a target with bright points transported to a distance multiplied by the factor a and with frequency characteristics simultaneously divided by the same factor(Fig.4). The dilations (8) can be combined with space translations of the reflectors. This leads to a representation of the two-dimensional affine group of the form:

$$H(f) \longrightarrow H'(f) = e^{-4i\pi(f/c).\delta x}\, H(af) \qquad (9)$$

In the distance-frequency half-plane this group must be represented by:

$$D(x,f) \longrightarrow D'(x,f) = a^{-1}\, D(xa^{-1} - \delta x,\ af) \qquad (10)$$

It can be verified that this transformation preserves the relation:

$$\int D(x,f)\, dx = |H(f)|^2$$

Expressions for the function D(x,f) may be obtained by simple adaptations of forms built for time-frequency representations of signals[8)9)10)]. When applied to the reflection coefficient:

$$H(f) = \begin{cases} e^{-4i\pi f(x_0/c)} & \text{for } f>0 \\ 0 & \text{for } f<0 \end{cases}$$

they lead to the safisfactory result:

$$D(x,f) = \delta(x - x_0)$$

A positive regularized form exists and example of application is given on Fig.5. It concerns the observation of a model of aircraft. For comparison the regularized form of the Wigner function corresponding to the same target is given on Fig.6.

4. CONCLUSION.

The above developments have shown the importance of the translation-dilation group for reflectivity description of one-dimensional targets. To progress towards higher dimension we have to consider general space translations, general dilations and rotations. This suggests that the euclidean group with dilations can play a central role in general imaging.

REFERENCES

1) Ausherman, D.A., Kozma, A., Walker, J.L., Jones, H.M. and Poggio, E.C. "Developments in Radar Imaging", IEEE-AES 20, no5, 363-400 (1984).
2) Mensa, D.L. "High Resolution Radar Imaging" Artech House 1981.
3) Wehner, D.R. "High Resolution Radar" Artech House 1987.
4) Langenberg, K.J. "Applied Inverse Problems for Acoustic, Electromagnetic and Elastic Wave Scattering" in "Tomography and Inverse Problems" Ed. P.C. Sabatier, Adam Hilger (1987)

5) Bertrand, P. and Fugier-Garrel, C. "Formulation de la théorie de la commmunication dans le plan temps-fréquence - Aspects pratiques" Colloque GRETSI (1981).

6) Norton, S. J. "Reconstruction of a two-dimensional reflecting medium over a circular domain: Exact solution" J. Acoust. Soc. Am. 67(4), 1266-1273 (1980).

7) Bertrand, P. "Méthodes et limitations en imagerie radar des cibles" Revue du CETHEDEC no 76, 107-118 (1983).

8) Bertrand, J. and Bertrand, P., C.R. Acad. Sc. Paris 299, Série I 635-638 (1984).

9) Bertrand, P. and Tardivel, P. "Représentation temps-fréquence des signaux et transformations affines sur le temps. Application à un problème d'imagerie radar" Colloque GRETSI (1985).

10) Bertrand, J. and Bertrand, P. "Time-frequency representations of broad-band signals" International Conference on Acoustics, Speech and Signal Processing, New-York (1988).

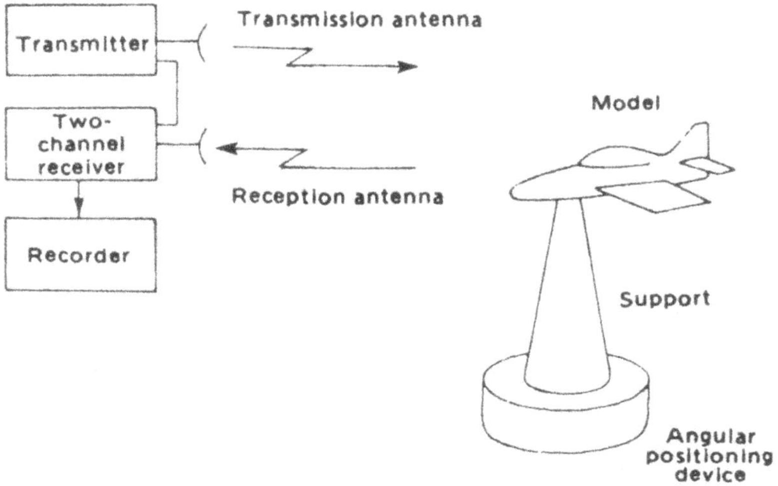

Fig.1-Data acquisition in laboratory.

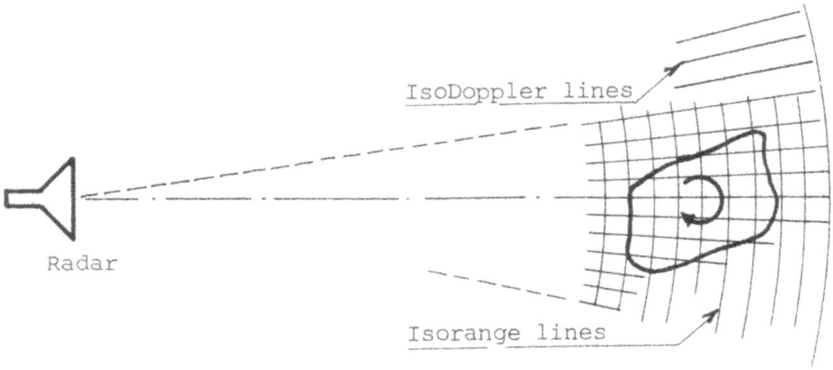

Fig.2-Principle of range-Doppler imaging.

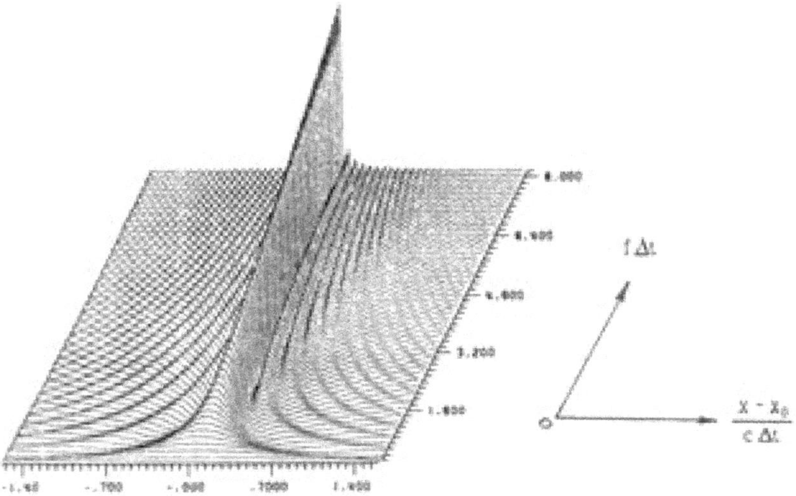

Fig.3-Wigner distance-colour representation of a perfect mirror.

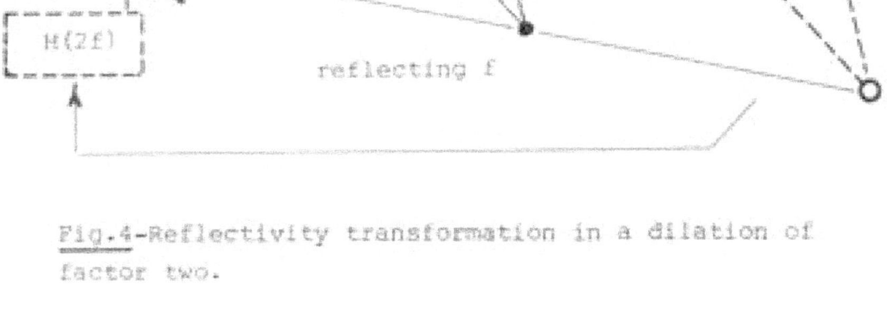

Fig.4-Reflectivity transformation in a dilation of factor two.

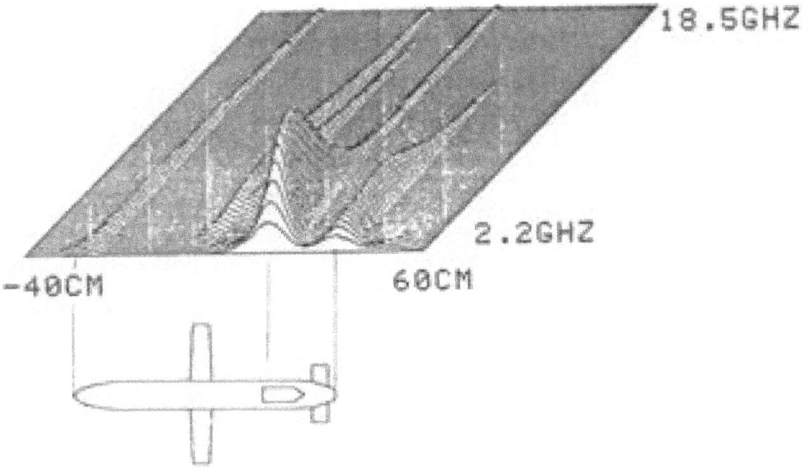

Fig.5-Distance-colour imaging by an affine-displaced window.

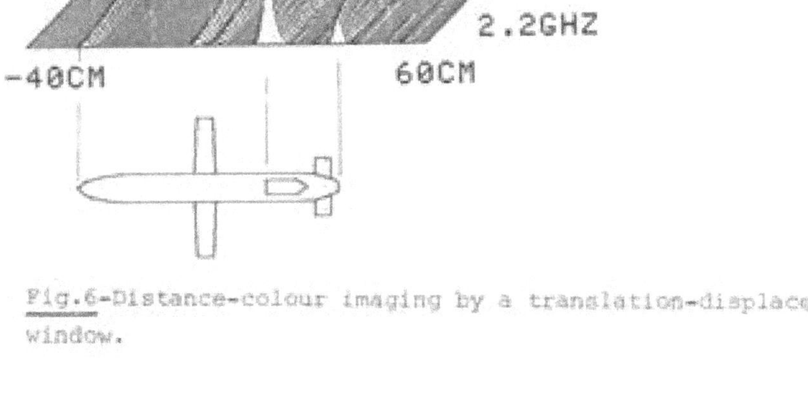

Fig.6-Distance-colour imaging by a translation-displaced window.

MEASURE OF THE SPECTRA OF THE TURBULENCE OF A PLASMA BY COHERENT SCATTERING

Jack OLIVAIN
CEN, Cadarache
13108 SAINT PAUL LEZ DURANCE

No written contribution received.

PROPAGATION IN RANDOM AND INHOMOGENEOUS MEDIA

ELECTROMAGNETIC AND ACOUSTIC WAVES IN RANDOM MEDIA FROM PROPAGATION TO ANDERSON LOCALIZATION

Bernard SOUILLARD
Centre de Physique Théorique
Ecole Polytechnique
91128 PALAISEAU

No written contribution received

WEAK DISORDER : HOMOGENEISATION METHOD AND PROPAGATION

Claude BARDOS
Centre de Mathématiques Appliquées
Ecole Normale Supérieure
Rue d'Ulm
75005 PARIS

No written contribution received

COHERENT PROPAGATION OF SURFACE ACOUSTIC WAVES IN QUASI-PERIODIC AND RANDOM ARRAYS OF GROOVES.

Louis MACON, Didier SORNETTE and Jean COSTE
Laboratoire de Physique de la Matière condensée
CNRS UA 190, Faculté des Sciences, Parc Valrose
06034 NICE Cedex, FRANCE

ABSTRACT : We present new experiments on the propagation of waves in 1D-inhomogeneous media, namely surface acoustic waves propagating through quasiperiodic or random arrays of parallel grooves (etched at the surface of piezoelectric crystals). Two competing coherent phenomena are involved : i) the interferences between all the multiply scattered waves at the corrugated surface of the crystals, which are responsible of the pure 1D-transport properties, ii) the coherent leakage due to surface-to-bulk radiation at each groove. After a brief exposition of a transfer matrix approach which takes into account both effects, we recover the anihilation of coherent losses at the frequency band-edges of periodic arrays and discuss the competition between localization and these losses in random arrays. Then, we show experimentaly, on a first quasicrystal of grooves, that the interesting features of the quasi-periodic systems can be experimentaly attainable. Finally, we show that quantitative predictions concerning a well-studied quasiperiodic system are confirmed and, notably, the regular structure of a quasi-localized proper-mode appearing in a very small frequency pass-band.

1-INTRODUCTION

Quasi-periodic or strongly disordered arrays of parallel grooves, etched at the surface of piezo-electric crystals and excited by surface acoustic waves (SAW) provide today one of the best experimental model-systems for studying the transport properties of such structured 1D-lattices (for 2D-studies, grooves could be replaced by dots).

Although the propagation of waves in random media has been extensively studied in the theoretical litterature [1-4], few experimental evidence is available concerning some coherent features of energy transport in these systems; for instance, the properties related to the "Anderson localization regime". This non intuitive regime relies on the subsistance of coherent phenomena, involving interference effects between multiply scattered waves by the inhomogeneities of quenched strongly disordered systems. These effects, generally neglected, were primarily proposed by Anderson [2] to explain anomalous conductivity in dirty metals and the existence of a metal-insulator transition in 3D-systems, depending on the amount of disorder.

Unfortunately, in the area of experimental solid-state physics (electronic waves) many competing effects are implied (Coulomb, spin-spin interactions...) and the main characteristic quantities are not directly attainable; thus, the expected phenomena have been only indirectly exhibited. Since Anderson localization has essentally an undulatory origin, people turned to classical contexts (acoustics, optics...). Among all experimental alternatives using "classical waves", SAW experiments appear to be good candidates in one and even two dimensions (figure 1-a). They allow to reach very accurately all the interesting physical properties : spectral dependance of the reflection and transmission coefficients (figure 1-b, and also figure 3), local structure of the proper-modes in the pass-bands and the temporal impulse and white noise responses (for instance, modal structure is obtained directly by a laser diffraction experiment shown in figure 1-c : a laser beam is scanned along the array and is locally diffracted by the harmonic deformations of the crystal surface due to the harmonic travelling SAW); other crucial advantages of these experiments are the possibility of engraving with accuracy a great number of scatterers (up to 10000 grooves) and the relatively weak dissipation and non linear effects.

For strongly disordered systems, we expect pure-point spectra (pass-bands), with localized associated proper-modes (i.e. roughly exponentialy decaying from some origin inside the system with a fractal fine structure), long temporal impulse responses and 1/f colouring of white phase noise [4].

Otherwise, there was recently a reviving interest in the quasi-periodic systems with, in particular, the discovery of "fifth-order symmetry crystals" by Shechtman et al. [5]. These systems can be characterized by the coexistence of several periodic properties with incommensurate periods. A great variety of spectral, modal and temporal behaviours are expected and some of these systems can be considered, from many respects, as intermediate between periodic and random ones [6-12].

Experiments on random arrays of grooves are developped at the present time, but for the moment and for sake of simplicity, we have begun our experiments on "quasiperiodic" arrays of grooves and already obtained some stricking results. A preliminary presentation of these experiments is the goal of this paper.

In such systems, two types of coherent effects are involved : i) the interferences between all the multiply scattered waves at the corrugated surface of the crystals, which are responsible of the pure 1D-transport properties and ii) the coherent leakage due to surface-to-bulk radiation at each groove as it does occur in all guided situation in presence of inhomogeneities. First, we present a synthetic transfer matrix formalism [13] which integrates all these coherent effects (section 2-a), then we apply it to the classical case of periodic arrays of grooves and recover the important effect of the anihilation of the coherent losses near the band-edges of a periodic system (section 2-b), which explains the great quality factors of the SAW resonators. We also discuss the competition between Anderson localization and the coherent leakage in a disordered array (section 2-c); this discussion covers the less critical case of quasiperiodic systems. It shows that near the band-edges of a periodic system and in presence of disorder, the "attenuation length" due to the leakage is much larger than the localization length, allowing then the observation of the Anderson localization phenomenon in these frequency regions. In section 3, we discuss some experimental results which fully confirm that the most spectacular localization effects, which take place in the finest pass-bands of quasiperiodic lattices, are fully preserved in these SAW modelling systems. Section 4 shows, briefly, that all quantitative predictions concerning a well-studied quasiperiodic system have been confirmed and, notably, the regular structure of a quasi-localized proper-mode appearing in a very fine frequency pass-band. In conclusion (section 5), we number some possible future applications of such quasiperiodic or disordered systems.

2 THEORETICAL CONSIDERATIONS.

We address here the general problem of the competition between 1D-wave propagation and coherent leakage in the bulk. Experimentally, no real 1D-system exists rigorously; 1D-scattering is often bound to lead to a coupling with bulk radiation modes in addition to the pure 1D-reflection. The interest in this problem is also suggested by the practical fact that guided waves are used to develop devices with many industrial applications [14].

We have studied a macroscopic model and shown how to incorporate the specific coherent nature of the guided wave to bulk radiation leakage in the usual transfer matrix formalism of 1D-wave propagation. From space reversal symmetry and a local energy flux conservation, we obtain the general form of the single symmetric scatterer transfer matrix in terms of four independent real parameters (compared to two parameters for symplectic matrices describing space-symmetrical scatterers without leakage corresponding to the pure one dimensional (1D) case).

We present in 2-1 the transfer matrix formalism, apply it to periodic systems in 2-2 and discuss the competition between coherent leakage and Anderson localization in 2-3.

2-1 Transfer Matrix Formalism.

Since 1D-systems are always topologically ordered [1], the equations for wave propagation in any 1D-system at any energy or frequency can be cast into essentially the same transfer matrix formalism [1]. To be concrete, let us consider a wave amplitude at point x_0 of the general form:

$$Y_1(x) = Y_1^+ e^{ik(x-x_0)} + Y_1^- e^{-ik(x-x_0)} \quad \text{before the scatterer,}$$

and :

$$Y_2(x) = Y_2^+ e^{ik(x-x_0)} + Y_2^- e^{-ik(x-x_0)} \quad \text{after the scatterer.}$$

We look for the general expression of the matrix T defined by :

$$\begin{pmatrix} Y_2^+ \\ Y_2^- \end{pmatrix} = T \begin{pmatrix} Y_1^+ \\ Y_1^- \end{pmatrix} = \begin{bmatrix} a & b \\ c & d \end{bmatrix} \begin{pmatrix} Y_1^+ \\ Y_1^- \end{pmatrix} \tag{1}$$

In absence of leakage, symmetry under time reversal imposes $d=a^*$ and $c=b^*$ [15]. In our case, this symmetry does not hold anymore. In the following, we assume that the scatterers have a center of symmetry. This leads to $c=-b$ and Det $T=1$ which implies $d=(1-b^2)/a$.

The bulk wave radiated in the direction characterized by the polar angle θ is described by the amplitude $Z(\theta)$ [13]. Due to the linearity of the equation of propagation, we write, without loss of generality, the following linear relation between $Z(\theta)$, Y_1^+ and Y_2^- :

$$Z(\theta) = e(\theta)\, Y_1^+ + e'(\theta)\, Y_2^- \tag{2}$$

Eq.(2) is the new ingredient characteristic of the leakage problem which writes that the total bulk radiation is the sum of the scattering of the two 1D-waves, Y_1^+ and Y_2^-, which are incident on the scatterer. Eq.(2) completely encaptures the role of interference effects on the coherent radiation of 1D-waves since it involves waves amplitudes.

Writting the conservation of energy, i.e. the energy flux incident upon a scatterer is equal to the energy flux outgoing plus the total radiated energy flux, we obtain a matrix equation on T in terms of the two parameters describing the leakage of the guided wave into the bulk :

$$p_1 = \int_0^{\pi} |e(\theta)|^2 d\theta \qquad\qquad p_2 = \int_0^{\pi} e(\theta) e^*(\pi-\theta) d\theta$$

With two complex parameters a, b and two real numbers p_1 and p_2, the matrix equation is found to degenerate into two real equations which determine completely p_1 and p_2 in terms of the coefficients of the transfer matrix (say a and b). Therefore, space symmetry and the conservation of energy flux let free four independent real parameters (a,b).

An additional constraint (inequalities) apply on the four free parameters since one must write that the bulk radiation energy is also positive. This leads to two cases :

1) the parameters a and b are not related by the expression :

$$a = (1+b)\, e^{i\phi} \qquad (3)$$

Then, it is easy to show that the energy radiated in the bulk is always positive and never vanishes. In this case, the energy loss per scatterer is of order p_1 in all situations : this is similar to a wave propagation in presence of a dissipation.

2) A more interesting case happens when eq.(3) holds. Then, ϕ is real (it is interesting to note that this recovers the case studied in ref. [16] within a coupled mode theory). Eq.(3) implies:

$$p_1 = p_2 = -(b+b^*)/|1-b|^2 = -2\xi/|1-b|^2 \text{ and } d = (1-(\pm)b)\, e^{-i\phi}$$

and we have three free parameters $b = \xi + i\mu$ and ϕ. In a periodic lattice of scatterers, ϕ can integrate the dephasing over one spatial period. Typically, one has $\xi \sim p \sim \mu^2$, where μ is the amplitude reflection coefficient per scatterer.

2-2 Case of a Periodic Array of Scatterers.

Consider a periodic lattice of such identical scatterers with period Λ. The general wave structure in the nth unit cell is [17] :

$$Y_n(x) = \{Y_0^+ e^{ik(x-n\Lambda)} + Y_0^- e^{-ik(x-n\Lambda)}\}\, e^{iK(x-n\Lambda)}\, e^{-iKx}$$

with K given by $s=e^{\pm iK\Lambda}$ and s verifies the secular equation :

$$s^2 - (a+d)\, s + 1 = 0$$

leading to : $\quad K\Lambda = \cos^{-1}\{(a+d)/2\}.$

The Bloch equations lead to $Y_0^+ = b$ and $Y_0^- = s - a$. In general, K is complex (and not a pure imaginary) which reflects the attenuation of the 1D-wave due to leakage in the bulk. However, for the particular values $\phi = n\pi$ with n integer, one finds $s = (-1)^n$. This corresponds to the "inferior" band-edges at which the modulus of the eigenvalue s raising to unity reflects a vanishing leakage. This striking result can be recovered explicitely by computing the wave energy reflection coefficient of a wave incident from the left upon a semi-infinite periodic lattice which gives R = 1 for $s = \pm 1$. The vanishing of the leakage at the band-edges demonstrate that leakage and dissipation are completely different phenomena. Dissipation is not sensitive to coherence whereas leakage is controlled by the coherent structure of the 1D-wave. Physically, this coherent effect results from the Bragg phase condition which insures that the guided standing wave has the period of the reflective array but is out of phase with it. This explains the high quality factors $\approx 10^5$ of SAW resonators made of two distributed reflective arrays forming an effective "cavity" surrounding a transducer, whereas the radiation loss is more than 20 times the reflection for one groove excited from only one side. We have also analysed the competition between leakage and reflection in the stop (forbidden) bands and in the pass-bands (see ref.[13]).

2-3 Competition between Localization and Coherent Leakage.

Let us now briefly discuss the influence of disorder. The disorder can enter in the position of the scatterers or in their reflective or radiative power. Consider for simplicity the case of a position disorder around a periodic frame, parametrized by $\eta = k\delta\Lambda$ and characterized by a zero average $<\eta> = 0$ and a r.m.s. amplitude $<\eta^2> = \sigma^2/\mu$ (the notation follows from a generalized central limit theorem [18]).

The T-matrix formalism is still valid. It leads to a problem of products of random matrices which has been well studied in the litterature [1,2]. It is known that even in the pass-bands, the waves are completely reflected ("localization" regime) by a semi-infinite system. iK is replaced by a real lyapunov exponent $\gamma = \xi_{loc}{}^{-1}$ inverse of the localization length which is always positive for any non-vanishing disorder σ. It remains of the order of μ near the center of the stop-band whereas, in the "old" pass-bands, it is no more vanishing and is approximately :

$$\gamma \sim \mu^2\sigma^2 / \sin^2(2\pi d/\lambda)$$

The leakage loss appears to be of the same order as the localisation effect in the old pass-bands as discussed in [13]. In concrete cases such as the SAW propagation on grooves, it can even dominate, since $p \sim 20\mu^2$.

However, a non-trivial effect occurs when one approaches an "inferior" band-edge at which the leakage is vanishing in absence of disorder. In this region, it is known [18-19] that the lyapunov exponent has a non-analytic dependance upon the disorder :

$$\gamma \sim \sigma^{2/3} \tag{4}$$

Both facts therefore suggest that Anderson localization should dominate over the leakage. This expectation has been made more precise in ref.[13]. Let us briefly give a feeling of the argument. The scaling (4) can be traced to the fact that at the band-edge, the problem becomes invariant under a dilatation of the lattice spacing. In our case, noting $a = (1+b)\, e^{i\eta} \sim e^{i((\eta+\mu)}$, the problem is invariant under a dilatation on $\mu \rightarrow \mu/\rho$ if the variance of the noise is changed according to $\sigma \rightarrow \sigma/\rho^{3/2}$. This implies $\rho \sim \sigma^{2/3}$ and since the lyapunov exponent γ scales as μ and therefore as ρ, one recovers eq.(4). Now, radiative losses are governed by the real part ξ of b. For ξ of order μ^2, we obtain a typical loss scaling as $\sigma^{4/3}$. This result offers a non-trivial signature of the competition between the two <u>coherent</u> effects : i) Anderson localization and ii) guided to bulk mode conversion.

3- ATTENUATION DUE TO LEAKAGE IN QUASIPERIODIC LATTICES.

We have studied experimentally the propagation of a SAW propagating at the free surface of corrugated crystals. One of the system studied is constituted of a lattice of N = 1000 grooves of identical width w=5µm, depth h=0.3µm and identical well characterized profile like an inverse plateau. The lateral scale of each groove (the so-called opening) is E=2150µm. The total length of the 1000 grooves system is L=15984µm corresponding to an average period a~16µm. The set of spaces between the grooves is constructed according the Fibonacci sequence which is built recursively as :

$$S_{j+1} = \{S_{j-1}, S_j\} \tag{5}$$

with S_0={s} and S_1={c}, where s and c are two chosen distances which constitute the two elementary tiles used for creating the quasicrystal. Since a finite number of elementary tiles (here two) are used to tile the system quasiperiodically, the structure can be termed a quasicrystal. With the recursion (5), one has S_2={sc}, S_3={csc}, S_4={sccsc} and so forth. In a system S_j built from the above recurrence up to j, there are F_j grooves where F_j is a Fibonacci number given recursively as $F_{j+1}=F_j+F_{j-1}$ with $F_0=F_1=1$. We have chosen s=11.6±0.1µm and c=18.7±0.1µm as the values of the two distances between the grooves found in the Fibonacci sequence, yielding s/c=0.620±0.005 near of the inverse golden mean $t=(\sqrt{5}-1)/2=0.6180...$

In this system, we expect [12] that the proper modes are critical in the sense that they are neither localized nor extended but exhibit a self-similar pattern. More precisely, the absolute value of a wave corresponding to a proper mode should be a product of periodic functions, the period of each function being the approximate period of the almost periodic problem obtained by cutting off the continued fraction representation of at/λ at a particular stage, where λ is the wavelength. Figure 2 constitutes the first direct experimental test of the theoretical models of wave propagation in quasi-periodic systems.

It is interesting to determine the influence of the wave attenuation on the observed spatial structure of the proper modes. On figure 2 we present the spatial structure of a mode (a) whose frequency lies inside a passing band. The spatial structure should therefore be uniform and flat. The intensity decreases exponentially with distance along the system with a characteristic length l_a=3180±40μm corresponding to a time scale l_a/c=0.98±0.05μs. The main source of dissipation is due in this case to the conversion of the guided SAW into the leaky bulk waves. This corresponds to an attenuation in energy of $p{=}a/l_a{\approx}2.3\ 10^{-3}$ per groove leading to $p{\approx}29\ \mu^2$. In this case, the energy loss due to SAW leakage per groove is about thirty times the energy reflection! This result would imply that attenuation due to leakage should always dominate the coherent 1D-propagation along the quasicrystal. Indeed, the attenuation length l_a measured on figure 2(a) should make extremely difficult the observation of the critical modes. However, we observe them well for particular frequencies. This implies that for these frequencies and their corresponding spatial structure, the SAW-to-bulk leakage is inhibited. The comparison of curve (a) and (b) in figure 2 constitutes a striking first verification of the effect discussed in 2-2 and 2-3. On curve (b), the rate of decrease of the SAW energy along the quasicrystal is strikingly less than in curve (a).
This can be explained as follows. The spectrum of a quasiperiodic system is characterized by the fact that as one increases the order of the rational approximation towards the irrational incommensurability, the pass-bands are cut by new forbidden bands and so on until only band-edges remain leading to pure point or singular continuous spectrum. This is precisely at band-edges that the leakage can be inhibited coherently[13]. More informations will be presented elsewhere[12].

4- LOCALIZATION IN A QUASIPERIODIC SYSTEM.

The structure of this system is depicted just below. Let P and Q be two relatively prime numbers. This lattice of grooves is the simple superposition of two periodic lattices of period proportionnal to 1/P and 1/Q. The groove density function is therefore :

$$n(x) = \Sigma_m\ \delta(x-\alpha m/P) + \Sigma_n\ \delta(x-\alpha n/Q) \qquad (6)$$

It is periodic and the elementary cell contains P+Q-1 grooves. The system under study is made of one such elementary cell, with $P{=}F_{10}{=}89$, $Q{=}F_{11}{=}144$ and the total number of grooves is $P{+}Q{=}F_{12}{=}233$. The smallest spacing between two successive grooves is d_{min}=4μm which constitutes the unit length of the problem; the largest spacing is d_{max}=358.5μm and the mean spacing a=221μm. The size of the system is $L{=}PQd_{min}$=89*144*4μm=51264μm. We can notice that the system is symmetric with respect to its center. The grooves have all the same well characterized profile (fig.1a), with w=2 μm width and h=0.3μm depth. The lateral scale of each groove (the so-called opening) is E=1860μm.

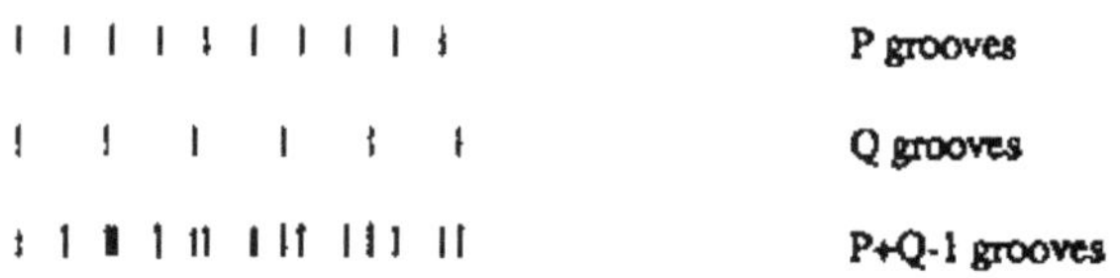

This system has been studied theoretically in details by Peyraud and Coste [10]. They predict a continuous spectrum of pass-bands separated by δ-like gaps, in the limit of quasi-periodicity (i.e. $P{=}F_j$, $Q{=}F_{j+1}$, with $j \rightarrow +\infty$) and in the vanishing reflection coefficient per groove limit. In these limits, the set of gaps is given by the Fourier transform of n(x) : which is simply the superposition of the Fourier transforms of the two sub-systems of P and Q grooves.The finite value of the reflection coefficient enlarges the gaps and makes them observable; in practice, these gaps always dominate the spectrum.

Figure 3 shows the measured transfer function of the system in transmission; the bell-like global shape corresponds to the transducers filtering and the peaks are the expected stop-bands of the system; as predicted, they constitute two periodic families of stop-bands with different intensities (see[10]).

Such a system presents some pairs of very close gaps with a narrow pass-band trapped inside (figure 4). The existence of these pass-bands is independant of the grooves depth; it is related to mathematical properties of irrationnal numbers. In such pass-bands, Peyraud and Coste have predicted the existence of quasi-localized modes. This can be well understood, since we always can expect special behaviours near a gap and so, particularly, between two neighboring gaps. The predicted structure of these modes is particularly simple. In figure 5, we compare predicted and measured shapes of a quasi-localized proper-mode along the array, in the small pass-band of figure 4. The concordance is specially striking, since the theoretical shape, given by [10], does not take into accont for dissipation and coherent losses. This mode is not localized in the usual sense, since, it has not a finite norm when P/Q tends to t, but it contains a non-vanishingly stored energy per unit length associated to an infinite sojourn time [1]. The relatively smooth shape of this mode is due to the little number of grooves of this device and their weak depth. Very sharp "localizations" are predicted in [10] for larger arrays and now we can expect to observe them easily.

5- CONCLUSION.

These preliminary experiments show that SAW devices are very appropriate to study accurately all the important transport features (spectral, temporal and modal) of these exotic systems; in particular we are able to measure directly the precise shape of the proper-modes. Moreover, we think that these experiments can be extended to 2D-arrays, by replacing the grooves by dots.

It is perhaps too early to give a full insight on the possible future applications of quasi-periodic or disordered structures; but it is quite evident that the former ones bring up a great number of well-controlled new properties which contrast with the relative poverty of the properties of periodic systems. Most likely applications are in the domain of signal processing (multi-band filters, tunable oscillators, memories relying on the existence of localized proper-modes...); random devices could be developped to perform random samplings or to understand experimentaly the features of the direct or inverse responses of stratified media (geophysical, medical probing...). From a general point of view, these experiments aim to link a large background of theoretical speculations with future real applications.

We aknowledge financial support from DRET under contract n°.86/177 for the research program "Propagation acoustique en milieux aléatoires" and THOMSON-SINTRA at Sophia-Antipolis, FRANCE for providing experimental facilities.

REFERENCES:

[1]: ZIMAN J.M., "Models of Disorder", Cambridge University Press (1979).
ISHII K., Suppl. Progr. Theor. Phys. 53, (1973), p.77.
ANDERSON P.W., THOULESS D.J., ABRAHAMS E. and FISHER D.S., Phys. Rev. B 22, p. 3519, (1980).
SOUILLARD B., "Waves and Electrons in Homogeneous Media", Proc.of the Summer School : "Chance and Matter", Les Houches, (1986), (North Holland, Amsterdam, 1987).

[2]: ANDERSON P.W., Phys. Rev. 109, p. 1492, (1958).

[3]: SORNETTE D., "Acoustic Waves in Random Media"
(I) : Weak Disorder Regime, Acustica 68 (1), (1989).
(II) : Coherent Effects and Strong Disorder Regime, Acustica 68 (2), (1989).
(III) : Experimental Situations, Acustica 68 (3), (1989).

[4]: SORNETTE D., "1/f Colouring of White Noise by Anderson Localization" Europhys. Lett. 4 (10), p. 1157, (1987).

[5]: SHECHTMAN D., BLECH I., GRATIAS D. and CAHN J.W. Phys. Rev. Lett. 53,, p.1951, (1984).

[6]: SOKOLOFF J.B., "Unusual Band Structure, Wave Functions and Electrical Conductance in Crystals with Incommensurate Periodic Potentials", Phys. Rep. 126 (4), (1985).

[7]: LUCK J.M. and PETRITIS D., "Phonon Spectra in One-Dimensional Quasicrystals" J. Stat. Phys., 42, , p.289, (1986).
[8]: KOHMOTO M., "Localization Problem and Mapping of One-Dimensional Wave Equations in Random and Quasiperiodic Media", Phys. Rev. Lett. 51, p. 1198, (1983).

[9]: KOHMOTO M., SUTHERLAND B. and IGUCHI K.,"Localization in Optics : Quasiperiodic Media", Phys. Rev. Lett. 58 (23), p. 2436, (1987).
[10]: PEYRAUD J., COSTE J., "Wave Propagation in a One-Dimensional Quasiperiodic Medium : Localization and Intermittency", Phys. Rev. B 37 , (March 1988).
[11]: KOHMOTO M., SUTHERLAND B. and TANG C. "Critical Wave Functions and a Cantor-Set Spectrum of a One-Dimensional Quasicrystal Model" Phys. Rev. B 35 (3), p. 1020, (1987).
[12]: MACON L., SORNETTE D., "Localization of Surface Acoustic Waves in Quasiperiodic Media" To be published.
[13]: SORNETTE D., MACON L. and COSTE J., "Transfer Matrix Theory of Leaky Guided Waves" J. Phys.49, (Oct. 1988).
[14]: E.A.ASH and E.G.S.PAIGE, "Rayleigh-wave theory and aplication", Springer Series on Wave Phenomena, Springer-Verlag (1985).
A.A.OLINER, "Acoustic surface waves", in Topics in Applied physics, vol.24, Springer-Verlag (1978).
R.F.WALLIS and G.I.STEGEMAN,"Electromagnetic surface excitations", Springer-Verlag (1986).
H.S.TUAN and J.P.PAREKH, IEEE Trans.Son.Ultrason. SU24, p.384, (1977).
S.D.WU and H.S.TUAN, J.Appl.Phys., 50, p.73, (1979).
O.W.OTTO, IEEE Trans.Son.Ultrason. SU22, p.251, (1975).
N.E.GLASS and A.A.MARADUDIN, J.Appl.Phys., 54, p.796, (1983).
E.L.CAMBIAGGIO and F.C.CUOZZO,Trans.Son.Ultrason. SU26, p.340, (1979).
A.RONNEKLEIV and J.SOUQUET, Ultrasonics Symp.Proc.IEEE.Cat.75, CHO 994-4SU (1975) and J.Appl.Phys.47, p.4422, (1976).
J.MENNGAILIS, R.C.WILLIAMSON, J.HOLTHAM and R.C.M.LI, Wave Electronics, 2, p.177, (1976).
E.BARON, Thèse de Troisième Cycle, Paris 7, (1986).
E.BARON,M. de BILLY and G.QUENTIN, Revue Phys.Appl.20, p.369, (1985).
S.DATTA and B.J.HUNSINGER, J.Appl.Phys., 49(7), p.3769, (1978).
[15]: P.A.MELLO, Phys.Rev.lett.60, p.1089, (1988).
[16]: U.SIVAN and A.SA'AR, Euro.Phys.Lett.5, p.139, (1988).
[17]: A.YARIV and P.YEH, "Optical waves in crystals", John Wiley.
[18]: J.P.BOUCHAUD and P.LE DOUSSAL, J.Phys. A19, p.797, (1986).
J.P.BOUCHAUD, A.GEORGES and P.LE DOUSSAL, "Fluctuations of the lyapunov exponent and intermittency in dynamical and disordered systems : th example of 1D localization", lecture given at the meeting on dynamicl systems-Rome-Octobe (1986).
[19]: B.DERRIDA and E.GARDNER, J.Physique, 45, p.1283, (1984).

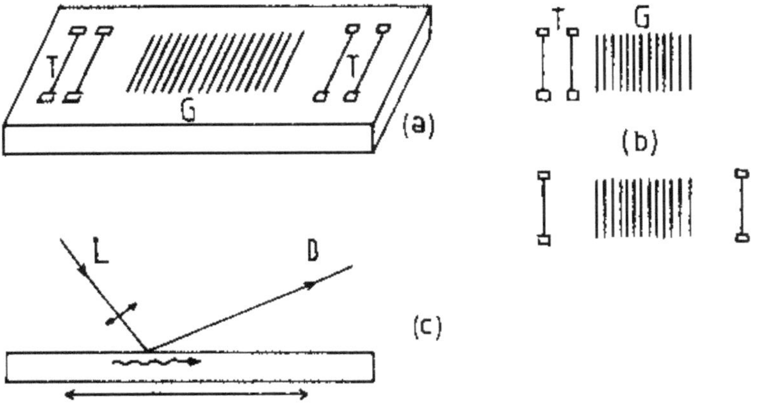

FIGURE 1 : Top view of a SAW device (figure 1-a) : a grooved array (G), etched at the surface of a YZ-$LiNbO_3$ crystal, is surrounded by electro-acoustical transducers (T) allowing SAW generation (harmonic or pulsed excitations) and then reflection or transmission measurements (fig. 1-b). The device can be translated (fig.1-c) while a laser beam is locally diffracted by the acoustic deformations of the surface. Thus one can detect the acoustic field along the array, when it is harmonicaly excited.

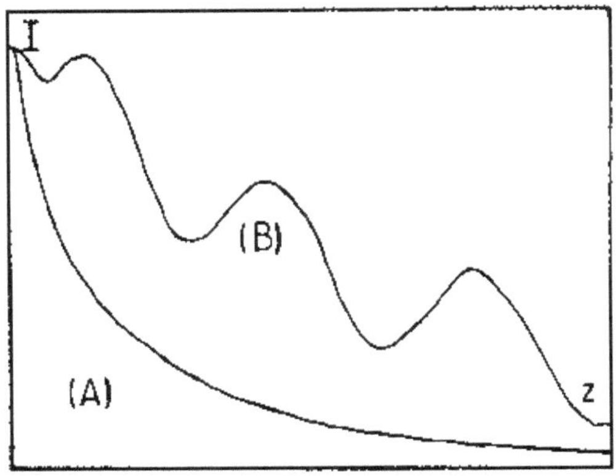

FIGURE 2 : Proper modes of a quasi-periodically corrugated YZ-$LiNbO_3$ crystal. The structure of the modes is obtained by the local diffraction of a laser beam by the SAW deformations of the surface. Curve(a) corresponds to a mode in a pass-band and the exponential decrease reflects the attenuation due to surface-to-bulk wave conversion. Curve (b) depicts a critical mode with three lobes. The present resolution is not sufficient to distinguish the inner structure of the mode. Note the much less rapid decrease of the envelop of mode (b): this is due to a much less efficient surface-to-bulk wave conversion resulting from destructive interference effects as discussed in §2.

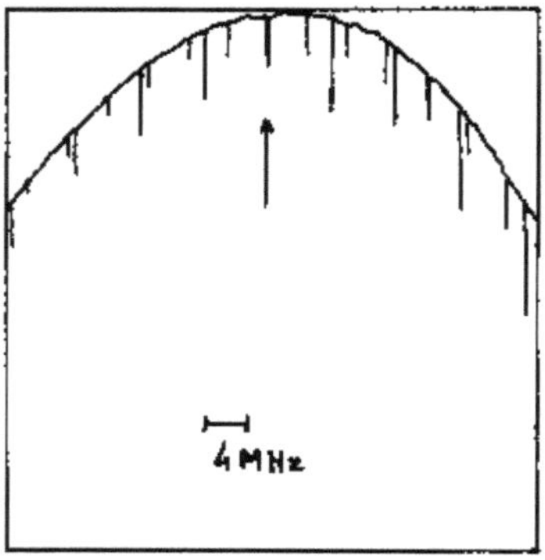

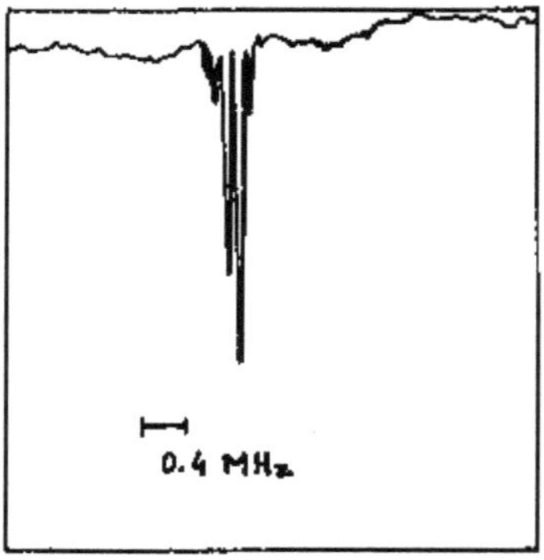

FIGURE 3 : Portion of the transmission transfer function of the quasiperiodic lattice of grooves depicted in section 4. Frequency resolution is 3.2 kHz over the 40 MHz band swept. Attenuation is given in dB versus the frequency. The peaks are the forbidden bands of the system and are distributed in two periodic families of different intensities. The global increase in intensity comes from the dependance of the reflection coefficient per groove on frequency. At about 166MHz (middle of the picture) two stop-bands seem to overlay each other.

FIGURE 4 : Magnification of the region around 166MHz; the stop-bands are well separated and we expect in the very small pass-band trapped between them a quasilocalized proper-mode (see figure 5).

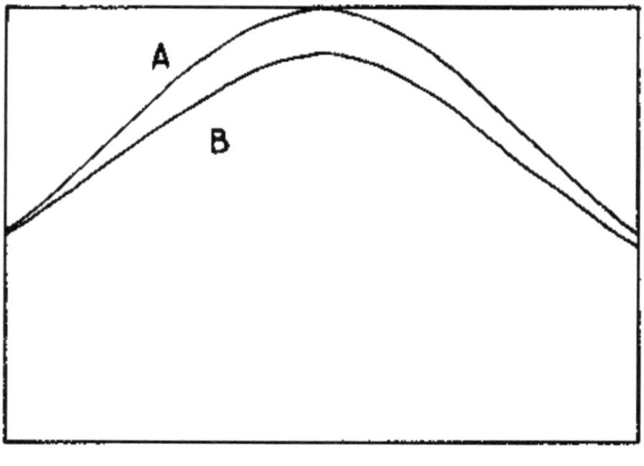

FIGURE 5 : Proper-mode envelope of the quasiperiodic system described in section 4. Intensity is given versus the position along the array. (A) : Theoretical shape. (B) : Experimental shape (obtained from the laser diffraction experiment).

ON ASYMPTOTIC BEHAVIOUR OF WAVE STATES IN RANDOM MEDIA

F.Bentosela R.Rodriguez

Centre de Physique Théorique
C.N.R.S. Luminy Case 907
F-13288 Marseille Cedex 09
France

Abstract: The asymptotic behaviour of wave states in inhomogeneous random potentials is deduced from perturbative computation of invariant measures in action-angle variables. Power law growth rate , Lyapunov exponents in magnetized plasmas and optical fibers are obtained as applications.

1. INTRODUCTION

Many problems in wave propagation in random media are in essence one dimensional, so they can be reduced to the study of a differential equation with random coefficients.It may be useful to know the asymptotic behaviour of the corresponding solution. It has been proven that for almost all the choices of the random coefficients in a certain class , the asymptotic behaviour is independent of such a choice. For some special models this behaviour is explicitly known but in most of the cases we can only hope to get expansions in terms of a parameter which caracterizes the disorder strength. Here we want to get such kind of results for second order differential equation of the type

$$\frac{d^2 \phi}{dx^2} + \left[E - V(x) - \alpha\, C(x)\, v(x)\right] \phi = 0$$

E is a constant and α a parameter, V(x) and C(x) being given deterministic, sufficiently smooth functions and v(x) is a stochastic process.

The asymptotic behaviour is found to be of the type exponential or power law, the exponential one appearing as a particular case. Some examples are given concerning waves in magnetized plasmas with particle densities fluctuating around smooth profiles and transmission of light waves in optical fibers with geometrical inhomogeneities.

2. FORMULATION OF THE PROBLEM AND MAIN RESULTS

To conform to usual probability notations, we will write t instead of x and denote v_t the stochastic process. The Schrödinger type equation becomes

$$-\frac{d^2\phi}{dt^2} + \left[V(t) + \alpha\, C(t)v_t\right]\phi = E\phi \tag{2.1}$$

It can be written in matrix form :

$$\frac{d\Phi}{dt} = A(E,\alpha,t)\Phi \tag{2.2}$$

$$\Phi(0) = \Phi_0 \qquad (\Phi_0 \text{ being given}) \tag{2.3}$$

where $\Phi = \begin{pmatrix} \sqrt{E}\phi \\ \dot{\phi} \end{pmatrix}$ and $A(E,\alpha,t) = \begin{pmatrix} 0 & \sqrt{E} \\ -\sqrt{E}+\frac{1}{\sqrt{E}}(V(t)+\alpha C(t)v_t) & 0 \end{pmatrix}$

Let $U^\alpha(t)$ be the flow associated with (2.1), then

$$\frac{dU^\alpha(t)}{dt} = A(E,\alpha,t)U^\alpha(t) \quad \text{with } U^\alpha(0) = 1$$

Let us denote : $W(t) = U^o(t)^{-1}\, U^\alpha(t)$. Define the action angles variables (I,φ) as :

$$X(t) = \sqrt{2I}\ \sin\varphi, \quad Y(t) = \sqrt{2I}\ \cos\varphi \tag{2.4}$$

where $$\begin{pmatrix} X(t) \\ Y(t) \end{pmatrix} = W(t)\begin{pmatrix} \sqrt{E}\phi(0) \\ \dot{\phi}(o) \end{pmatrix}$$

They satisfy the following equations :

$$\dot{I} = \frac{\alpha}{\sqrt{E}} I\, C(t)\, v_t\, f(\varphi,t) \tag{2.5}$$

$$\dot{\varphi} = \frac{\alpha}{\sqrt{E}} C(t)\, v_t\, g(\varphi,t) \tag{2.6}$$

where f and g are given by :

$$f(\varphi,t) = 2\, u^0_{12}(t)\, u^0_{11}(t) \cos 2\varphi \;+\; (u^0_{11}(t)^2 - u^0_{12}(t)^2) \sin 2\varphi \tag{2.7}$$

$$g(\varphi,t) = -\left[\, u^0_{12}(t) \cos\varphi + u^0_{11}(t) \sin\varphi \right]^2 \tag{2.8}$$

Here, $U^o(t) = \left\{ u^0_{ij}(t) \right\}_{i,j=1,2}$ and $\det U^o(t) = 1$.

The stochastic process v_t will be chosen in the following manner : let η_t be a diffusion process on S^1 (the unit circle) or whatever compact manifold M, the density of probability transition of which is called $h(s,x;t,y)$. By the ergodic theorem $h(s,x;\, t,y) \to \rho(y)$ as $t \to \infty$. If $\rho(y)$ is taken as the initial distribution, then η_t becomes a stationary Markov process.

Now let $H : M \to R$ be a continuous function on M.Then

$$v_t = H(\eta_t) \tag{2.9}$$

Furthermore, we take H such that

$$\int_M H(y)\, \rho(y)\, dy = 0 \tag{2.10}$$

Let us define the "process" $\tau_t = \tau + t$, $t > 0$, $\tau = \tau_0 \in R$; the process $\xi_t = \{\eta_t, \tau_t, \varphi_t\}$ is a homogeneous Markov process. The semi-group of shifts acting on $C^1(R)$ has $\frac{\partial}{\partial\tau}$ as generator, so the generator of ξ_t is :

$$K_\alpha = G + \frac{\partial}{\partial\tau} + \frac{\alpha}{\sqrt{E}} C(\tau)\, H(\eta)\, g(\varphi,\tau) \frac{\partial}{\partial\varphi} \tag{2.11}$$

where G is the self-adjoint generator of η_t. The invariant measure associated to the process $\{\eta_t,\tau_t\}$ is μ given by

$$\mu(q) = \lim_{T\to\infty} \frac{1}{T}\int_0^T d\tau \int_M dy\; q(y,\tau)\; \rho(y). \tag{2.12}$$

Let us assume that the potential V in (2.1) is such that the functions f and g in (2.7), (2.8) are bounded with respect to t. This will be the case for potentials which tend to zero at infinity and are sufficiently regular. The calculation of an invariant measure for the process ξ_t may be solved by a perturbation method. We construct a solution of

$$K_\alpha^* \, p = 0 \tag{2.13}$$

of the form

$$p(\eta,\tau,\varphi) = \sum_{k=0}^{\infty} \alpha^k \, v^{(k)}(\eta,\tau,\varphi). \tag{2.14}$$

where $v^{(k)}$ are distributions.(Remark : criteria for (weak) convergence of such series are given in [4].)Injecting this expression in (2.13) and equating the coefficients of the same order in α, we obtain at the lowest order

$$-\frac{\partial}{\partial\tau} v^{(0)}(\eta,\tau,\varphi) + G\, v^{(0)}(\eta,\tau,\varphi) = 0$$

the solution of which is the density of the following measure :

$$\tilde{v}^{(0)}(q) = \frac{1}{2\pi}\int_{S^1} d\varphi \lim_{T\to\infty} \frac{1}{T}\int_0^T d\tau \int_M dy\; \rho(y)\; q(y,\tau,\varphi) \tag{2.15}$$

At the kth order, one has

$$-\frac{\partial}{\partial\tau} v^{(k)} + G v^{(k)} = \Gamma^{(k-1)} \tag{2.16}$$

where $\Gamma^{(k-1)}(\eta,\tau,\varphi) = \dfrac{H(\eta)}{\sqrt{E}} \dfrac{\partial}{\partial\varphi}\left(C(\tau) g(\varphi,\tau)\, v^{(k-1)}(\eta,\tau,\varphi)\right).$

The solution of (2.16) may be obtained for each k, but we shall give it only for the first term. We want to solve :

$$(G - \frac{\partial}{\partial \tau}) v^{(1)} = \frac{H(\eta)}{\sqrt{E}} \; v^{(0)}(\eta,\tau) \; \frac{\partial}{\partial \varphi} g(\varphi,\tau)$$

Using the fact that formally

$$\left(G - \frac{\partial}{\partial \tau}\right)^{-1} = - \int_0^{\infty} du \, \exp(uG) \; \exp(- u \frac{\partial}{\partial \tau})$$

and $$[\exp(uG)F](\eta) = \int_M h(0,y \,;\, u,\eta) \, F(y) \, dy$$

we get:

$$v^{(1)}(\eta,\tau\varphi) = - \int_0^{\infty} d u \int_M dy' \; h(0,y' \,;\, u, \eta) \; \frac{H(y')}{\sqrt{E}} v^{(0)}(y',\tau\text{-}u) \; C(\tau\text{-}u) \frac{\partial}{\partial \varphi} \; g(\varphi,\tau\text{-}u) \quad (2.17)$$

Using the first two terms in the invariant measure expansion and the ergodic theorem, we have :

$$\lim_{t \to \infty} \frac{1}{t} \; \mathrm{Log} \frac{I_t}{I_0} \overset{a.s}{=} \frac{\alpha}{\sqrt{E}} \int d\eta \, d\tau \, d\varphi \; C(\tau) \; H(\eta) f(\varphi,\tau) \big(v^{(o)}(\eta,\tau,\varphi) + \alpha v^{(1)}(\eta,\tau,\varphi) + \ldots\big) \quad (2.18)$$

The first term of the expression is equal to zero by (2.10) ; so, at the lowest order in α, one obtains :

$$\lim_{t \to \infty} \frac{1}{t} \mathrm{Log} \frac{I_t}{I_0} \overset{a.s}{=} \frac{- \alpha^2}{2\pi E} \; \lim_{T \to \infty} \frac{1}{T} \int_0^T d\tau \; \int_{S^1} d\varphi \; \int_0^{\infty} du \; C(\tau\text{-}u) \; C(\tau) R(u) f(\varphi,\tau) \frac{\partial}{\partial \varphi} g(\varphi,\tau\text{-}u) \quad (2.19)$$

where R(u) is the correlation function of the process v_t, f and g being given by (2.7) and (2.8). It may be useful for applications, to consider situations where the funcitons f and g have a simple form. This will be the case when the so called quasi classical solutions of (2.1) for $\alpha = 0$ are sufficiently good approximated solutions, namely when $\sqrt{E} \frac{|n'_t|}{n^2} \ll 1$, where $n^2(t) = \left(1 - \frac{V(t)}{E}\right)$.

As is well known, these solutions are such that $u_{11}(t)$ and $u_{12}(t)$ have the following form :

$$u_{11}(t) = \frac{\gamma(t)}{\gamma(0)} \cos\alpha(t) \quad ; \quad u_{12}(t) = \frac{\gamma(t)}{\gamma(0)} \sin\alpha(t)$$

where $\gamma(t) = (E-V(t))^{-1/4}$ and $\alpha(t) = \int_0^t (E - V(\tau))^{1/2}\, d\tau$.

Let us suppose V(t) is an even function and let us consider E high enough in such a way $\gamma(t)$ and $\alpha(t)$ are real quantities. Then :

$$f(\varphi,t) = \sin 2(\varphi+\alpha(t)) \frac{\sqrt{E}}{\alpha'(t)}$$

$$g(\varphi,t) = -\sin^2(\varphi+\alpha(t)) \frac{\sqrt{E}}{\alpha'(t)} \tag{2.20}$$

and one obtains from the general formula (2.19), at lowest order in α :

$$\mathrm{Log}\ \frac{I_1}{I_0} \overset{a.s}{=} \frac{\alpha^2}{2} \int_0^t \int_0^\infty C(\tau-u)\, C(\tau)\, R(u) \frac{\cos 2(\alpha(\tau)-\alpha(\tau-u))}{\alpha'(\tau)\,\alpha'(\tau-u)}\, du\, d\tau + 0(t) \tag{2.21}$$

We shall give, in what follows some examples of application of (2.21).

3. APPLICATIONS

Ex.1 : when $V(t) = 0$, one has $\alpha(\tau) = \sqrt{E}\tau$. From (2.21) with $C(\tau) = 1$, one has the usual formula for the Lyapunov exponent for a correlated noise with correlation function R(u)

$$\lambda = \frac{\alpha^2}{2E} \int_0^\infty R(u) \cos(2\sqrt{E}\,u)\, du + 0(\alpha^3) \quad \text{(see e.g. [4])}$$

Ex.2 : the second example may be taken in plasma physics where high frequency waves enter a magnetized plasma. Let us take the incident wave vector orthogonal to the resident magnetic field. For the so called extraordinary wave, the transverse component of the electric field E_x verifies the equation $\frac{d^2E_x}{dz^2} + \frac{\omega^2}{c^2}\varepsilon(z)E_x = 0$ where the permittivity is $\varepsilon(z) = \frac{\nu(z)-(1-q(z))^2}{\nu(z)-(1-q(z))}$ (see [5]) ; $\nu(z) = \frac{\omega^2{}_B(z)}{\omega^2}$ and $q(z) = \frac{\omega^2{}_p(z)}{\omega^2}$ where $\omega_B(z)$ is the cyclotronic frequency and $\omega_p(z)$ is the plasma frequency which depends on the density. Both quantities may fluctuate due to fluctuations of the density and the magnetic field. Thus ε may be of the form $\varepsilon(z)+\alpha\varepsilon_z$ with ε_z some stochastic process. Taking $C(\tau) = 1$ and white noise perturbations of the type $\langle\varepsilon_z\, \varepsilon_{z'}\rangle = D\delta(z-z')$, equation (2.21) becomes

$$\mathrm{Log}\frac{I_z}{I_o} \overset{a.s}{=} \frac{\alpha^2 c^2 D}{2\omega^2}\int_0^z \frac{v(y)-(1-q(y))}{v(y)-(1-q(y))^2}\, dy + 0(z)$$

For a flat profile and a non zero constant magnetic field v and q are constant (called respectively v_o and q_o) if we except the small fluctuations. For white noise perturbations, one obtains the Lyapunov exponent : $\gamma = \frac{\alpha^2 c^2 D}{2\omega^2}\left(\frac{v_o-(1-q_o)}{v_o-(1-q_o)^2}\right)$. For $B = 0$, we get $\gamma_0 = \frac{\alpha^2 c^2 D}{2(\omega^2-\omega_p^2)}$, which can be compared with the results given in [6]. For a correlated noise , one has $\gamma_1 = \frac{\alpha^2}{2\mu^2}\int_0^\infty R(u)\cos(2\mu u)du$ where $\mu = \frac{\omega}{c}\left(\frac{v_o-\eta_o}{v_o-\eta_o^2}\right)$ In particular , for an exponential correlation of the form $R(s)=\Gamma e^{-s/\tau}$, the Lyapunov exponent takes the form $\gamma_1 = \frac{\alpha^2\Gamma\tau}{4\mu^2(1+4\mu^2\tau^2)}$.

Here we recover the same type of formula as in [7].

Ex.3 : Let us take $V = 0$ and $C(x) = \frac{1}{|x|^\varepsilon}$ $\varepsilon>0$ then we get with $\alpha(\tau) = \sqrt{E}\tau$, at the lowest order in α :

$$\mathrm{Log}\frac{I_x}{I_o} \overset{a.s}{=} \frac{\alpha^2}{2}\int_0^x\int_0^\infty \frac{R(y)\cos(\sqrt{E}y)}{E|z|^\varepsilon\,|z-y|^\varepsilon}\, dy\, dz + 0(x)$$

For white noise fluctuations such that $R(y)=D\delta(y)$, on obtains the following asymptotics for I_x : if $\varepsilon<1/2$ $I_x \approx I_0\exp(Cx^{-2\varepsilon+1})$ where C is some constant depending on α ,E and D (See [1] and [3] for the derivation of such a behaviour by different methods ; see also [2]). If $\varepsilon \geq 1/2$, $\log(I_x / I_0)=0(x)$.

Ex.4 : Optical fibers. Among the various sources of attenuation of electromagnetic waves propagating in fibers, one may consider the one caused by irregularities in their geometrical structure.

The wave equation can be written as a system of second order coupled differential equations with random coefficients. It can be proven that attenuation is given by the smallest positive Lyapunov exponent. In some cases, this Lyapunov exponent is well approximated looking only at one equation i.e. neglecting the coupling of this equation with all the others. Once we are reduced to a random oscillator type equation and we can use the same formula as in example 1.

For instance when the fiber is random bended and the correlation function for the bends is $\alpha^2 \exp(-z^2/L^2{}_c)$ we get the following approximated formula for their attenuation :

$$\gamma = 2(\pi^{5/2}) \frac{\alpha^4}{\lambda^2} n \ (\Delta n)^2 L_c \exp(-4\pi^2 n^2 L_c{}^2/\lambda^2)$$

where λ is the wave length, n the core index, Δn the difference of refractive indexes between core and cladding [8].

REFERENCES

[1] Delyon F. , Simon B. ,Souillard B., Ann. Inst. H. Poincaré, 42 , n°3 283-309 (1985)

[2] Souillard B., Waves and Electrons in Non Homogeneous Media.
Proceedings of : Les Houches Summer School "Chance and Matter"
Souletie J. , Vannimenus J.,Stora R. Ed. North Holland.(1986).

[3] Kotani S.,Ushiroya N., Comm. Math. Phys. 115, 247-266 (1988).

[4] Arnold L., Papanicolaou G., Wihstutz V., SIAM J. Appl. Math. 46, n°3, 427-450 (1986).

[5] GinzburgV.L., The propagation of electromagnetic waves in plasma, Pergamon Press (1970).

[6] Escande D.F. , Souillard B., Phys. Rev. Letters. 52 n°15 1296-1299 (1984).

[7] Hansel D. , Luciani J.F., On diffusion equations for dynamical systems driven by noise. Preprint CPT Ecole Polytechnique A 800 - 1187 (Nov. 87).

[8] Bentosela F., Piccoli P., An expression for attenuation in deformed optical waveguides. Preprint CPT/88/P.2117, Marseille.(mai 1988)

AN ATTEMPT OF OPTIMUM DESIGN FOR REACTIVE MUFFLERS

J. HARDY and J. KERGOMARD

Laboratoire d'Acoustique de l'Université du Maine

UA 1101 CNRS

Route de Laval BP 535

72017 LE MANS Cedex (FRANCE)

1. ABSTRACT

A geometrical method is given to interpret sound propagation in one dimensional systems. In the complex plane the acoustical impedance of the noise source moves when new portions of waveguide are added at the output. The character of this motion determines the behaviour of sound propagation in the muffler. Applications to random and deterministic arrangements of expansion chambers is given as well as the first experimental results, although other applications areas seam to be possible.

The efficiency of reactive mufflers comes from reflection effects of incident waves by the system and dissipative effects are not essential. We focus our attention on low frequencies range (100-500 Hz) where usually dissipative mufflers do not work very efficiently, in this case propagation of sound waves can be modelized by products of 2x2 matrices, which greatly simplify the problem.

We first give a geometrical interpretation for such produces in term of motion of the acoustical impedance in the hyperbolic or Lobachevsky plane. It is shown that a riemannian metrix gives the reflection coefficient for incident waves ; then, in order to obtain much attenuation, one has to take an arrangement of the waveduct giving fast motion (with respect to this metrix) of the acoustical impedance.

As an application of our method, it is shown that periodic waveducts are fairly bad candidates for sound muffling. We then investigate "random" waveducts: the point standing for the acoustical impedance moves randomly in the hyperbolic plane to each new matrix product, and because of the negative curvature, trajectory escapes to infinity with a velocity (non-zero as in the random walk on the usual plane) which is obviously not the greatest possible.

We show that direct methods in order to get much attenuation are almost impossible even with help of computers. Then we apply our method for an arrangement of expansion chambers which is one of the most commonly used for reactive mufflers and the first experimental results are given.

2. INTERPRETATION OF PRODUCTS OF MATRICES IN TERMS OF MOTION IN THE HYPERBOLIC PLANE AND ITS APPLICATION TO PERIODIC MEDIA

We consider a sound wave propagating in a guide ; for sufficiently low frequencies the wave equation can be

$$\frac{\partial^2\varphi}{\partial t^2} - c^2 \frac{\partial^2\varphi}{\partial x^2} = 0 \qquad (1.1)$$

interpreted in terms of two-port network matrices, each portion of waveguide transforms linearly the vector $\begin{pmatrix} p \\ v \end{pmatrix}$ where p is the acoustical pressure and v the velocity.

$$\begin{pmatrix} p_1 \\ v_1 \end{pmatrix} = \begin{pmatrix} a & b \\ c & d \end{pmatrix} \begin{pmatrix} p_2 \\ v_2 \end{pmatrix} \qquad (1.2)$$

The matrix depends on the wave frequency and without loss of generality one can suppose it unitary (ad-bc = 1). The basic idea of our paper is to associate to this transfer matrix a homographic transformation for the acoustical impedance [1,2] z = p/v :

$$M = \begin{pmatrix} a & b \\ c & d \end{pmatrix} \qquad T_M(z) = \frac{az+b}{cz+d} \qquad (1.3)$$

This transformation has the property of replacing matrix products by iterations :

$$T_{M_1M_2}(z) = T_{M_1}(T_{M_2}(z)) \qquad (1.4)$$

Then for a product of n matrices one obtains, iterating (1.4) :

$$T_{M_1M_2\,\ldots\,M_n}(z) = T_{M_1}(T_{M_2}(\;\ldots\;(T_{M_n}(z))\;\ldots)) \quad (1.5)$$

Such iteration aroises in theory of continued fractions, but in this case (a,b,c,d) are integers. Homographic transformations have special properties ; we shall briefly describe the most important of them.

From (1.4) one see that iteration of two homographic transformations is itself an homographic transformation ; it can be shown that they form a group. The matrices encountered in wave propagation in sound guides belong to a subgroup which leads invariant a particular circle. Modulo a conformal transform, this circle can be chosen as we want.

Although another choice is possible (as an example in transmission lines theory the Smith [3] chart corresponds to the choice of keeping the disc $r^2 \leq 1$ invariant), we use the Lobachevsky version. In this case a,b,c and d in (1.3) are real and the homographic transformation in this subgroup are the isometries of the remannian metrix defined by :

$$dS^2 = \frac{dx^2+dy^2}{y^2} \quad (1.6)$$

In other terms they can be interpreted as displacements on the half upper plane, known as hyperbolic or Lobachevsky plane. The geodesics between two points z_1 and z_2 are the lines minimizing the integral $\int_{z_1}^{z_2} dS$: it can be shown that they are circles orthogonal to the real line. The points on the real line are infinitly distant from every other points, and points (x,y) with small y are very far.

We now suppose that our waveguide repeats itself periodically [5]. The property of propagation depends in this case of only one transformation. This transformation has two fixed points O_1 and O_2 :

$$O_{1,2} = \frac{a\,O_{1,2} + b}{c\,O_{1,2} + d} = T(O_{1,2}) \tag{1.7}$$

We define then a reflection function (its modulus represents the reflection coefficient) as :

$$g(z) = \frac{z-O_1}{z-O_2} \tag{1.8}$$

and reporting (1.3) in (1.8) one gets :

$$g(T(z)) = K\,g(z) \tag{1.9}$$

where $K = \dfrac{a-c\,O_1}{a-c\,O_2}$.

The coefficient K is called the multiplier of the transformation ; its value determines the character of the transformation. For a periodic arrangement of the waveguide one has

$$g(T_M^n(z)) = K\,g(T_M^{n-1}(z)) = K^n\,g(z) \tag{1.10}$$

and then

$$T_M^n(z) = g^{-1}(K^n\,g(z)) \tag{1.11}$$

Three cases exist :

- <u>the elliptic transformation</u> :

If $|a+d| < 2$, then $K = e^{i\theta}$ with $\theta \neq 2n\pi$. On Fig.2 an example is shown each point in a black square is transformed into a point in a white square in the direction of the arrays ; for obvious reasons of clarity a special case is represented ($\theta = \pi/12$), but usually θ/π is irrationnal and the black squares get mixed up with the white ones after n products with $n\theta > 2\pi$. The two fixed points are stable points and the transformation is a rotation around these points of an angle θ ; in fact it is easy to obtain :

$$d(O_1, T_z) = d(O_1, z) \qquad (1.12)$$

the distance d being defined by (1.6)

An example the transformation which corresponds to a tube of length ℓ and with a section area S :

$$\begin{pmatrix} a & b \\ c & d \end{pmatrix} = \begin{pmatrix} \cos k\ell & \frac{\rho c}{S} \sin k\ell \\ -\frac{\rho c}{S} \sin k\ell & \cos k\ell \end{pmatrix} \qquad (1.13)$$

k is the wave number, ρ is density and c the sound velocity ; in this case $\theta = k\ell$.

- **<u>the parabolic transformation</u>**

If $|a+d| = 2$ then $K = 1$ and there is one fixed point at infinity ; usually this transformation exists for one frequency known as cutoff frequency, but the Helmholtz resonator is an example of parabolic transformation.

For the Helmholtz resonator one has :

$$\begin{pmatrix} a & b \\ c & d \end{pmatrix} = \begin{pmatrix} 1 & 0 \\ \dfrac{1}{L\omega - \dfrac{1}{c\omega}} & 1 \end{pmatrix} \qquad (1.14)$$

- **the hyperbolic transformation**

If $|a+d| > 2$ then K is real number: there are two fixed points at infinity (on the real axis), one is stable and the other unstable.

An example of hyperbolic transformation is given by change of section in a waveguide or horns ; we give not it explicitly because to much explanation is required.

Suppose now that a part of waveguide is infinitely repeated, with an output impedance (at infinity) is z_o, then for each addition of a portion of waveguide the input impedance

- remains at a finite distance of a fixed point if the transformation is elliptic

- goes to infinity with a null velocity if the transformation is parabolic

- goes to infinity with a positive velocity which is called the Liapunov exponent of the transformation.

We shall see with more details the asymptotic properties of the impedance after a lot of matrix produces.

3. ASYMPTOTIC LAWS AND RANDOM WAVEGUIDES

A very large amount of litterature [6-9] exists on asymptotic laws of products of random matrices. Because originally applied to solid state physics, the results must be used with some modifications for the design of performant mufflers.

The major differences comes from the number and nature of matrices : in acoustics the matrices used commonly can be very different each other and they depend drastically of a parameter - the sound frequency - although their number in the product is not so important say (5-10). For that reason the weak disorder approximations cannot be made in our case, we may use some applications of [10-11] which are also valid for deterministic produces. The most useful result for our purpose can be found in [11] p.41 and, translated in terms of acoustician vocabulary it gives :

Suppose you have a waveguide composed with n portions numbered from the source to the output, the output impedance being the radiation impedance then the input impedance converges to a point on the real axis when $n \longrightarrow \infty$ (the waveguide is changed by adding a new portion of waveguide at the output). The value of this impedance depends mostly on the part of the waveguide which is near the source and weakly on the part which is far from it and on the load impedance at the output. Secondly the distance between the input and output impedance gives roughly speaking the attenuation of the muffler (this law is only true for large n number of matrices).

From these results, the only periodic muffler with attenuation proportional to the length of the waveguide corresponds to an hyperbolic transformation. If one try to realize practically such transformations, one see that it is possible only in a range of frequencies, and repeating periodically this waveguide one obtains successive pass band an stop bands frequencies ; for that reason periodic mufflers give attenuation on relatively narrow range of frequencies and such arrangement is not suitable, as suggested in [12].

At the opposite of periodic media, random arrangement gives attenuation for all frequencies (although its importance can depend on frequency). This result is well known in localization theory, and has been developped in [13-14].

Although a random waveguide is a good idea in itself, its practical realization is almost impossible because a muffler is usually composed of resonators, tubes and expansion chambers whose the number is rarely more than ten and asymptotic behaviour is not fully obtained. We show an example in a next section with an arrangement of expansion chambers.

If one considers the impedance z_M at the input of the mufflers and that new portions chosen at random are added at the output, with z_o for radiation impedance, one has :

$$z_n = T_1(T_2(\dots T_n(z_o) \dots)) \qquad (2.1)$$

The homographic transformation keeps invariant the distance defined in (1.6) then

$$d(z_n, z_{n-1}) = d(T_n(z_o),\ z_o) \qquad (2.2)$$

Eq.(2.2) suggests that the point z_n has a trajectory as n grows defined by a Markov chain. At each matrix M_n are associated two angles α_n and β_n and a length λ_n : suppose the matrix n is chosen (by a random generator process) then z_n is obtained from z_{n-1} by rotating of an angle α_n then translating of a length

$\lambda_n = d(T_n(z_o), z_o)$ and then rotating of an angle β_n, as shown on Fig.6. Roughly speaking the point z_n (as n increases) has a trajectory which looks a zig-zag or random walk. Obviously this process does not give the optimum in escaping the departure point z_o, and then not give the optimum for wave attenuation.

4. OPTIMIZATION TECHNIQUES

Since the attenuation depends on the hyperbolic distance between input and output impedances it could be thought that the faster way to escape is to follow a geodesic ; in this context the better muffler composed with N>M portions of waveguide is made with the better muffler for M portions plus N-M added portions. Then it is easy with a computer to find iteratively the good choice for muffling at the best without to enter in the details of trajectories of impedance. Athough this direct method should be only valid for a given frequency, the following example will show that even in this case it does not work so simply.

Suppose we want to realize a muffler composed with successive expansion chambers. The homographic transformations corresponding to a length of tube ℓ and an expansion chamber of length ℓ are rotations of angle $k\ell$ (k is the wavenumber) around two points on the imaginary axis, as shown on Fig. 7. The optimization problem reduces in this case to the choice of the lengths ℓ_1, ℓ_2... etc of tubes and expansion chambers. It is easy to see that as soon as the impedance reaches a point on the imaginary axis under the two centers it cannot escape under the action of any of the two rotations ; the optimizations gives $k\ell = 2m\pi$, m integer or zero. This is just that one obtains directly by computing : paradoxally one gets the lower attenuation for the frequency for which the larger attenuation was wanted.

We avoid this difficulty by considering attenuation globally in a range of frequencies rather than a precise frequency. Suppose the sound source has a noise with spectrum $f(\omega)$ (ω is frequency), we say that a muffler is better if

$$\int_{\omega_1}^{\omega_2} f(\omega)\gamma(\omega)\, d\omega \quad \text{(where } \gamma(\omega) \text{ is the attenuation at frequency } \omega\text{)}$$

is maximized . This does mean that muffler is better at every frequency ; we think this requirement is too hard to be realized in practrice. In the example given in the next section we take $f(\omega) = 1$ and $\gamma(\omega) \sim d(z_e(\omega), z_n(\omega))$ which is only asymtotically exact. The point which is very important is to add new portions of waveguides near the output, it corresponds to the product

$$M_1 M_2 \dots M_n \tag{3.1}$$

rather than adding near the source which corresponds to products taken as in usual order

$$M_n M_{n-1} \dots M_2 M_1 \tag{3.2}$$

In this case the input impedance fluctuates very quickly and the optimization is no more valid.

5. NUMERICAL AND EXPERIMENTAL RESULTS

We apply the above method to an arrangement of expansion chambers in the frequency range 150-500 Hz. The low frequency 150 Hz corresponds to an arbitrary limit of length of 1 m for the expansion chambers ($kl \sim \pi$). To get efficient muffling for lower frequencies a greater length is required.

Although the optimization has been made without to take into account dissipative effects due to viscosity and to thermal conductivity, after the choice we enter it in the equations. The important result is that much attenuation is obtained, especially in the case of low attenuation without dissipation but the main shape of attenuation versus frequency remains the same. In any case we suppose that there is no fluid flow in the waveguide and we neglect the higher modes effects. Of course this is not realistic, but in the case of expansion chamber it is shown in [15] that the mean flow effect is not important and in [16-17] that for the dimensions of chambers in question it is possible to neglect the higher modes effects.

The attenuation appearing in the following curves are not the insertion loss as usual [18] but another quantity which seens to us to be more significant of the muffling effect. Following the idea of Fukuda [19] we use the quotient of radiating power with and without muffler . An hypothesis is made for the noise source : [20] we suppose that acoustic velocity of the source is constant. Experimental results are obtained with a woofer as a noise source ; the radiated power being proportional to the pressure for large distances we obtain two transfer fonctions P_o/V_1 with and without muffler, the attenuation figuring in our curves are their quotient. The acoustic pressure p_o is mesured at 20 cm of the output (the output diameter of the muffler is 3 cm) and the input velocity at the input is measured with an intensimeter METRAVIB.

On Fig. 8 one shows the results obtained for a "random" arrangement of expansion chambers by computer simulation.

On Fig. 9 the same result for an optimized waveguide is shown and on Fig. 10 the corresponding pratical realization. The Fig. 11 is the same arrangement turned back : although the results are interesting they are a little weaker than in Fig. 9.

6. CONCLUSION

The geometrical method exposed in this paper can be considered as a generalisation of the Smith chart. It can be applied to any one dimensional wave propagation problems as well as design of electric filters.

We come to the conclusion that considerable progress in the design of performant muffler can be made. It is shown that periodic mufflers are not good to provide a wide band attenuation and random arrangement of waveguide gives better results. We will try to apply this method to waveguides with variable sections (like in horns) : the attenuation is expected to be more important.

REFERENCES

[1] L.R. FORD, Automorphic functions . Chelsea Pub. Co. (1951).

[2] M.E. GERTSENSHTEIN and V.B. VASIL'EV, Waveguides with random inhomogeneities and brownian motion in the Lobachevsky plane. Theory Prob. ApplicationsIV n°4 (1959).

[3] R.A. CHIPMAN, Transmission lines. Shaum outlines series Mc Graw Hill Co. (1968).

[4] V.I. ARNOLD et A. AVEZ, Problèmes ergodiques de la mécanique classique. Paris, Gauthier-Villars (1967).

[5] L. BRILLOUIN et M. PARODI, Propagation dans les milieux périodiques. Paris, Masson-Dunod (1956).

[6] P.W. ANDERSON, Absence of diffusion in certain random lattices. Phys. Rev. 109, p. 1492-1505 (1958).

[7] V.N. TUTUBALIN, On limit theorems for a product of random matrice. Theory of probability Appl. , 10 n°1, p.25-27(1965).

[8] H. FURSTENBERG, Non commuting random products. T.A.M.S., 108 p. 377-428 (1963).

[9] S. ALEXANDER, J. BERNASCONI, W.R. SCHNEIDER, and R. ORBACH Excitation dynamics in random one-dimensional systems. Rev. of modern Phys., 53, n°2, p 175-198 (1981).

[10] Y. GUIVARC'H and A. RAUGI, Products of random matrices : convergence theorems contemporary math. 50, p 31-53 (1986).

[11] P. BOUGEROL and S. LACROIX, Products of random matrices with applications to Schrödinger operators, Progress in proba and stat,8, Birkhaüser (1985).

[12] J.S. LAMANCUSA, The transmission locs of double expansion chamber muffler with unequal size chambers. Appl. Acoustics, 24, p. 15-32 (1988).

[13] J.A. FAWCETT, Numerical simulations and analytic approximations of transmission through a random medium. Wave motion, 10, p. 99-114 (1988).

[14] D. DEPOLLIER, J. KEGOMARD, F. LALOE, Localisation d'Anderson des ondes dans les réseaux acoustiques unidimensionnels aléatoires. Ann. Phys. Fr., 11, p. 457-492 (1986).

[15] T.F.W. EMBLETON, Mufflers, in noise and vibration control L. BERANEK Ed. Mc Graw-Hill, N.Y. (1971).

[16] L.J. ERIKSSON, Higher order mode effects in circular ducts and expansion chambers. J.Acoust.Soc.Am., 68, p.545-550 (1980).

[17] J.KERGOMARD and A. GARCIA, Simple discontinuities in acoustic waveguides at low frequencies : critical analysis and formulae. Journal of sound Vibration,114, p. 465-479 (1987).

[18] M.L. MUNJAL, Acoustics of ducts and mufflers, Wiley and sons, N.Y. (1987).

[19] M. FUKUDA, A study on characteristics of cavity-type muffler Bull. of J.S.M.E., 12, p. 333-349 (1969).

[20] M.L. MUNJAL and A.G. DOIGE, On uniqueness, transfer and combination of acoustic sources in one-dimensional systems. J. Sound. Vib., 121, p. 25-35 (1988).

FIGURES CAPTIONS

Fig. 1: Geometry of the waveguide

Fig. 2: The elliptic transformation

Fig. 3: The parabolic transformation

Fig. 4: The geometry of the Helmholtz resonator

Fig. 5: The hyperbolic transformation

Fig. 6: Zig-Zag trajectory of the acoustical impedance

Fig. 7: Optimization of an arrangement of expansion chambers for a precise frequency cannot be made directly

Fig. 8: A random arrangement of expansion chamber : a computer simulation

Fig. 9: An optimized arrangement of expansion chamber : a computer simulation

Fig. 10: The same attenuation curve from experiment

Fig. 11: Experimental results for the same arrangement than in Fig. 8 but turned back (the source is at the output and the input of the waveguide radiates sound waves).

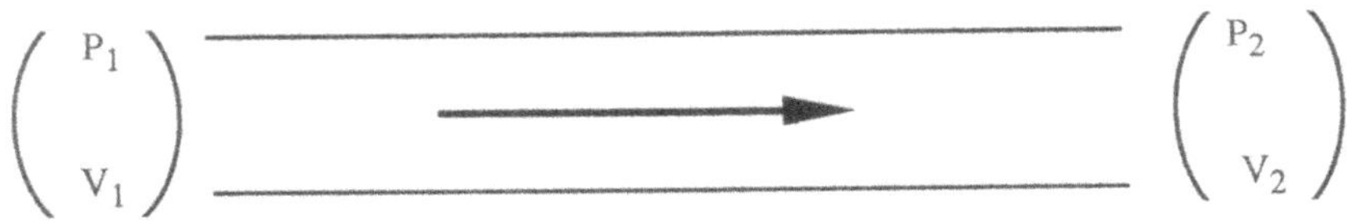

Fig. 1

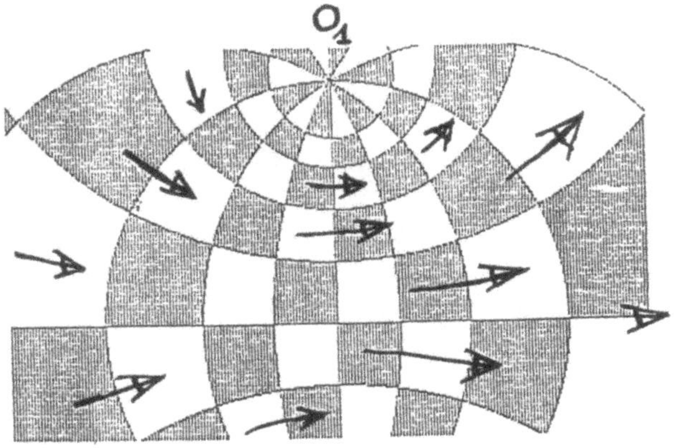

Fig. 2

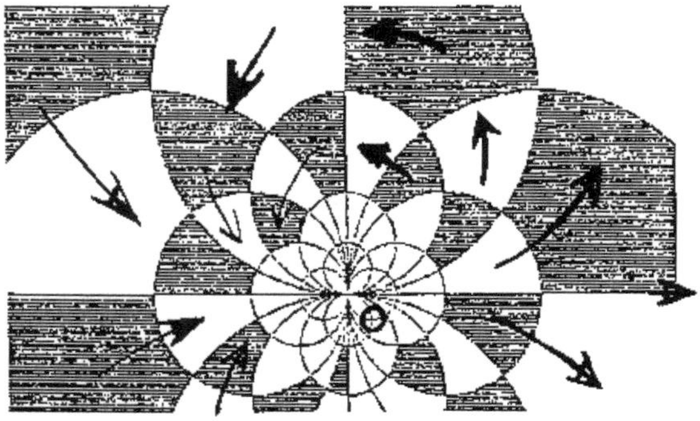

Fig. 3

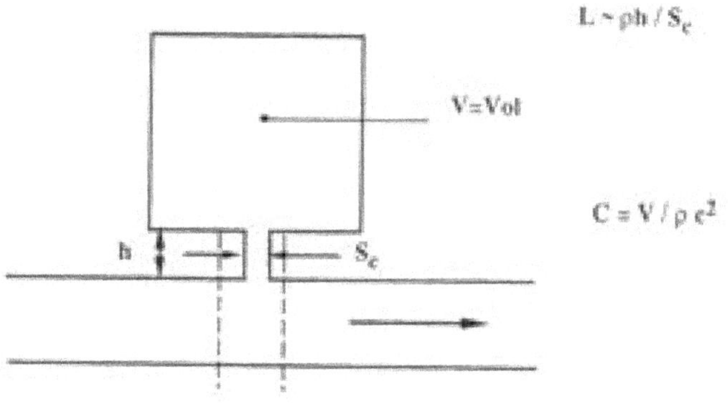

Fig. 4

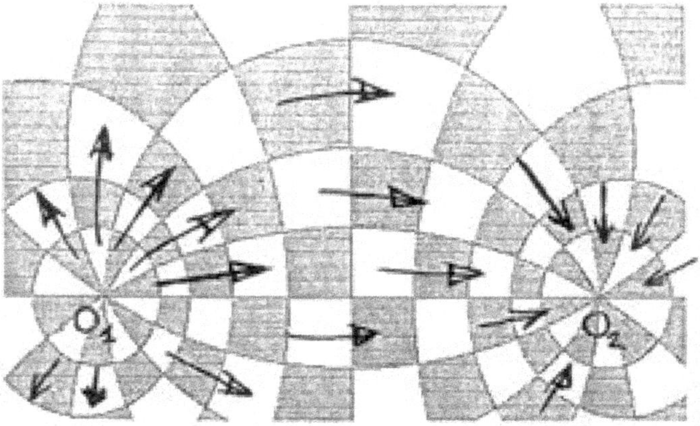

Fig. 5

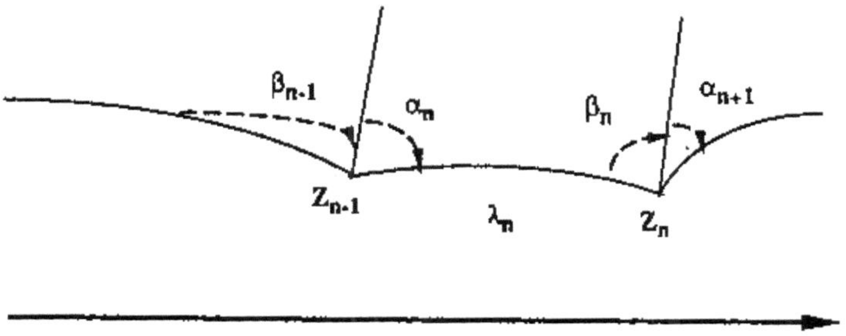

Fig. 6

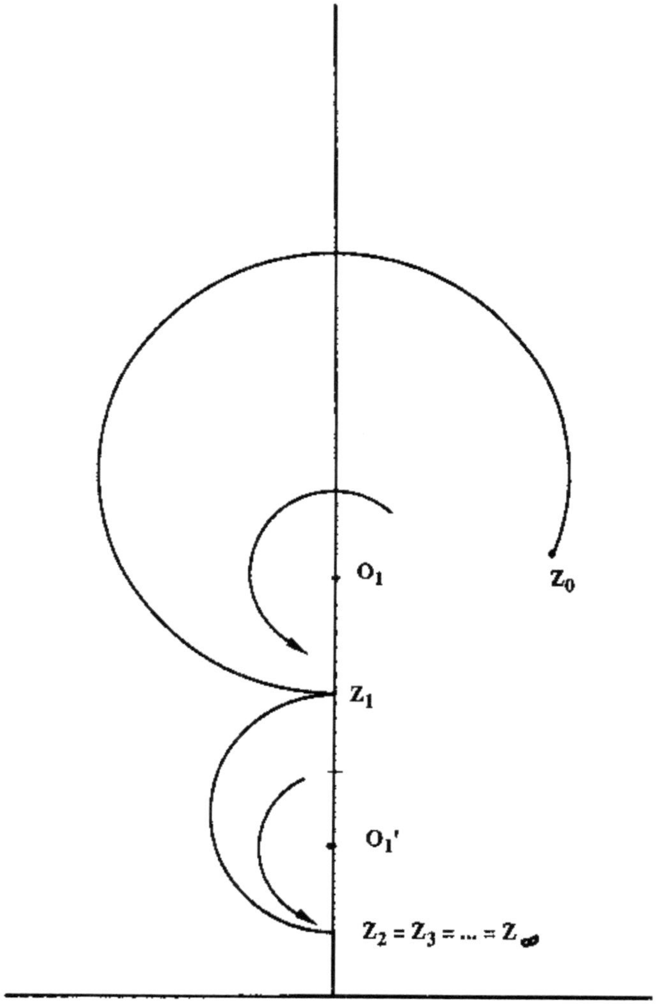

Fig. 7

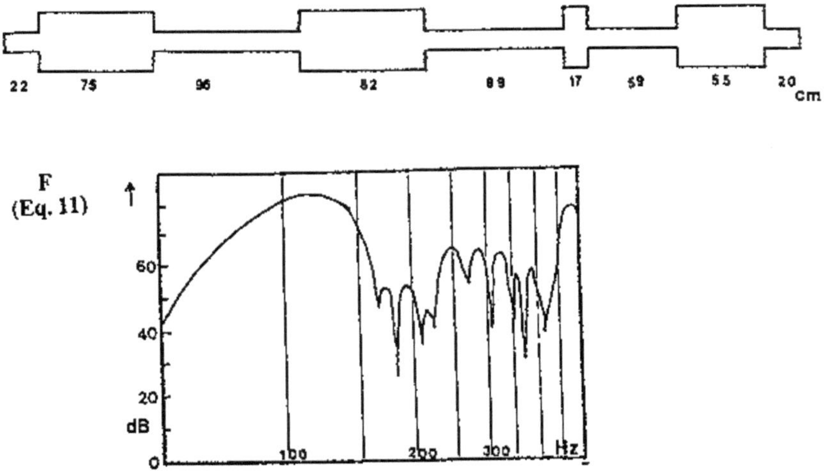

Fig. 8

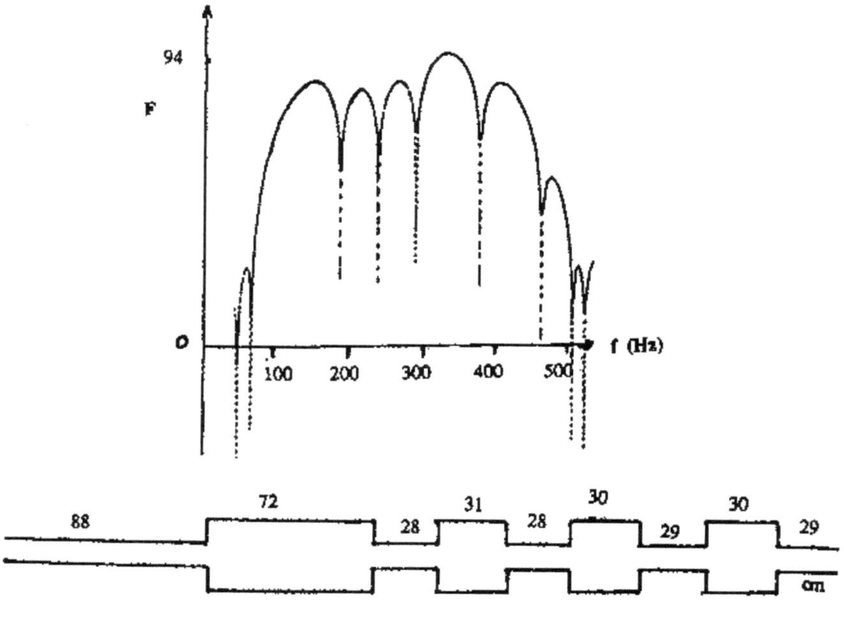

Fig. 9

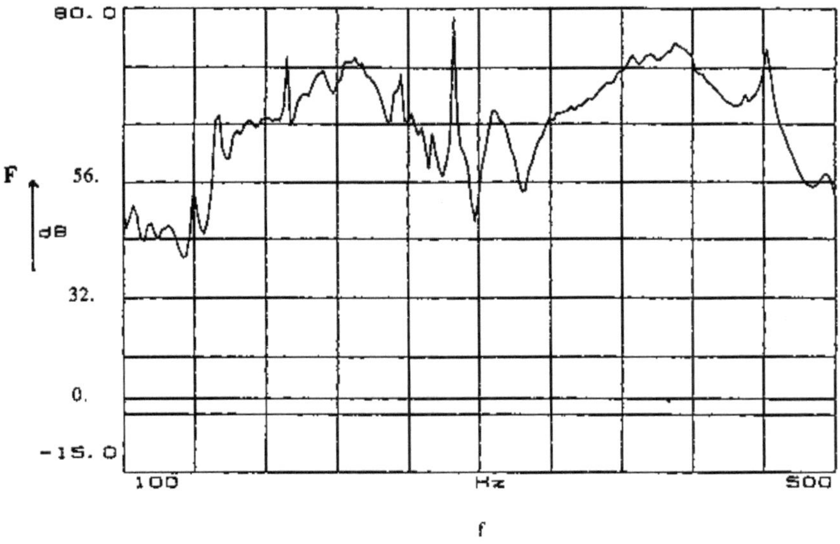

Fig. 10

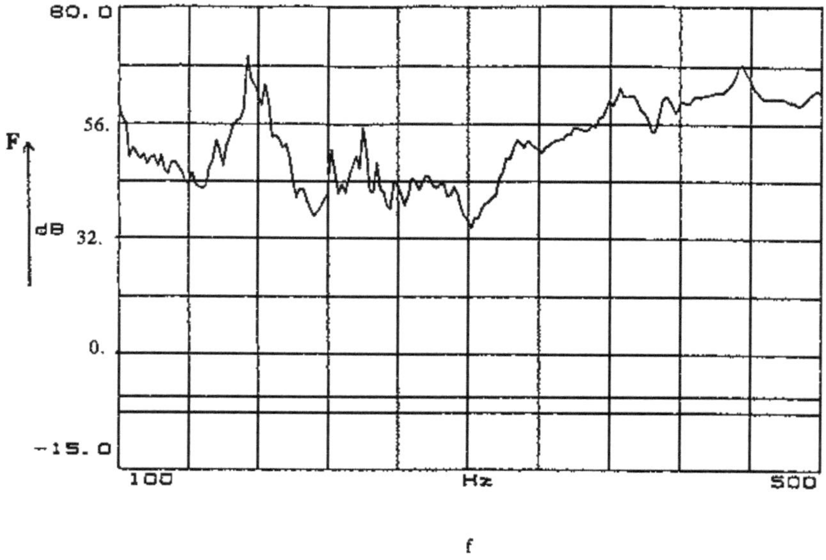

Fig. 11

NONSTATIONARY STIMULATED BRILLOUIN BACKSCATTERING

Carlos MONTES and Olivier LEGRAND

Laboratoire de Physique de la Matière Condensée, Université de Nice, Parc Valrose, 06034 NICE Cedex
and Observatoire de Nice, B.P. 139, 06003 NICE Cedex, FRANCE.

ABSTRACT

Here we present recent issues concerning dynamical features of stimulated Brillouin backscattering through the coherent 3-wave description. Nonstationary conditions are envisaged under which long time behaviors are theoretically as well as numerically investigated. Comparison is achieved with results from a parallel experiment in a fiber ring laser. Amplification and compression of the backscattered Stokes pulse are limited by nonlinear optical Kerr effect which we include into the model. Arguments are given in favor of the possible generation of quasi-solitons in a ring resonator configuration.

RESUME

Nous présentons ici quelques résultats récents sur la dynamique de la rétro-diffusion Brillouin stimulée au travers du modèle à 3 ondes cohérentes. On considère des conditions non stationnaires pour lesquelles sont étudiées, théoriquement et numériquement, les comportements aux temps longs. Ces résultats sont comparés à ceux issus d'une expérience dans un laser à fibre optique en anneau. L'amplification et la compression de l'impulsion Stokes rétrodiffusée sont limités par l'effet Kerr optique non linéaire que nous incluons dans le modèle. Des arguments sont avancés en faveur de la possibilité d'engendrer des quasi-solitons dans une configuration de résonateur en anneau.

I. INTRODUCTION

Stimulated Brillouin backscattering (SBS) is one of the most important nonlinear mechanisms responsible for the limitation of electromagnetic propagation of a pulse in nonlinear optical media. Actually, in optical fibers, appearing at low threshold values, it is the dominant stimulated scattering process. In this parametric instability, a forward propagating pump wave (at frequency $\omega_1 = k_1 c/n$) stimulates a backscattered Stokes wave (at frequency $\omega_2 = k_2 c/n$) in resonant interaction ($\omega_1 = \omega_2 + \omega_s$) with a sound wave (at frequency $\omega_s = k_s c_s \simeq 2c_s\omega_1 n/c$); counterpropagation yields for the phase matching condition: $k_s = k_1 + k_2 \simeq 2k_1$. Recent theoretical and experimental works on nonstationary SBS [1] [2] [3] demonstrate the possibility of achieving efficient compression and amplification of the backscattered Stokes pulse in the presence of a long pump wave.

Within the frame of the standard resonant 3-wave model [2] [3], we outline typical dynamical behaviors arising in the long time evolution of SBS interaction. Effective compression and amplification of the Stokes pulse are obtained by numerically solving the coherent 3-wave SBS equations which yield the following main dynamical features. Starting with a spatially bounded fluctuation, the envelope of the backscattered wave grows while depleting the counterpropagating pump wave. For long interaction times, the backward envelope displays a series of alternating peaks in phase opposition, the amplitudes of which spatially decrease as the distance from the leading edge of the Stokes envelope gets larger. The first peak intensity grows as t^2 while its width shrinks as t^{-1}. We stress the importance of the inertial response of the medium which is properly described by taking the dynamics of the sound wave into account, thus relaxing the usual "strongly damped approximation" which proves to be irrelevant in the nonstationary cases.

Here we wish to present some recent results concerning SBS under essentially nonstationary conditions even for the pump wave input. In order to validate the coherent 3-wave description, comparisons were made between results provided by numerical simulations of the model equations and experimental results obtained in fiber ring configurations [4]. The experimental set-up was a single-mode optical fiber ring laser of length L = 83 m pumped by an Ar-ion CW laser. The numerical investigations of the SBS equations allowed major improvement in the experimental configuration, in particular through a periodic interruption of tbe pump input, thus preventing the accumulation of the sound waves over a short distance from the input end of the fiber, which accumulation would otherwise shorten the effective length of the nonlinear interaction due to a strong and abrupt depletion of the pump wave at its entry. In return, precise measurements of the finite width of the Stokes pulse in the actual experiment led the authors of Ref. [4] to include nonlinear optical Kerr effect into the model, thus accounting for the limitations in the compression supposedly due to phase and amplitude modulation.

In the next section we briefly introduce the coherent 3-wave SBS equations used in the numerical simulations, as well as the two principal categories of asymptotic solutions which analysis yields in the nondissipative case. (The main behaviors still hold in the dissipative case). In the last section, we present experimental results from Ref.[4] in connection with corresponding numerical results obtained through modified SBS equations where Kerr nonlinearities are included. Finally, on the basis of theoretical and numerical considerations, we suggest the possibility of generating soliton-like structures in a particular ring configuration.

II. THE COHERENT SBS EQUATIONS : ASYMPTOTIC SOLUTIONS

Coherent SBS nonlinear evolution equations couple a forward propagating pump wave $\widetilde{E}_1 = E_1 \exp[i(k_1 x - \omega_1 t)]$ to a backscattered Stokes wave $\widetilde{E}_2 = E_2 \exp[-i(k_2 x + \omega_2 t)]$ through a forward acoustic wave $\widetilde{E}_s = E_s \exp[i(k_s x - \omega_s t)]$. Within the "slowly varying amplitude" approximation, the amplitudes of the wave envelopes obey the following system of partial differential equations (for a complete derivation, see for instance Ref.[4]):

$$\begin{aligned}(\partial_t + \frac{c}{n}\partial_x + \gamma_e)E_1 &= -KE_2E_s\\ (\partial_t - \frac{c}{n}\partial_x + \gamma_e)E_2 &= KE_1E_s^*\\ (\partial_t + c_s\partial_x + \gamma_s)E_s &= KE_1E_2^*.\end{aligned} \tag{1a}$$

We write Eqs.(1a) in dimensionless form after introducing time $\tau = 1/KE_p$ and length $\Lambda = c\tau/n$ scales, where E_p is the input amplitude of the pump E_1, and $K = [(\epsilon_0 cn^7)/(2\rho_0 c_s)]^{1/2}\pi p_{12}/\lambda$, is the SBS coupling constant for optical materials[4]. For an optical fiber of fused silica with a refractive index $n = 1.46$, an elasto-optic coefficient $p_{12} = 0.286$, a sound velocity $c_s = 5.96 \times 10^3$ m s^{-1}, an unperturbed fiber density $\rho_0 = 2.21 \times 10^3$ kg m^{-3}, and a pump wavelength $\lambda = 5145\AA$, the SBS coupling value is $K = 65.9$ m s^{-1} V^{-1}. Changing ($E_i/E_p \longrightarrow E_i$; $t/\tau = tKE_p \longrightarrow t$; $x/\Lambda = xnKE_p/c \longrightarrow x$), normalizing the damping rates ($\mu = \gamma_s\tau = \gamma_s/KE_p$; $\mu_e = \gamma_{1,2}\tau = \gamma_{1,2}/KE_p$), and neglecting the sound wave velocity in the last Eq.(1a) (because $nc_s/c \sim 10^{-5}$ in silica), Eqs.(1a) yield the dimensionless form:

$$(\partial_t + \partial_x + \mu_e)E_1 = -E_2 E_s$$
$$(\partial_t - \partial_x + \mu_e)E_2 = E_1 E_s^* \tag{1b}$$
$$(\partial_t + \mu)E_s = E_1 E_2^*.$$

In a first stage, consider the simple case $\mu = \mu_e = 0$. As shown in Ref.[2], the SBS equations leave the phases unchanged in time except for sudden π shifts occuring when fields vanish. Hence we may restrict our attention to the real parts of the E_i's, leading to the modified system :

$$(\partial_t + \partial_x)E_1 = -E_2 E_s$$
$$(\partial_t - \partial_x)E_2 = E_1 E_s \tag{2}$$
$$\partial_t E_s = E_1 E_2.$$

When looking for asymptotic solutions of system (2), an approximate self-similar one can be derived in the case of an initially bounded Stokes envelope counter-propagating with respect to a constant pump input. Following Ref.[2], approximate asymptotics ($t \longrightarrow \infty$) assume the form :

$$E_1 = F_1(\xi)$$
$$E_2 = tF_2(\xi) \tag{3a}$$
$$E_s = F_s(\xi),$$

where $\xi = (x + t)t$ and the F_i's obey the following system of ordinary differential equations :

$$F_2 + \xi \frac{dF_2}{d\xi} = F_1 F_s$$
$$2\frac{dF_1}{d\xi} = -F_2 F_s \tag{3b}$$
$$\frac{dF_s}{d\xi} = F_1 F_2.$$

One should note that the validity of this approximation is limited by the fact that the condition $F_2(\xi = 0) = 0$ (where $x + t = 0$ is the leading edge point abcissa of the Stokes envelope in its rest frame) cannot be satisfied except for the trivial solution $F_2(\xi) = 0$. Hence one should consider the solution of system (3b) as an approximate solution of system (2) as $t \longrightarrow \infty$ only over a semi-infinite $x + t$-domain excluding the origin. Moreover, as pointed in Ref.[2], this self-similar behavior seems to be inherited from the behavior of the solution for the linearized version of system (2) (set $E_1 = 1$) whose domain of validity lies within a decreasingly small neighborhood of $x + t = 0$ and which also exhibits a self-similar dependence in ξ, thus providing a link between both solutions as $t \longrightarrow \infty$. The general solution of Eqs.(3b) depend upon three parameters. Following Ref.[2] we set the integral of motion $F_1^2 + F_s^2/2$ equal to unity, and defining: $F_1(\xi) = \cos[\Phi(\xi)/2]$, $F_2(\xi) = (1/\sqrt{2})[d\Phi(\xi)/d\xi]$, and $F_s(\xi) = \sqrt{2}\sin[\Phi(\xi)/2]$, Eqs.(3b) yield the following ordinary differential equation:

$$\xi \frac{d^2\Phi}{d\xi^2} + \frac{d\Phi}{d\xi} = \sin\Phi \tag{4}$$

The self-similar solution of Eqs.(3b) or Eq.(4) (hereafter called "π-pulse" after Lamb [5] who derived this last equation for the study of self-induced transparency in a two-level inverted medium) fits very well the numerical solution of Eqs.(2) for large values of time except in a vanishingly small neighborhood of the origin $\xi = 0$, where we can use[2] the linearized solution of Eqs.(2) around $E_1 = 1$. Fig.1 shows the numerical solution of Eqs.(2) as function of space in the frame of the backscattered envelope for a large value of time ($t = 40$). We suspect that the "π-pulse" has something to do with the "radiation" spectrum or continuum involved in the inverse scattering theory (IST) of the 3-wave equation (as described for instance by Chiu[6]) in spite of the nondecaying (even singular) behavior of its narrowing leading peak as $t \longrightarrow \infty$.

Now, we shall introduce the other part of the alternative in the IST approach, that is to say the solitary wave solutions or solitons (travelling waves), which are special cases of a broad class of soliton solutions fully described by Chiu[6] (and references therein). Wishing to avoid the rather lengthy calculations involved in the IST approach, we shall use the more heuristic one formerly used by Armstrong et al.[7] and by Nozaki et al.[8].

If we proceed to the following change of frame ($x \longrightarrow x + vt$, $t \longrightarrow t$), Eqs.(2) become:

$$\begin{aligned} [\partial_t + (1+v)\partial_x]\, E_1 &= -E_2 E_s \\ [\partial_t + (v-1)\partial_x]\, E_2 &= E_1 E_s \\ [\partial_t + v\partial_x]\, E_s &= E_1 E_2 . \end{aligned} \tag{5}$$

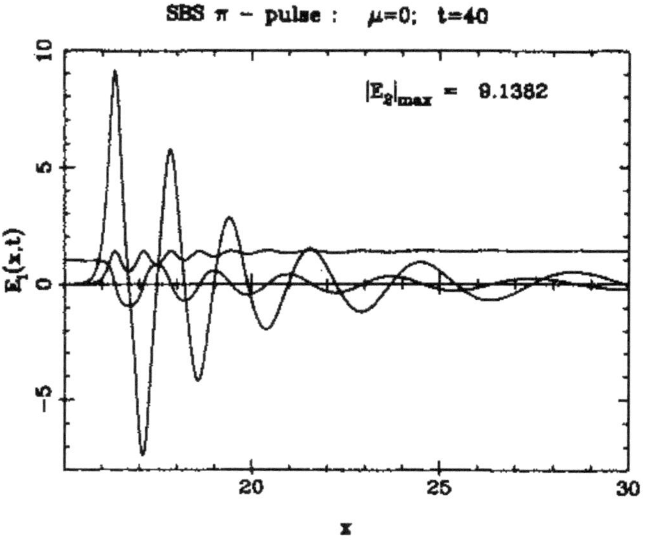

Fig.1 Amplitude spatial distribution for the π-pulse in the frame of the backscattered envelope.

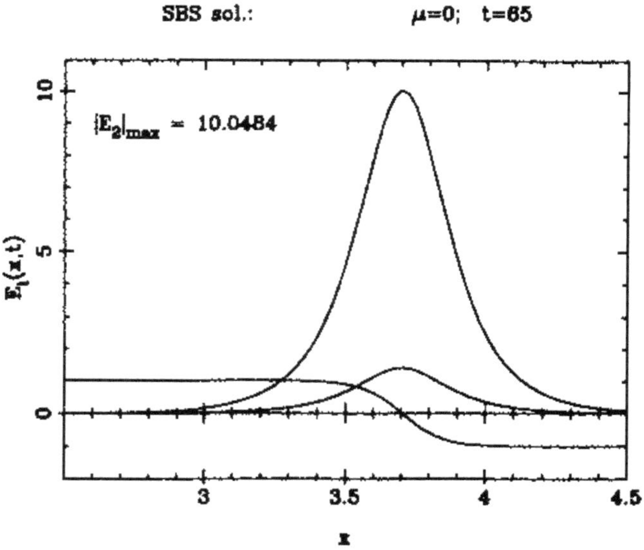

Fig.2 Triple solitary waves solution given by formulas (6,8) for $v = 1.01$ and $A = (1 + v)^{1/2}$.

Then, by defining the A_i's fields as

$$\begin{aligned} A_1 &= |1+v|^{1/2} E_1 \\ A_2 &= |v-1|^{1/2} E_2 \\ A_3 &= |v|^{1/2} E_s \end{aligned} \tag{6}$$

and looking for stationary solutions in the new frame, we have :

$$\begin{aligned} \partial_X A_1 &= -s_1 A_2 A_3 \\ \partial_X A_2 &= s_2 A_1 A_3 \\ \partial_X A_3 &= s_3 A_1 A_2 \end{aligned} \tag{7}$$

where $X = x/|(v-1)(1+v)v|^{1/2}$ and $s_1 = sgn(1+v)$, $s_2 = sgn(v-1)$, $s_3 = sgn(v)$. If one requires $A_1(x=-\infty) = const.$ and $A_{2,3}(x=-\infty) = 0$ (it just means that A_1 is the pump), then one must have $(s_1, s_2, s_3) = \pm(1,1,1)$ or equivalently $|v| > 1$ [see Ref.[9]]. Taking arguments of stability into account[6], we shall be concerned only with the solutions for which $v > 1$ (the $v < -1$ case only intervenes in a transient stage of the nonlinear interaction and is asymptotically unstable). The most general solutions to Eqs.(7) involve Jacobi elliptic functions but if we restrict ourselves to infinite period solutions, they read:

$$\begin{aligned} A_1 &= \mp A \tanh AX \\ A_2 &= \pm A_3 = A\, \mathrm{sech} AX. \end{aligned} \tag{8}$$

This triple solitary waves solution is shown in the original variables in Fig.2. Figs.3(a,b) show the counterpropagation of an initial profile for E_2 given by (8) with $A = (1+v)^{1/2}$ and $v = 1.02$, with respect to a constant pump $E_1 = 1$ and a vanishing initial sound amplitude $E_s = 0$. For long interaction times, a "superluminous" ($v = 1.02$) triple solitary waves structure of the form given in Eqs.(8) leaves the radiative envelope (π-pulse structure) which moves at light velocity ($v = 1$). However, this "superluminous" generation of solitary waves will become impeded as soon as the E_2 background in front of its leading edge gets vanishingly small. Then, the self-similar amplification, which causes a "superluminous" displacement, can no longer take place and causality imposes a further evolution of the π-pulse type. Figs.3(a,b) show the transformation of the "superluminons" quasi-soliton into the "luminous" π-pulse envelope. We stress the importance of an extended E_2 background for a self-similar "superluminous" amplification of the

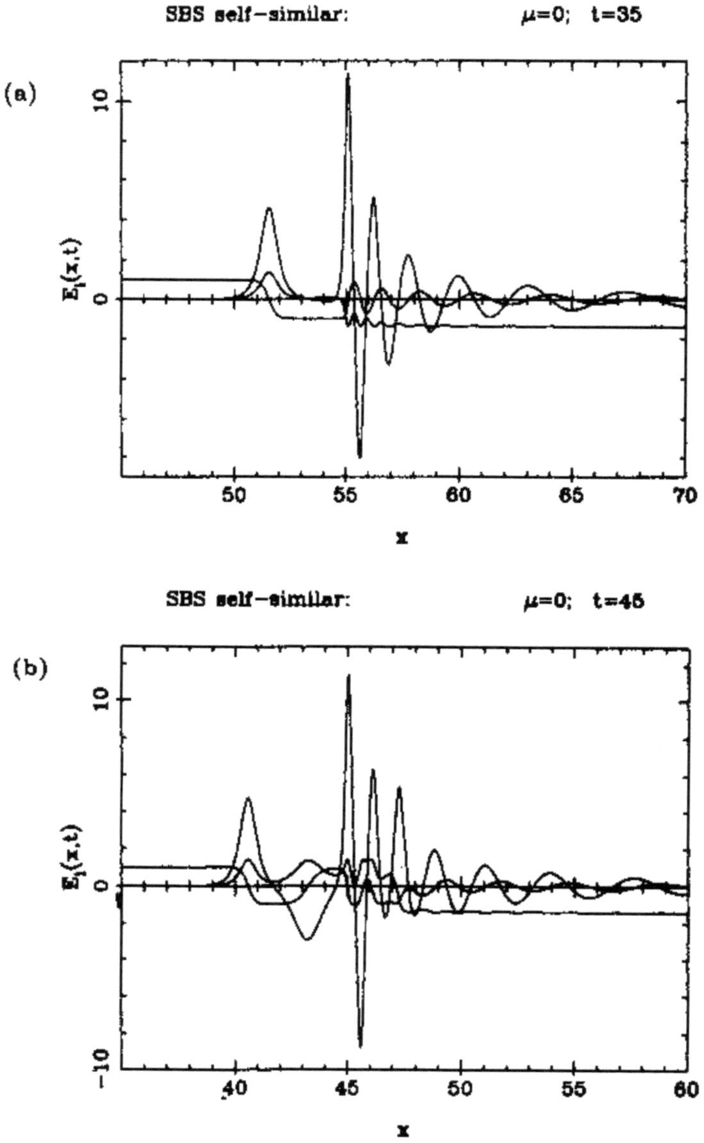

Figs.3 Transformation of the solitary wave structure moving at "superluminous" velocity $v = 1.02$ (a) into the π-pulse structure moving at light velocity $v = 1$ (b).

solitary type to occur. This can be obtained via a ring configuration where the leading E_2 pulse propagates over a permanently excited E_2 background.

III. SBS IN A REAL FIBER-RING CONFIGURATION

First of all, in a real optical experiment, dissipation cannot be neglected. However, it is of little consequence for the type of asymptotic solutions presented above. Sound damping does not affect amplification nor compression of the leading backscattered Stokes pulse in the π-pulse evolution[2]. Moreover, dissipation plays a stabilizing rôle in the solitary wave evolution, imposing maximum amplitude and velocity and smearing out the leaving radiative tail. Numerical simulations in the presence of dissipation ($\mu_e \neq 0, \mu \neq 0$) display a high stability of the quasi-soliton solution.

When the SBS amplification and compression of the backscattered Stokes pulse becomes important we must look for saturation mechanisms in order to interpret the finite spatial width observed in the actual experiment. For a time width of the order of γ_s^{-1}, i.e. for a few nanoseconds, the dispersion is not efficient. However, self-phase modulation due to nonlinear optical Kerr effect[10] seems to be a good candidate because it couples phases and amplitudes, producing self-amplitude modulation and spreading the Stokes pulse. The refractive index of the fiber medium is

$$n = n_0 + n_2|\tilde{E}|^2/2, \tag{9}$$

where n_0 $(= 1.46)$ is the unperturbed refractive index, n_2 $(= 1.27 \times 10^{-22}\ m^2\ V^{-2}$ for silica) is the Kerr coefficient and $\tilde{E} = \tilde{E}_1 + \tilde{E}_2$ is the total electromagnetic field. Here, we only retain the zero frequency terms of the nonlinear Kerr effect which describe a uniform bias of the nonlinear index. The other terms are oscillating at frequencies different from those involved in SBS, namely at $exp[3i(kx \pm \omega t)]$, $exp[i(3kx \pm \omega t)]$, and $exp[i(kx \pm 3\omega t)]$. Those terms can be responsible for Raman instabilities, but with lower gains than the SBS process here considered. Nevertheless, the bias terms of the refraction index are resonant with SBS and will be responsible for self-phase and self- amplitude modulation which considerably modify the profile and width of the backscattered amplified Stokes pulse. Introducing the nonlinear optical Kerr effect in Eqs.(1a) and performing the same change of variables as for Eq.(1b), we obtain the dimensionless SBS-K equations

$$\begin{aligned}(\partial_t + \partial_x + \mu_e)E_1 &= -E_2E_s + iK_r[|E_1|^2 + 2|E_2|^2]E_1,\\ (\partial_t - \partial_x + \mu_e)E_2 &= E_1E_s^* + iK_r[2|E_1|^2 + |E_2|^2]E_2,\\ (\partial_t + \mu)E_s &= E_1E_2^*,\end{aligned} \tag{10}$$

where $K_r = n_2\omega E_p/2K$ is the dimensionless coefficient comparing nonlinear optical Kerr effect to SBS effect, K being the SBS coupling coefficient of Eqs.(1a). The validity domain of Eqs.(10) is restricted to perturbative Kerr contribution.

Here we shall present numerical results corresponding to a parallel experiment[4)] performed with an optical fiber-ring cavity, pumped by a single mode Ar-ion CW laser beam of wavelength $\lambda = 5145\AA$. The dimensionless length L/Λ = 40 corresponds to a maximum input pump power of 50 mW in a silica fiber of length L = 83 m and $3\mu m$ core diameter. The characteristic SBS gain length Λ is about 2 m and the feedback efficiency $\rho \equiv |E_2(x = L,t)/E_2(x = 0,t)|^2 \simeq 0.56$. Fig.4 shows a stable train of SBS Stokes pulses, compressed to 10 ns, which has been obtaiued by strong modulation of the laser beam. Interruption of the pump action at each round-trip time t_r = L$n/c \simeq 415$ ns, allows spontaneous damping of the excited acoustic waves which otherwise accumulate at the input end of

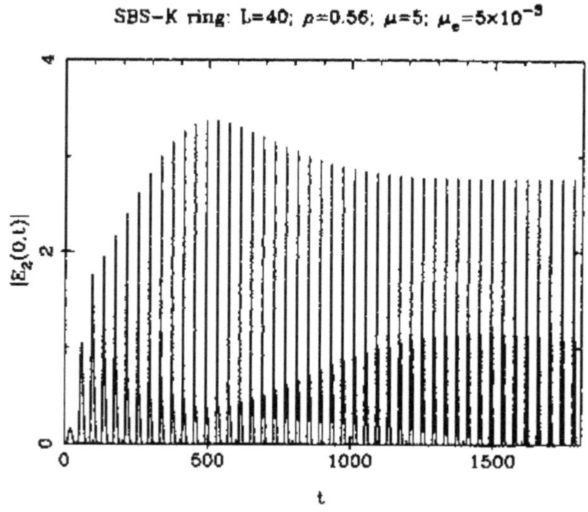

Fig.4 Time evolution of the backscattered amplitude $|E_2(x = 0,t)|$ in the intra-cavity modulated ring configuration obtained from Eqs.(10) [after [4)]].

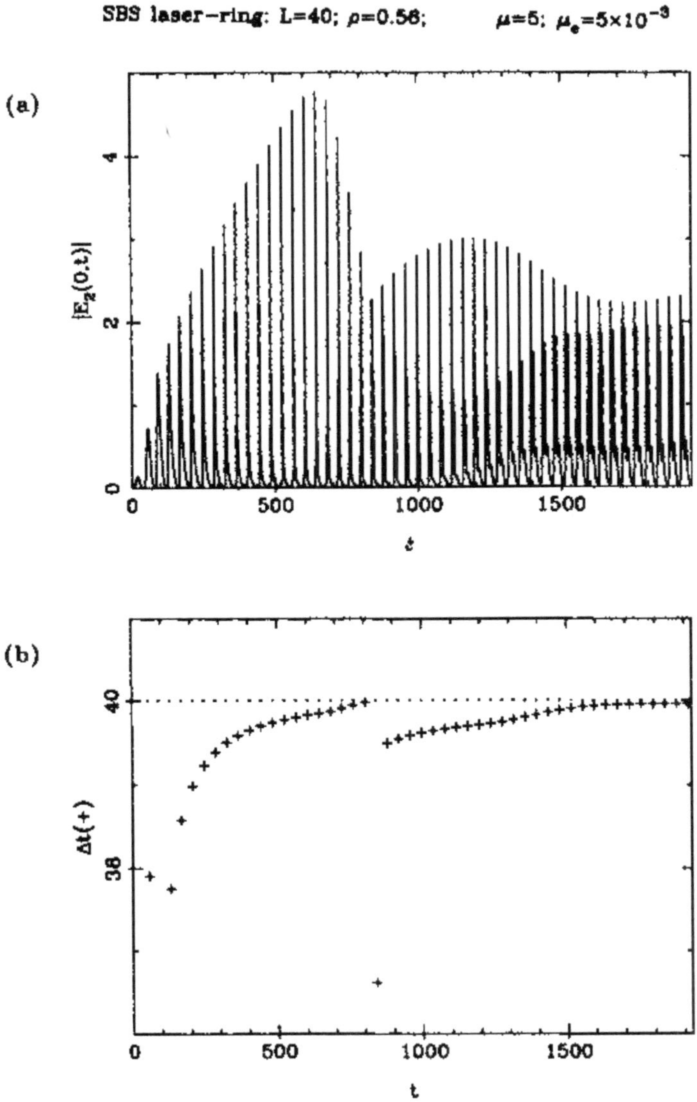

Figs.5 (a) Time evolution of the backscattered amplitude $|E_2(x=0,t)|$ in the ring configuration, where only the pump is modulated, obtained from Eqs.(2).
(b) Time intervals between two pulses (pseudo-periods) versus time. $\Delta t < 40$ represent "superluminous" pseudo-periods.

SBS ring: $L=40$; $\rho=0.56$; $\mu=5$; $\mu_e=5\times10^{-3}$

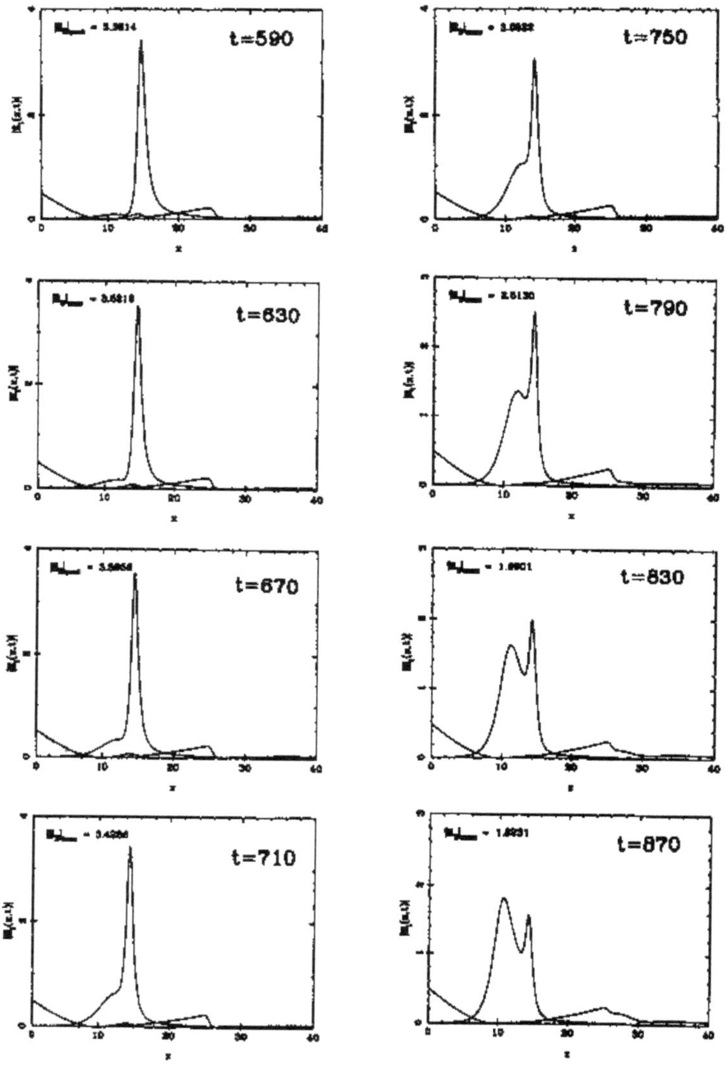

Fig.6 Time sequence of amplitude spatial distribution. Generation of the quasi-soliton structure for E_2 as a self-similar amplification of the Stokes background. From t $\simeq$ 800 on the sech function [cf. formula(8)] fits very well the leading profile of the Stokes pulse.

the fiber and shorten the effective length of the nonlinear interaction medium by depleting the pump just at its entry. Pump modulation is therefore a good mechanism to use the maximum fiber depth. In Ref.[4], the modulation is introduced through an intra-cavity acousto-optic modulator which modulates not only the pump but also the backscattered Stokes envelope and therefore reduces its background width. The train of Stokes pulses presents the round-trip period t_r corresponding to a propagation at the light velocity c/n in the fiber : radiative-like behavior.

Now, if we only modulate the pump and recouple the whole Stokes envelope, we allow the Stokes background to fill the total length of the fiber. In Figs.5(a,b) we present the results of a numerical simulation of Eqs.(2) for the ring configuration in the latter case. Fig.5a shows the time evolution of the backscattered envelope $|E_2(x = 0, t)|$ versus time. After a transient amplification stage, an almost periodic regime of Stokes pulses sets in, where the leading edges are separated by pseudo-periods lower than t_r (cf. Fig.5b), clearly displaying the "superluminous" behavior of the solitary wave type. Figs.6 show the spatial distribution of the envelope amplitudes inside the fiber at consecutive periods, before and after the amplitude saturation shown in Fig.5a. One can directly see the generation of the quasi-soliton structure of E_2 as a self-similar amplification of the Stokes background.

1) Hon, D.T., Opt. Lett. 5, 516 (1980)

2) Coste, J., and Montes, C., Phys. Rev. A 34, 3940 (1986)

3) Gorbunov, V.A., Sov. J. Quant. Electron. 14, 1066 (1984)

4) Botineau, J., Leycuras, C., Montes, C. and Picholle, E., "Stabilization of stimulated Brillouin fiber ring laser by strong pump modulation", J.O.S.A. B to be published

5) Lamb Jr., G.L., Phys. Lett. 29A, 507 (1969); Rev. Mod. Phys. 43, 99 (1971)

6) Chiu, S.C., J. Math. Phys. 19, 168 (1978)

7) Armstrong, J.A., Sudhanshu, S.J. and Shiren, N.S., J. Quant. Elect. $QE-6$, 591 (1974)

8) Nozaki, K. and Taniuti, T., J. Phys. Soc. Jpn. 34, 796 (1973); Oshawa, Y. and Nozaki, K., J. Phys. Soc. Jpn. 36, 591 (1974)

9) McKinstrie, C.J., Phys. Fluids 31, 288 (1988)

10) Stolen, R.H. and Lin, C., Phys. Rev. A17, 1448 (1978)

A STUDY OF ACOUSTIC TRANSMISSION OF TRANSIENT SIGNAL IN AN INHOMOGENEOUS MEDIUM WITH THE HELP OF A WAVELET TRANSFORM. APPLICATION TO AN AIR-WATER PLANE INTERFACE.

Saracco G.*, Tchamitchian Ph.**

*Laboratoire de Mécanique et d'Acoustique, équipe d'Acoustique sous-marine et Ultrasons, C.N.R.S. 31 chemin J. Aiguier 13402 Marseille Cedex 09 FRANCE.
**C.P.T. et Faculté des Sciences et Techniques de Saint-Jérôme 13397 Marseille Cedex 13, FRANCE.

1. Introduction.

The analysis of acoustic propagation in an inhomogeneous fluid medium, generated by an impulsive point source (fluid-fluid or fluid-solid interface) has been discussed innumerable times [1-6]. Most of these works use the Cagniard-De Hoop method which consists essentially in an application of a Fourier transform in the space variables along the plane interface and of a Laplace transform in the time variable. The resolution of the propagation equation is very easy in the transformed domain, but requires elaborate tools for the return to the original space-time domain. An exact analytic solution, hard to interpret physically, can be found only in certain cases.

More recently, it has been shown that one can use also a Fourier transform in the time variable (so-called impulse method [7]). The solution is then easier to interpret physically; however the problem then becomes to find an explicit causal solution.

Our aim here is the analysis and solution of an acoustic propagation problem in which we preserve at every scale the time and frequency characteristics of different waves of the transmitted acoustic field [8]. A recent time-and-scale method adapted to the study of rapid transitions is the wavelet transform. This method can furthermore decompose an arbitrary signal into a sum of elementary contributions, each well localized both in time and in frequency domain.

The properties - on the one hand - of the analyzing function called wavelet and - on the other hand - of the transform, will allow us to study selectively and with arbitrary precision the different contributions or "waves" that appear in the transmitted acoustic field. These contributions are calculated exactly, without the use of high-frequency asymptotic expansions. The separation of the acoustic field into three contributions comes about because of the existence of branch-points in the expression for the acoustic potential. The total field can then be obtained as a sum of these three contributions, due to the linearity of the transform.

We shall see later, by analogy with the existing formulas for inversion of the transform, how one can obtain a reconstruction of the time behaviour of the source. This one is obtained from the knowledge of the total acoustic pressure at all point under a given point at the interface (integration over all depths). The concrete interface problem to be studied is the air-water interface.

The point source, located in a medium of propagation velocity c_1 (air), at a finite height h above the interface, is described by an arbitrary function of time. The observation point is in the second medium, of propagation velocity $c_2 > c_1$. These two fluids are supposed to be homogeneous and isotropic. In an arbitrary point of space and time, we are interested in the behaviour of the acoustic pressure.

A preliminary experiment performed in laboratory will be shown.

2. What is a wavelet transform?

The wavelet transform is a method that allows the decomposition of an arbitrary signal into a sum of elementary contributions called wavelets. They all have the same shape and are obtained by dilation (contraction) D^a and translation T^b from an original wavelet, called analyzing wavelet. They form a two-parameter family $g^{(b,a)}(p)$ (p being a space or time variable), where (b,a) belongs to the open half-plane a>0.

$$g^{(b,a)}(p) = D^a\, T^b\, [g(p)] = a^{-1/2}\, g((t-p)/a).$$

There are many possible choices of wavelets g(p). However, they have to satisfy a - not very restrictive - admissibility condition:

$$(1) \qquad \int |\hat{g}(\omega)|^2/|\omega| \, d\omega < \infty \; ,$$

where $\hat{g}$ is the Fourier transform of g. In practice, this means $\hat{g}(0)=0$, ie: g is of zero mean.

We can also impose additional restrictions on g, depending on the problem to be solved. We shall require here the wavelet to be causal in frequency space (or progressive) and to be well localized in time and in frequency. The progressivity of the wavelet allows us to define correctly the phase of the transform. The wavelet used here will be of Morlet type [9], i.e. a modulated gaussian.

$$(2) \qquad g(p) = \exp(i\, \omega_0\, p) \; \exp(-p^2/2\, \sigma) \; .$$

We have seen that the wavelet has to be admissible (1). Numerically, this will hold if ω_0 is greater than or equal to 5.5 when $\sigma = 1$, if we suppose that the analyzing wavelet is known with a precision of10^{-5}.

The wavelet transform of a signal s(p) at a point (b,a) of the half-plane is defined as the scalar product of this signal and of the wavelet
$a^{-1/2}\, \bar{g}\; ((p-b)/a)$:

$$(3) \qquad S(b,a) = a^{-1/2} \int_{\sup(g)} s(p)\; \bar{g}((p-b)/a)\; dp$$

Here $\bar{g}$ is the complex conjugate of g.

If p is a time variable, then the transformation is the transform of an analytic signal [10]. This is due to the progressivity of the wavelet, and can be seen if the definition (3) is written in frequency space:

$$(4) \qquad S(b,a) = a^{1/2} \int_{\sup(g)} s(\omega)\; \hat{\bar{g}}(a\omega)\; \exp(-ib\omega)\; d\omega$$

The admissibility condition is necessary, since it gives rise to a transform that is isometric, in the following sense:

There exists a constant C(g), depending only on the wavelet g, such that for every signal s(p) one has:

$$|s(p)|^2 dp = c(g) \int |S(b,a)|^2 db\, da/a^2 .$$

In other words, one has conservation of signal energy. There exists then an inversion formula that allows us to reconstitute the signal. It is:

$$S(p) = R_e[\, C(g) \int S(b,a) a^{1/2} g((p-b)/a)\, da\, db/a^2 \,]$$

where R_e denotes the real part.

This is not the only possible inversion formula. Another one, which will be used below, is given by a one - dimensional integral over the scale parameter:

$$S(p) = R_e[\, K^{-1}(g) \int S(p,a)\, da/a \,]$$

where K(g) is defined by:

$$K(g) = (2\pi)^{1/2} \int \hat{g}(\omega)/\,|\omega|\, d\omega ,$$

and is required to be different from zero.

The wavelet is then assumed to satisfy:

$$\int \hat{g}(\omega)/\,|\omega|\, d\omega < \infty$$

We should make the following remark: Independently of the choice of the wavelet, all the information contained in the signal can be found in the wavelet transform. However, when we want to study some special characteristics of the signal, we shall choose an analyzing wavelet that is appropriate to the phenomena to be observed.

We do not need any "a priori" knowledge on the signal in order to apply the transform, since it is a non-parametric method. On the other hand, and this is the main point, the linearity of the transform will allow us to study separately the various contributions of the refracted field. The total field is obtained by summation of the contributions. This transformation, applied to wave propagation, allows us to analyze the signal while keeping track of time and frequency characteristics.

3. Applications of wavelet transform to the propagator: formulas.

The study of acoustic propagation through a plane interface shows that the acoustic field can be written in the form:

$$(6)\quad \hat{\Phi}(r,z,\omega) = \frac{i\ \mathrm{sgn}(\omega)}{c1}\sqrt{\frac{2}{\pi}}\,\hat{F}(\omega)\,\omega\int_0^{\infty} J_0\!\left(\frac{u\omega r}{c1}\right) N(\omega,u,z)\,\frac{1}{D(u)}\,u\,du\,,$$

where $\hat{\Phi}(r,z,\omega)$ is the Fourier transform with respect to time of the scalar acoustic potential and where:

$$\hat{F}(\omega) = \frac{1}{\sqrt{2\pi}}\int e^{i\omega t}F(t)dt \quad ;$$

$$N(\omega,u,z) = \exp\{\ i\ \mathrm{sgn}(\omega)\,\frac{\omega}{c1}\,[h\sqrt{1-u^2} + z\sqrt{n^2-u^2}]\ \};$$

$$D(u) = m\sqrt{1-u^2} + \sqrt{n^2-u^2}\ .$$

For $u > n$, the expression $\sqrt{n^2-u^2}$ is defined to mean $i\mathrm{sgn}(\omega)\sqrt{|(n^2-u^2)|}$.

Similarly, for $u > 1$, we define $\sqrt{1-u^2}$ as $i\ \mathrm{sgn}(\omega)\sqrt{|(1-u^2)|}$.

J_0 represents the Bessel function.

c_1, n are respectively the sound velocity in air and the refraction index for water: one has $n < 1$; m is the ratio of densities of the two media: one has $m > 1$.

This integral can be decomposed into three contributions due to the existence of branch-points . We can then write:

$$\Phi = \Phi_1 + \Phi_2 + \Phi_3$$

These contributions will be called respectively "geometric", where the variable u goes from 0 to n, "lateral" or "surface" contribution for $n<u<1$ and "evanescent" for $u>1$. These name are chosen by analogy with the terminology used in the case of monochromatic source [11].

With the help of the expression of the wavelet transform in (3), we can study separately the behaviour of every contribution, with respect to the analyzing wavelet g(t) described in formula (2). So, we have:

$$(7) \quad (S\Phi)(b,a) = \sqrt{a}\int_0^{\infty} \hat{\Phi}_j(r,z,\omega)\, \hat{g}(a\omega)\, e^{ib\omega} d\omega. \quad (j = 1..3)$$

$$\hat{g}(\omega) = \exp\left(-\frac{(\omega - 5.5)^2}{2}\right)$$

As we have seen, this wavelet is admissible and progressive. In our analytic and numerical evaluation, we have calculated these expressions in the case where $\hat{F}(\omega) = \text{const}$, i.e. where the source emits a δ-function pulse.

Since the problem is linear, we can say in this case ($F(t) = \delta(t)$) that it is equivalent to decompose the signal measured at the hydrophone (transmitted signal) into wavelet contributions and to assume that the source is a wavelet g(t). One can then see which coefficients of wavelet give rise to the three familiar contributions described above.

We will not make any use of asymptotic expansions in the evaluation of these contributions.

4. Applications of wavelet transform to the propagator: numerical simulations.

The numerical evaluation of the propagator and its wavelet transform have been calculated by algorithms of the Romberg type, adapted to the somewhat singular behaviour of the integrand at the boundary. The values of the refraction index n and the ratio of the densities have been chosen, respectively, 0.2266 and 800.

The analysis of signals is perfomed with a range of scale parameter of 6 octaves. Each octave is linearly decomposed into five intervals, called voices, associated to five values of the scale parameter. The first dilation parameter is equal to 0.15. The horizontal axis describes the translation parameters and the vertical axis, the dilation parameters. The signal is considered as a function of time.

The graphic representation are displayed in level lines. The modulus of the transform is shown in logarithmic scale, with a dynamic span of 40db, while the phase representation is linear, between $-\pi$ and π. We go from high to low level lines progressively in decreasing order from black to white color.

The figures 1a and 1b visualize, respectively, the modulus and the phase of the wavelet transform of the "geometric" contribution Φ_1. The observation point is directly under the source (i.e $r= 0$), at a finite depth z under the interface ($z/h = 0.2$). The wavelet transform of the acoustic potential approximates clearly the wavelet transform of a δ-function. The moment of arrival of this contribution can be seen as the vertical line of constant phase on Fig 1b, corresponding to the abscissa a little over 1.0. The same abscissa appears as a line of maximum modulus on any horizontal line on Figure 1a. This result is a very clear indication of the arrival of a discontinuity. Quantitatively, the position of this discontinuity corresponds exactly to the calculated time of arrival.

The figure 2 corresponds to the modulus of the wavelet transform of the "lateral" contribution Φ_2 at the same observation point. In the same scale parameter, we can see the modulus corresponding to the sum of these two contributions (fig 3), due to the linearity of the transform.

1. a:

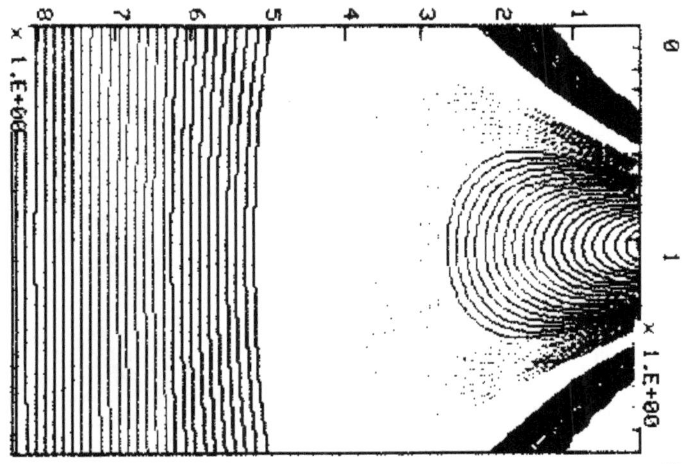

1. b:

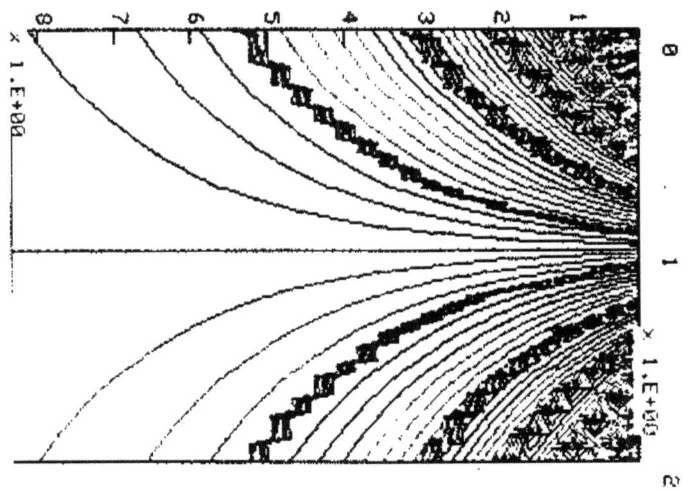

Figure 1.a: Modulus of the wavelet transform of the "geometric" contribution Φ_1. The observation point is under the source, at a depth of z/h=0.2. The smallest dilation parameter is equal to 0.15. The total number of voices is 30 for 5 voices per octave.

Figure 1.b: Phase of the wavelet transform of the "geometric" contribution at the same observation point and parameters (fig.1.a).

fig. 2:

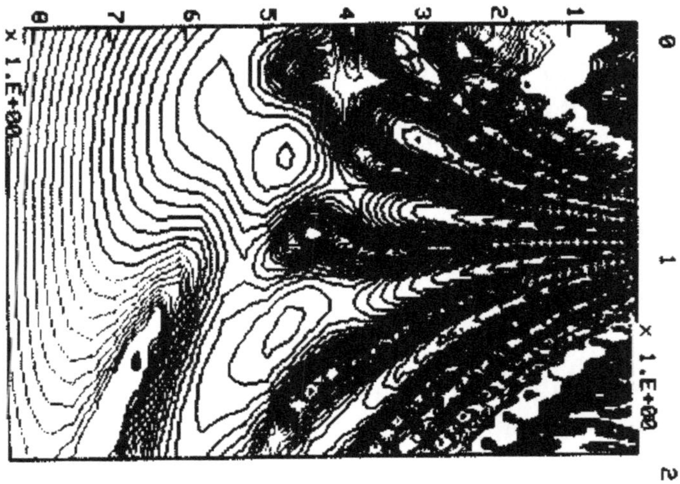

fig. 3:

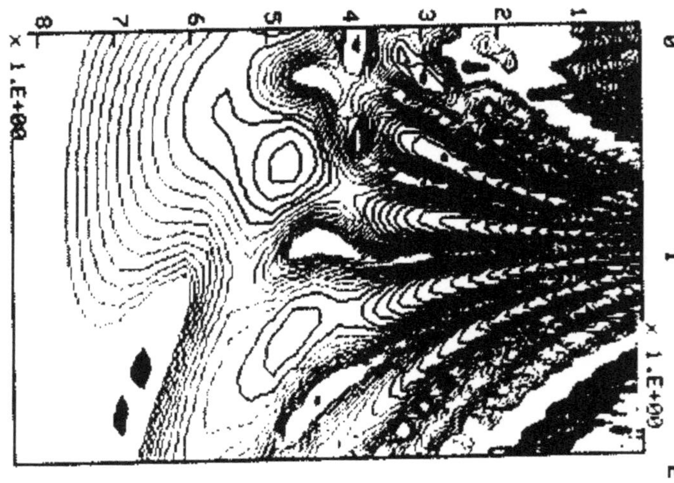

Figure 2: Modulus of the wavelet transform of the "lateral" contribution Φ_2 at the same observation point (r/h=0, z/h=0.2) and scale parameters of the fig. 1a.

Figure 3: Modulus of the wavelet transform of the sum of the two first contribution $\Phi_1 + \Phi_2$. (same conditions that fig. 1.a and 2).

In figures 4 - 8, the observation point is in a region where the contribution Φ_2/Φ_1 increases. It is at a shallow depth z/h = 0.1 and at a finite radial distance from the source r/h = 1, except for the fig. 8 where r/h = 2. In figure 4, one sees the modulus of the wavelet transform of the "geometric" contribution Φ_1, while the figure 5 displays the modulus of the wavelet transform of the "surface" contribution Φ_2. The third or "evanescent" contribution Φ_3 is represented in figure 6, and the total contribution in figure 8. It is very instructive to compare the first two pictures. One notices that the "geometric" contribution arrives before the "surface" contribution, in agreement with theory.

In other hand, as the radial distance from the source increases, we can see the "geometric" contribution split into two parts. The figure 7 shows the modulus of the wavelet transform of the contribution Φ_1.

Independently, we have also calculated the contributions to the propagator, for different coordinates of the observation point. The figures 9.a and 9.b show the results for the "lateral" propagator respectively at a radial distance r/h = 1 and r/h = 2. The wavelet transform applied to the propagator confirms the results already obtained. A more systematic and quantitative discussion is in preparation [12].

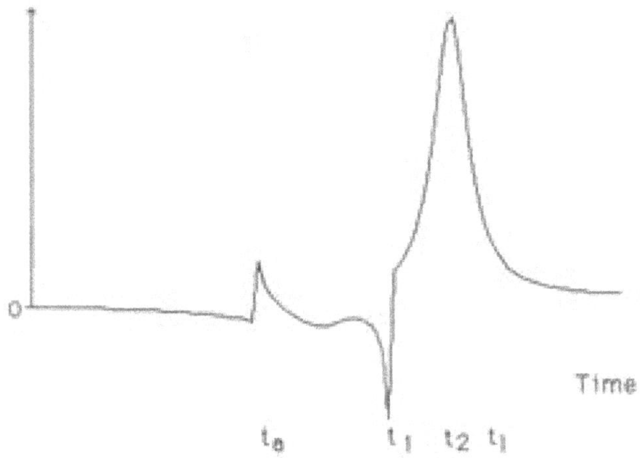

Figure 9.a

fig.4

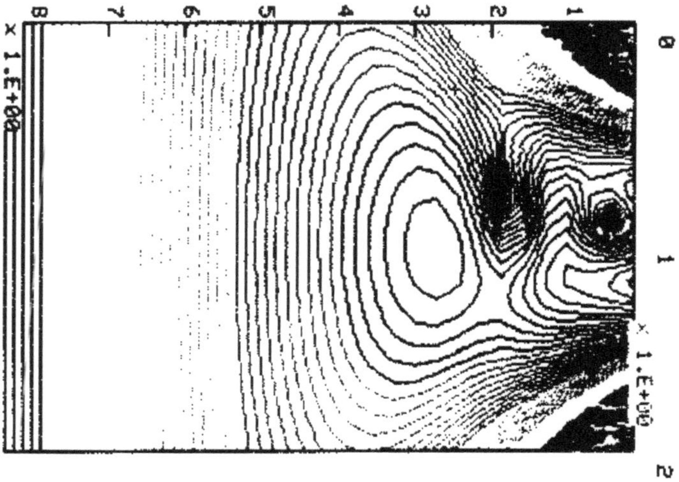

fig.5:

Figure 4: Modulus of the wavelet transform of the "geometric" contribution Φ_1 at a radial distance r/h = 1, and at a shallow depth z/h = 0.1.(same scale parameters of fig.1.a)

Figure 5: Modulus of the wavelet transform of the "lateral" contribution Φ_2 at the same observation point (r/h = 1 and z/h =0.1).

fig. 6:

fig. 7:

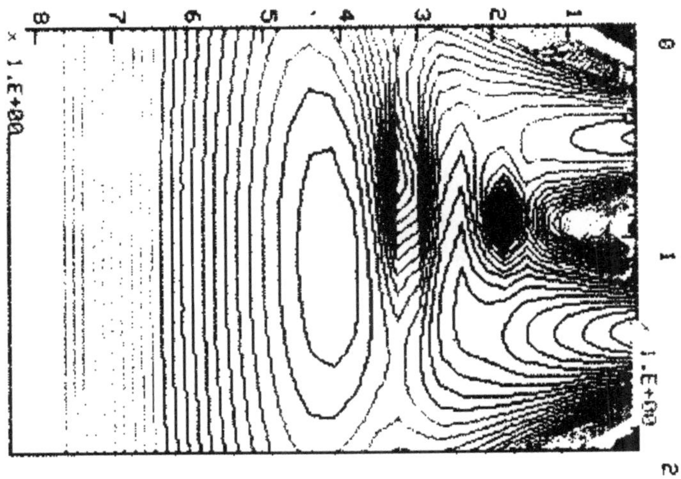

Figure6: Modulus of the wavelet transform of the "evanescent" contribution Φ_3 (r/h = 1 and z/h =0.1).

Figure 7: Modulus of the wavelet transform of the "geometric" contribution Φ_1 (r/h = 2 and z/h =0.1).

The scale parameters are the same of the figure 1.a.

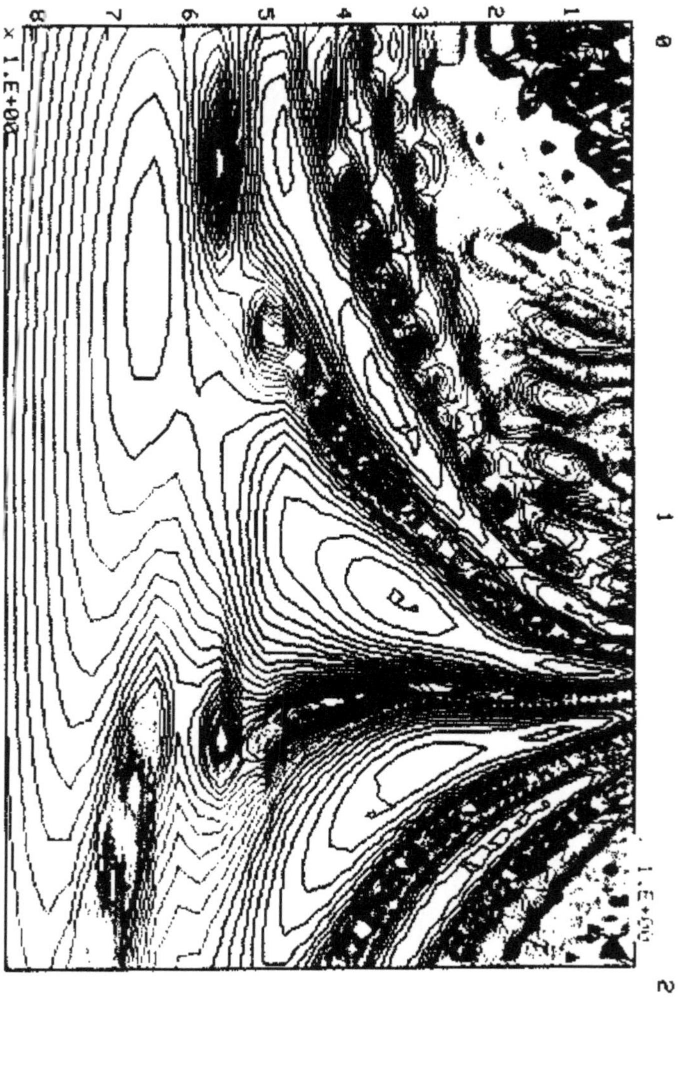

Figure 8: Modulus of the wavelet transform of the sum of the three contribution $\Phi_1 + \Phi_2 + \Phi_3$, at a radial distance $r/h = 1$ and at a depth $z/h = 0.1$. These are the same scale parameters of the fig.1.a.

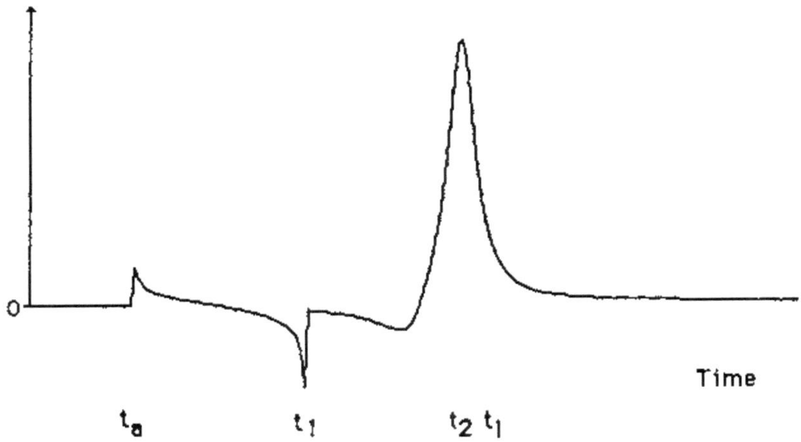

Figure 9.b

Figures 9.a,b: Real part of the lateral propagator (contribution of the "lateral" waves to the Green's function), for respectively r/h =1 and r/h = 2.

t_a = artefact due to the decomposition into "geometric" and "surface waves" contribution seen [12].

t_1 , t_2 = delays appearing in formula (8)

t_l = arrival time of the "surface" wave.

5. Applications of wavelet transform to the propagator: experimental results.

Preliminary experiments were performed in laboratory (acoustic basin) to measure the times of arrival and the behaviour of various contributions. We show some preliminary results obtained, nevertheless the reader will find more information in a paper which is in preparation (Corsain G., Gazanhes C., Grossmann A., Holtzer R., Léandre J., Saracco G., Tchamitchian P.).

The impulsive point source was simulated by a fire-cracker. The source signal measured at a microphone and the transmitted signal measured at a hydrophone were recorded synchronously.

The microphone was fixed at 1m. from the source and gave rise to a reference signal, while the hydrophone moved around in the space occupied by water. The temperature of the two media and air-pressure was measured and taken into account in the calibration of experimental data.

In this preliminary set of experiments we have recorded the transmitted signal at three positions: under the source (r = 0), at r = 1.m. and r = 1.45 m., at a depth of 0.1m. In the bandwidth between 500 - 9.10^3 hz the response of the source is almost linear with accurancy of ±6db and between 9.10^3 - 10^4 hz with accurancy of ±10db.

The analyzing wavelet oscillates at a mean frequency of 8130hz. The wavelet analysis has been made over 6 octaves at 5 voices per octave, starting with a= 0.8.

In figure 10, we see the modulus of the wavelet transform of the source. The modulus of the wavelet transform of transmitted acoustic pressure, just under the source (i.e. r=0) is shown figure 11. The behaviour agrees with the calculations.

The figure 12 displays the same quantities as Fig.11, but at a radial distance from the source r = 1m. We find the splitting describes in Sect. 4, with excellent quantitative agreement of the three times with calculations (t_1, t_2, t_l).

Finally, the figure 13 corresponds to a radial distance r = 1.45m. Here the arrival time t_1 is still in excellent agreement with our calculations but t_2 and t_l are with relative error of 7%. We are working on additional experiments.

6. An expression for the reconstruction of time dependence of the source.

We consider now, a point source with arbitrary time variation F(t). Our aim is to reconstruct F(t) from information obtained through "underwater listening". It turns out that there exists a simple reconstruction formula that involves the total acoustic potential evaluated at all depths for a fixed horizontal distance r. The notations being as above, define P(r,z,t) as the pressure obtained from the total acoustic potential. The formula to be written is valid asympotically as r/h -->∞. Consider $I(r,t) = \int P_2(r,z,t)dz$.

Define $t_2 = \dfrac{\sqrt{r^2+h^2}}{c_1}$ and $t_1 = \dfrac{h\sqrt{1-n^2}}{c_1} + \dfrac{r}{c_2}$.

Fig.10:

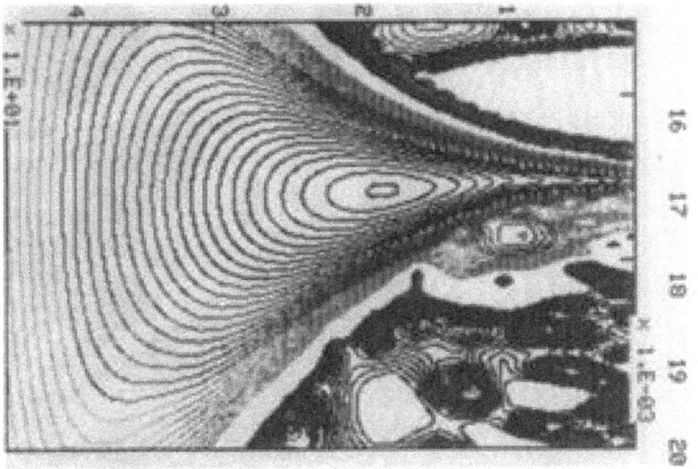

fig.11:

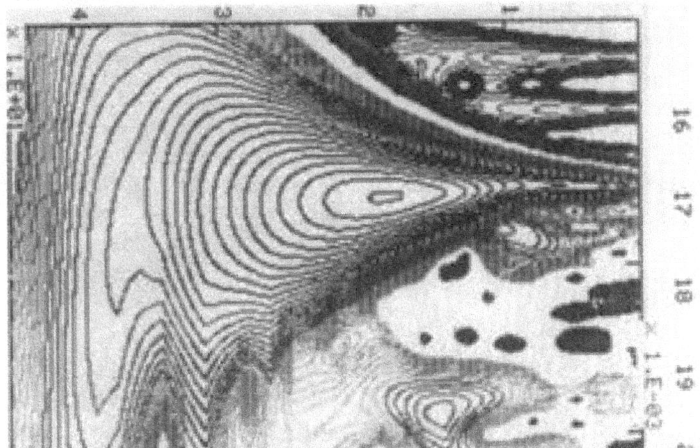

Figure 10: Modulus of the wavelet transform of the signal source $(F(t)=\delta(t))$. The analysis of the signal is performed over 6 octaves and five voices per octaves. The smallest dilation parameter is 0.8.

Figure 11 : Modulus of the wavelet transform of the transmitted acoustic signal (r=0.m and z=0.1m.)

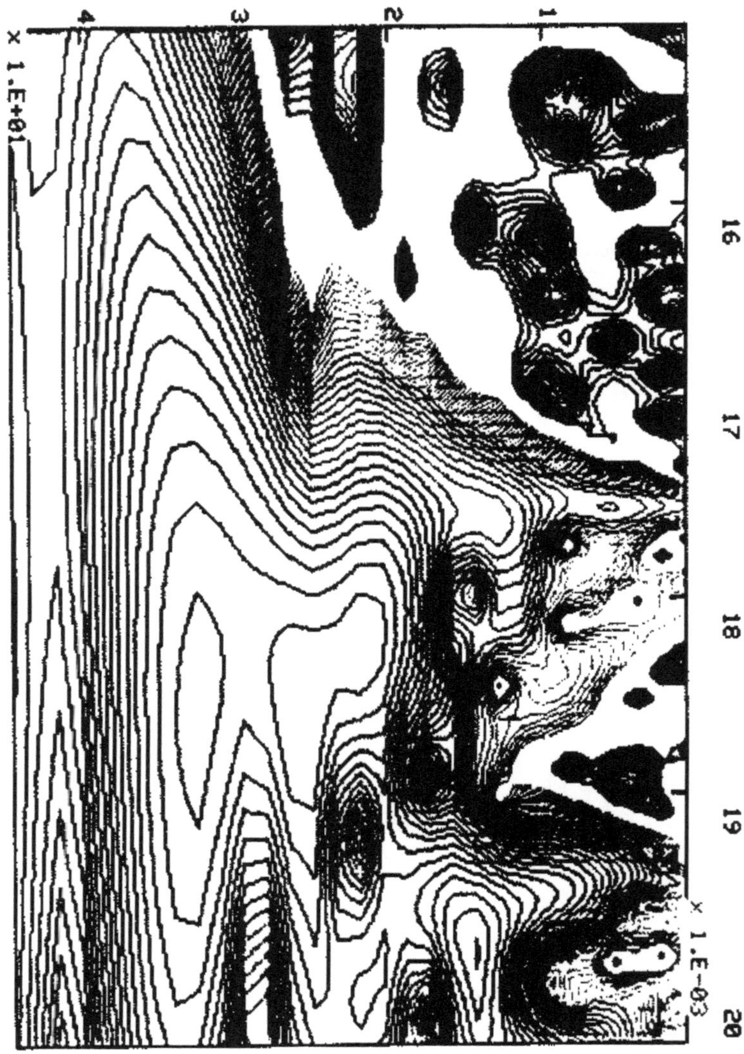

Figure 12: Modulus of the wavelet transform of the transmitted acoustic signal measured at a radial distance r=1.m. and at a depth z=0.1m. The same scale parameters of the fig.10 is used.

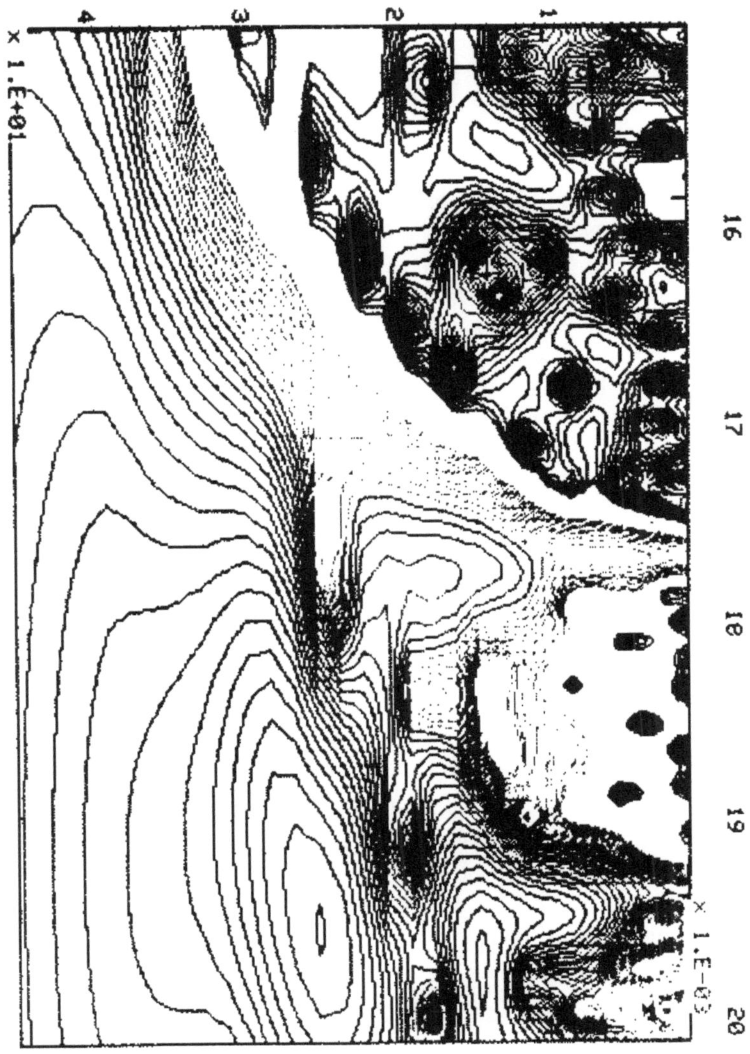

Figure 13: Modulus of the wavelet transform of the transmitted acoustic signal measured at a radial distance r=1.45.m. and at a depth z=0.1m. The same scale parameters of the fig.10 is used.

We have then:

(8) $I(r,t) \approx A(r,h)F(t-t_1) + B(r,h)\ F(t-t_2)$

where A(r,h) and B(r,h) can be explicitly calculated. The arrival time t_1 corresponds to the path of the wave at the critical angle of incidence given by the Fermat principle (extremum path principle). The arrival time t_2 corresponds to the direct path in the first medium. If t_2-t_1 is larger than the duration of the emitted sound F(t), we obtain the exact signal emitted in air repeated twice but with different amplitudes. We have not yet performed any numerical evaluations of (8).

Acknowledgements: One of the authors (S. G.) wishes to thank Dr C. Gazanhes for helpful advice and support. We wish also to express our thanks to M. G. Corsain, R. Holtzer, J. Léandre for their excellent experimental job. Special thanks are due to A. Grossmann for numerous enlightening discussions. This work has been partially supported by the company DIGILOG.

References.

[1] Cagniard, L., Réflexion et Réfraction des Ondes Séismiques Progressives, (Gauthier-Villars, Paris) 1939.

[2] Gottlieb, P., Sound Source near a Velocity Discontinuity, J. Acoust. Soc. Am. 32 (1960) 1117-1122.

[3] Towne, D. H., Pulse shapes of Spherical Waves Reflected and Refracted at a Plane Interface separating Two homogeneous Fluids, J. Acoust. Soc. Am. 44 (1968) 65-76.

[4] Ewing, M., Jardetzky, W., Press, F., Elastic Waves in Layered Media, (McGraw-Hill Book Company, Inc, New York)1957, pp. 74-121.

[5] De Hoop, A. T. and Van Der Hijden, J., Generation of Acoustic Waves by an Impulse point Source in a fluid/solid configuration with a Plane boundary, J. Acoust. Soc. Am. 75 (1984) 1709-1715.

[6] Van Der Hidjen, J., Propagation of transient elastic waves in stratified anisotropic medium, (North-Holland series in Applied Mathematics and Mechanics, Amsterdam) 1987 pp107-166.

[7] Casserau, D., Nouvelles Méthodes et Applications de la Propagation Transitoire dans les milieux Fluides et Solides, Thèse de l'U. E. R. Paris VII (1988).

[8] Saracco, G., Grossmann, A., Tchamitchian, P., Use of wavelet transform in the study of propagation of transient acoustic signals across a plane interface between two homogeneous media, "Wavelet, Time-Frequency methods and Phase Space" proceeding, Marseille, Dec. 1987.

[9] Grossmann, A., Morlet, J. Decomposition of Hardy functions into square integrable wavelets of constant shape. SIAM Journ. of Math. Analysis, 15 (1984) 723-736.

[10] Papoulis, A., The Fourier Integral and its Applications, (McGraw-Hill, New York) 1962.

[11] Saracco G., Transmission acoustique à travers le dioptre air-eau, J. Acoust. (1988) 1 71-79.

[12] Grossmann, A., Saracco, G., Tchamitchian, P., Study of propagation of transient acoustic signals across a plane interface between with the help of the wavelet transform, in preparation.

INVERSE PROBLEM

THREE-DIMENSIONAL IMPEDANCE SCATTERING THEORY

Pierre C. SABATIER
Laboratoire de Physique Mathématique - U.S.T.L.- MONTPELLIER (FRANCE)

ABSTRACT

This is a report on a three-dimensional Scattering Theory which takes into account altogether reflectors, i.e discontinuities of the impedance or its derivative, and the diffuse scattering by a potential. After the mathematical tools (Green function, etc), it is given the basic Integral Equation of Scattering, which yields the wave function from a knowledge of the "singular data" (reflection factors, slope factors, etc) and the diffuse potential. The scattering amplitude due to reflectors and the one due to diffuse scattering when reflectors are present are defined, together with the full scattering amplitude. More generally, most quantities and relations of interest in the "direct" study of Quantum Scattering Theory are generalized to this case, which encompasses as well acoustic or heat waves scattering.

We present several elements of a rigorous Scattering Theory related to the mixed Impedance-Potential equation. This equation is <u>not</u> equivalent to the Schrödinger equation if the impedance factor shows discontinuities, and this is the physical case. On the other hand, it can be reduced to the Schrödinger equation between discontinuities, so that a development of our Theory is parallell to that of Quantum Scattering Theory. We shall give results on the general three-dimensional case (easy to extend to n-dimensional cases), except in section II which contains results known only in the one-dimensional case up to now (however, a "simple" Born approximation is also known in the three-dimensional case and is given in another set of Proceedings). A more elaborate presentation of these results will be published in a Journal.

I. EQUIVALENCE THEOREMS

We are interested by the equation

$$[\alpha^{-2}\ \mathrm{div}\ \alpha^{2}\ \mathrm{grad} + k^{2} - V]\ \varphi(k,x) = 0$$

$$\varphi,\ \alpha^{2}\ \mathrm{grad}\ \varphi \text{ continuous} \tag{1}$$

where $V(x)$ is a real function, belonging to a class $\mathcal{V}$ of "potentials" for which the scattering problem described by (1) in the case $\alpha = 1$ is meaningful - for instance the Rollnick class :

$$\int dr \int dr'\ |V(r)\ V(r')|\ |r-r'|^{-2} < \infty \tag{2}$$

$k,\ x,\ \in \mathbb{R}^3$, $k = |k|$, the impedance factor α is everywhere twice differentiable with $\alpha^{-1}\ \Delta\ \alpha \in \mathcal{V}$, except at a finite number of disconnected surfaces $S_0,\ S_1, \dots, S_N$, which separate $\mathbb{R}^3$ into ordered shells Ω_i (between S_{i-1} and S_i), Ω_0 being the part of $\mathbb{R}^3$ beyond S_0, Ω_{N+1} that beyond S_N. We assume that each S_i is sufficiently smooth for our derivations, and, in particular, there is one unit normal vector at each point : $\nu(x)$ pointing into the + side. At a point x of a surface S_i, $\alpha(x)$ and $\nu.\mathrm{grad}\ \alpha(x)$ may be undefined but their limits from the (+) and (-) sides are defined and described by the "transmission factor" t, the "reflection factor", r, the "slope-factor", s, :

$$[t(x)]^{-1} = \frac{1}{2}\left[\frac{\alpha(x^{+})}{\alpha(x^{-})} + \frac{\alpha(x^{-})}{\alpha(x^{+})}\right] \tag{3a}$$

$$r(x)/t(x) = \frac{1}{2}\left[\frac{\alpha(x^{+})}{\alpha(x^{-})} - \frac{\alpha(x^{-})}{\alpha(x^{+})}\right] \tag{3b}$$

$$s(x)/t(x) = \frac{1}{2}\ \nu\ .\ \left[\frac{\mathrm{grad}\ \alpha(x^{-})}{\alpha(x^{+})} - \frac{\mathrm{grad}\ \alpha(x^{+})}{\alpha(x^{-})}\right] \tag{3c}$$

The first results we obtained in this three-dimensional case are summarized by the following theorem :

<u>Theorem I.</u> Let $\beta(x) > 0$, such that $\alpha(x)/\beta(x)$ is everywhere continuous, $\beta(x)$ is twice differentiable at each regular point, and

$$\frac{1}{2}\,[(\alpha \text{ grad } \beta - \beta \text{ grad } \alpha)\ .\ \nu]_{x^-}^{x^+} = \alpha(x^-)\ \beta(x^+)\ p(x) \qquad (4)$$

It can be proved[1,2] that the function $\Phi = \alpha\varphi/\beta$ satisfies the equation :

$$[\beta^{-2} \text{ div } \beta^2 \text{ grad} + k^2 - V + (\alpha\beta)^{-1} \text{ div } (\alpha \text{ grad } \beta - \beta \text{ grad } \alpha)]\Phi = 0 \quad (5)$$

at each regular point and that its continuity conditions across a singular surface at x are

$$\Phi(x^+) = \Phi(x^-) \qquad (6a)$$

$$[\nu.\beta^2 \text{ grad } \Psi]_{x^-}^{x^+} = -2\ p(x)\ \Phi(x)\ \beta(x^-)\ \beta(x^+) \qquad (6b)$$

The impedance factors α and β have the same transmission and reflection factors but not the same slope factor :

$$s_\beta(x) - s_\alpha(x) + p(x)\ t(x) = 0 \qquad (7)$$

By making $p(x) = 0$, we can use this theorem to transfer "twice differentiable parts" of the impedance factor into the potential. By introducing a function $\sigma(x)$ which is harmonic inside the shells and such that α/σ is everywhere continuous we can also reduce the problem to one which differs from the Schrödinger problem only by the substituting σ^{-2} div σ^2 grad to the laplacean, and by the slope discontinuities (6b). By choosing β such that $\alpha\,\Delta\beta - \beta\,\Delta\alpha$ vanishes, we can also construct a scattering problem which is completely equivalent, although the impedance factor β has replaced α, V remaining unchanged, in the central equation (1) : we call this the "standard equivalence result"[2].

II. PERTURBATION THEORY (One-dimensional case)

In the one-dimensional case, it is easy to derive from the

above equation an analytic perturbation theory of the impedance equation, starting from an unperturbed model (impedance factor and potential), at $k \neq 0$, and going to a perturbed model (perturbed impedance factor and perturbed potential) at the same value of k. This enabled us to write down the distorted wave Born approximation (first and second order) corresponding to a small modification of potential and impedance factors, including of course a small shift of singular surfaces and a modification of singular factors. We shall only write down the simple distorted wave Born approximation for the scattering data. These data are defined by using the Jost solutions $p_{\pm}(k,x)$ of the one-dimensional equation :

$$\left[\alpha^{-2} \frac{d}{dx} \alpha^2 \frac{d}{dx} + k^2 - V\right] p(k,x) = 0$$

$$p, \quad \alpha^2 \frac{dp}{dx} \quad \text{continuous} \tag{8}$$

i.e. the solutions respectively asymptotic to $\alpha^{-1} \exp[ikx]$ at $+\infty$ and $\alpha^{-1} \exp[-ikx]$ at $-\infty$. Like in the usual theory, the transmission and reflection coefficients couple p_+ and p_-, for instance

$$T(k)\ p_-(k,x) = p_+(-k,x) + R_+(k)\ p_+(k,x) \tag{9}$$

and they depend only on k. Now, if we use a tilde to denote the perturbed data, the simple distorted wave Born approximation reads[3]

$$\tilde{T}(k) = T(k) \tag{10}$$

$$2ik \frac{\tilde{R}_+(k) - R_-(k)}{T^2(k)} = - \int_{-\infty}^{+\infty} ds\ [p_-(k,s)]^2\ \Big\{\alpha^2(s)\ \delta\ V(s)$$

$$+ \sum_{x_n \in S} \alpha\left(x_n^+\right)\ \alpha\left(x_n^-\right)\ \omega_n(k)\ \delta(s-x_n)\Big\} + O(\epsilon^2) \tag{11}$$

where $\delta\ V(s)$ is defined at each point that is regular before and after perturbation :

$$\delta W(x) = V(x) + \alpha''(x)/\alpha(x) - \tilde{V}(x) - \tilde{\alpha}''(x)/\tilde{\alpha}(x) \qquad (12)$$

the points x_n are those which correspond to discontinuities either before or after the perturbation. They form the set S and the ω_n's are readily derived from the singular factors and the position discontinuities before and after the perturbation[3]. They are of order ϵ. ϵ is an upper measure of the differences $|r_n - \tilde{r}_n|$, $|s_n - \tilde{s}_n|$ (unshifted discontinuities), $\mathrm{Sup}\left(|x_{n-1} - x_n|, |r_{n-1} - \tilde{r}_n|, |s_{n-1} - \tilde{s}_n|\right)$ when x_{n-1} and x_n respectively correspond to an unperturbed and a perturbed case obtained by shifting the discontinuity. Notice that the approximation may be relevant only if $\epsilon \ll k$. This approximation makes possible modeling a distribution of small discontinuities by a continuous model or by a model which shows smoothly varying parts between large discontinuities. But if one tries to "wash out" a random distribution of small discontinuities, a short calculation shows the importance of their mean square deviation and it does not seem possible to homogeneize the layers to the point of using constant parameters.

We shall not recall here a very valuable tool of one-dimensional problems, the "Darboux transformations". They enable a complete and self contained[4] study of direct and inverse one-dimensional problems for large classes of scattering data e.g. rational coefficients. Up to now, we do not know *simple* generalizations to the three-dimensional problem

III. THE ADJOINT OPERATOR

Let $L = \alpha^{-2}\,\mathrm{div}\,\alpha^2\,\mathrm{grad}$ be the linear differential operator of (1). Its formal adjoint is $\tilde{L} = \mathrm{div}\,\alpha^2\,\mathrm{grad}\,\alpha^{-2}$ and one easily checks that

$$v\,L\,u - u\,\tilde{L}\,v = \mathrm{div}[v\,\mathrm{grad}\,u - u\,\mathrm{grad}\,v + 2uv\,\alpha^{-1}\,\mathrm{grad}\,\alpha] \qquad (13)$$

The adjoint impedance equation is

$$(\mathrm{div}\,\alpha^2\,\mathrm{grad}\,\alpha^{-2} + k^2 - V)\,\phi(k,x) = 0$$

$$\alpha^{-2}\phi\,,\ \alpha^2\,\mathrm{grad}\,(\alpha^{-2}\phi)\ \ \text{continuous} \qquad (14)$$

It is clear that if ψ is a solution of (14), the function $\alpha^2\,\psi$ is a solution of (1). Hence, a theorem similar to theorem A can be devised for the equation (14). If we construct for the equation (14) a Green's function satisfying Sommerfeld's conditions at infinity, we can use it to solve the inhomogeneous form of (1), or to derive integral equations for solutions of (1).

IV. CHAINS OF EQUATIONS

In the following sections, we shall not use the adjoint operator or another one *directly* related to the impedance equation. Rather, we shall keep in mind that setting $\varphi = \Phi/\alpha$ in (1) readily shows that φ satisfies inside each domain of regularity the Schrödinger equation :

$$\Delta\Phi + (k^2 - V - \alpha^{-1}\,\Delta\alpha)\Phi = 0 \tag{15}$$

whereas the "continuous" or "conserved" quantities through any discontinuity surface, say, S_k, read :

$$[\Phi(x)/\alpha(x)]_{x\in S_k^+} = [\Phi(x)/\alpha(x)]_{x\in S_k^-} \tag{16}$$

$$\left[\alpha(x)\,\frac{\partial\Phi(x)}{\partial\nu(x)} - \Phi(x)\,\frac{\partial\alpha(x)}{\partial\nu(x)}\right]_{x\in S_k^+} = \left[\alpha(x)\,\frac{\partial\Phi(x)}{\partial\nu(x)} - \Phi(x)\,\frac{\partial\alpha(x)}{\partial\nu(x)}\right]_{x\in S_k^-} \tag{17}$$

The formulas (1)-(17) define a *chain of Schrödinger equations* linked by special linear homogeneous conservation laws. We shall call it the *impedance Schrödinger chain.*

V. BASIC PROPERTY OF HOMOGENEOUS SCHRODINGER-HELMOLTZ CHAINS

From now on, (see Fig.1), we deal with a set of ordered domains Ω_i and surfaces S_i of class C^2, with $S = \cup_i S_i$, such that each domain is finite but Ω_{N+1}, which extends to infinity in all directions, and $i \neq j \Rightarrow S_i \cap S_j = \emptyset$. The "topological aspect" of such surfaces is that of sphere-like surfaces, each S_i enclosing a ball-like domain

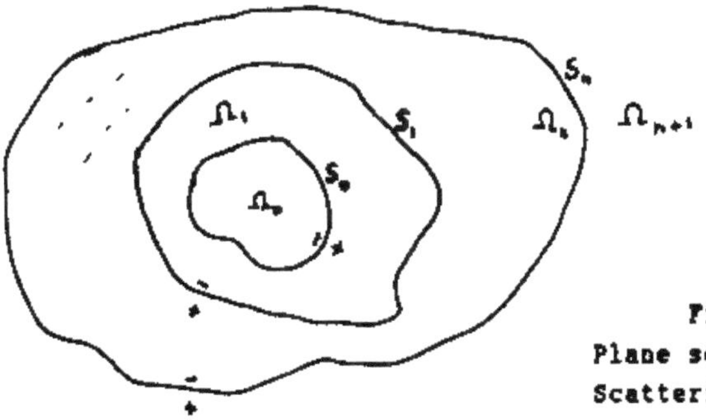

Figure 1.
Plane section of the Scattering system.

$\overline{\Omega}_i = \sum_{k=0}^{k=i} \Omega_k$. Now, we study $\Psi \in C^2\left(\mathbb{R}^3\setminus S\right)$ that satisfies the free wave Schrödinger (or Helmholtz) equation inside each Ω_i :

$$(\Delta + k^2)\,\Psi(x) = 0 \qquad \left(x \in \mathbb{R}^3\setminus S\right) \tag{18}$$

and the continuity conditions

$$[\Psi(x)/a(x)]_{x\in S_i^+} = \lambda_i\,[\Psi(x)/a(x)]_{x\in S_i^-} \tag{19}$$

$$\left[a(x)\,\frac{\partial\Psi(x)}{\partial\nu_x} - b(x)\,\Psi(x)\right]_{x\in S_i^+} = \mu_i\left[a(x)\,\frac{\partial\Psi(x)}{\partial\nu_x} - b(x)\,\Psi(x)\right]_{x\in S_i^-} \tag{20}$$

$$\langle x/|x|,\ \mathrm{grad}\,\Psi(x)\rangle - ik\,\Psi(x) = o(|x|^{-1}) \quad \text{as} \quad |x| \to \infty$$

$$\text{uniformly for all directions} \quad x/|x| \tag{21}$$

where λ_i and μ_i are real positive *numbers*, the functions $a(x) > 0$, $b(x)$ real, belong to $C\left(\mathbb{R}^3\setminus S\right)$ and can have jumps through the surfaces S_i. The "Sommerfeld condition" (21) (see Ref 5 p.70) implies in particular that $\Psi(x) = O(|x|^{-1})$ as $|x| \to \infty$. It physically means that $\Psi(x)$ is a pure "outgoing" solution.

The conditions (19) and (20) imply that the total current on the (+) side of a surface S_i is proportional to the total current on the (-) side :

$$\int_{S_i^+}\left(\Psi \frac{\partial \Psi}{\partial \nu} - \Psi \frac{\partial \overline{\Psi}}{\partial \nu}\right)ds = \lambda_i \ \mu_i \int_{S_i^-}\left(\Psi \frac{\partial \Psi}{\partial \nu} - \Psi \frac{\partial \overline{\Psi}}{\partial \nu}\right)ds \tag{22}$$

We claim that a *function $\Psi(x)$ that satisfies (18)-(19)-(20)-(21) must vanish.*

Proof. Inside each domain Ω_i, the Green's theorem (5 p.68) yields

$$\left(\int_{S_i^+} - \int_{S_{i+1}^-}\right)\left(\overline{\Psi} \frac{\partial \Psi}{\partial \nu} - \Psi \frac{\partial \overline{\Psi}}{\partial \nu}\right) ds(x) = \int_{\Omega_{i+1}}\left(\overline{\Psi}\ \Delta\Psi - \Psi\ \Delta\overline{\Psi}\right)dx = 0 \tag{23a}$$

A similar argument inside Ω_o implies that the total current on S_o^- vanishes. Using (22) and (23a) repeatedly it follows that

$$\int_{S_N^+}\left(\overline{\Psi} \frac{\partial \Psi}{\partial \nu} - \Psi \frac{\partial \overline{\Psi}}{\partial \nu}\right)ds = \mathrm{Im} \int_{S_N^+} \Psi \frac{\partial \overline{\Psi}}{\partial \nu}\ ds = 0 \tag{23b}$$

Remember $\overline{\Omega}_N = \sum_{k=o}^{N} \overline{\Omega}_k$. A known result (Ref.5) on the Helmholtz equation is that any function Ψ that satisfies (18) in $\mathbb{R}^3 \backslash \overline{\Omega}_N$ and the condition (21) for a given (real) k also satisfies

$$\lim_{R\to\infty} \int_{|y|=R}\left\{\left|\frac{\partial \Psi}{\partial \nu}\right|^2 + k^2\ |\Psi|^2 + 2k\ \mathrm{Im}\left(\Psi \frac{\partial \overline{\Psi}}{\partial \nu}\right)\right\}ds = 0 \tag{24}$$

Equation (23b) implies that the last term in (24) vanishes, and thus Equation (24) implies that $\lim_{R\to\infty} \int_{|y|=R} |\Psi|^2 ds = 0$. It follows from

Rellich lemma (Ref. 5 p.77) that Ψ identically vanishes in $\mathbb{R}^3 \setminus \bar{D}_N$. Hence ψ and $\frac{\partial \Psi}{\partial \nu}$ vanish on S_N^+, and thanks to (19) and (20), on S_N^-, so that Ψ satisfies (18) throughout $\mathbb{R}^3 \setminus \bar{D}_{N-1}$ and hence vanishes therein according to Rellich lemma. Continuing the process yields $\Psi = 0$. Q.E.D.

VI. A GREEN FUNCTION FOR THE IMPEDANCE CHAIN

VI.1 The Problem

The impedance factor $\alpha(x)$, the domains Ω_i and surfaces S_i being defined above, we seek $G(x,y)$ such that

$$x \longrightarrow G(x,y) \in C^2(\mathbb{R}^3 \setminus S,\ x \neq y) \tag{25}$$

$$\Delta\, G(x,y) + k^2\, G(x,y) = -\,\delta(x-y) \tag{26}$$

$$G(x,y)/\alpha(x) \quad \text{and} \quad G(x,y)\,\frac{\partial \alpha(x)}{\partial \nu(x)} - \alpha(x)\,\frac{\partial G(x,y)}{\partial \nu(x)} \quad \text{are}$$
continuous through S, and Sommerfeld condition (27)

These conditions obviously imply that for $x \neq y$, $x \to G(x,y)$ satisfies (18) and that if $\Phi(x,y)$ is the standard Helmholtz Green function (see Ref.5) :

$$\Phi(x,y) = [4\pi\ |x-y|]^{-1} \exp\,[i|k|\ |x-y|] \tag{28}$$

$x \to G(x,y) - \Phi(x,y)$ is a solution of Helmholtz equation (18) that satisfies the Sommerfeld condition. We recall by the way that two functions $g(x)$ and $h(x)$ which satisfy the Sommerfeld condition are such that

$$\lim_{R\to\infty} \int_{|x|=R} \left[g(x)\, \frac{\partial}{\partial \nu}\, h(x) - h(x)\, \frac{\partial}{\partial \nu}\, g(x) \right] ds = 0 \tag{29}$$

The function $G(x,y)$ is defined uniquely by (26)-(27) since the difference of two solutions would be a solution of the problem

(18)-(21). If it exists, it is symmetric with respect to the exchange x,y : using the solution of (26), completed by (27), and the solution $G(x,z)$ of the same equation, with z instead of y, completed by (27), we calculate by means of Green's formula

$$\int_{x\in\Sigma\Omega_i} [G(x,z)\ \Delta\ G(x,y) - G(x,y)\ \Delta\ G(x,z)]dx$$

$$= \lim_{R\to\infty} \int_{|x|=R} \left[G(x,y)\ \frac{\partial}{\partial\nu_x}\ G(x,y) - G(x,y)\ \frac{\partial}{\partial\nu_x}\ G(x,z)\right]ds_x \quad .$$

$$- \int_{S_N^+} \left[G(x,z)\ \frac{\partial}{\partial\nu_x}\ G(x,y) - G(x,y)\ \frac{\partial}{\partial\nu_x}\ G(x,z)\right]ds_x$$

$$+ \int_{S_N^-} \left[G(x,z)\ \frac{\partial}{\partial\nu_x}\ G(x,y) - G(x,y)\ \frac{\partial}{\partial\nu_x}\ G(x,z)\right]ds_x$$

$$+ \cdots \left(\int_{S_e^+} - \int_{S_e^-}\right) \left[G(x,z)\ \frac{\partial}{\partial\nu_x}\ G(x,y) - G(x,y)\ \frac{\partial}{\partial\nu_x}\ G(x,z)\right]ds_x = 0 \quad (30)$$

and it follows from (26) that the left-hand side is also equal to $-G(y,z) + G(z,y)$, Q.E.D. It remains to construct the solution of (26)-(27). However, we need for this purpose several generalizations of the operators usually introduced for Helmholtz problems, which deal with one surface only (Ref.5), and we need a number of results concerning these generalizations. Statements and simple proofs are given below, other proofs are in the appendix.

VI.2 Surface Operators

We first define the operators $S_{ij}, K_{ij}, K'_{ij}, T_{ij}$, by

$$(S_{ij}\ f)(x) = 2 \int_{S_j} ds(z)\ \Phi(x,z)\ f(z) \quad (x \in S_i) \qquad (31)$$

$$(K_{ij}\ f)(x) = 2 \int_{S_j} ds(z)\ \frac{\partial\ \Phi(x,z)}{\partial\ \nu_z}\ f(z) \quad (x \in S_i) \qquad (32)$$

$$(K'_{ij}\ f)(x) = 2\int_{S_j} ds(z)\ \frac{\partial\ \Phi(x,z)}{\partial\ \nu_x}\ f(z) \qquad (x \in S_i) \tag{33}$$

$$(T_{ij}\ f)(x) = 2\ \frac{\partial}{\partial\ \nu_x}\int_{S_j} ds(z)\ \frac{\partial\ \Phi(x,z)}{\partial\ \nu_z}\ f(z) \qquad (x \in S_i) \tag{34}$$

Each of these operators maps a space of functions defined on S_j, with values in $\mathbb{C}$, say $E(S_j)$, into a space of functions defined on S_i, with values in $\mathbb{C}$, say $F(S_i)$. Stating precisely the couples E, F must be done in a different way for $\underset{\sim}{S}, \underset{\sim}{K}, \underset{\sim}{K'}, \underset{\sim}{T}$, and may also depend whether $i = j$ or $i \neq j$. It can be said that in all cases, the operator with $i \neq j$ is more regular than the same with $i = j$. Since the surfaces are unconnected, a "non diagonal" operator is the value of a single or double layer potential, or its derivative, at a point that does not belong to the layer, and it is not difficult to prove by means of Ascoli theorem that, for instance, S_{ij} is a compact mapping from $C(S_j)$, K_{ij}, K'_{ij} from $C^{0,\alpha}(S_j)$, T_{ij} from $C^{1,\alpha}(S_j)$, to sets included in $C(S_i)$ (where $\alpha \in (0,1)$). As for the "diagonal operators", they are nothing but the "one surface" operators already studied in the litterature (Ref.5). Hence S_{ii}, K_{ii}, K'_{ii} are compact in $C(S_i)$ and $C^{0,\alpha}(S_i)$ for $0 < \alpha < 1$. K'_{ii} maps $C(S_i)$ into $C^{0,\alpha}(S_i)$, K_{ii} and S_{ii} map $C^{0,\alpha}(S_i)$ into $C^{1,\alpha}(S_i)$. But T_{ii} is an unbounded operator, defined on a set $\mathcal{N}(S_i)$ containing $C^{1,\alpha}(S_i)$, and which is mapped into $C(S_i)$. Hence the reader of this preliminary paper can make with us the choice $E = C^{1,\alpha}$, and state that not only the four operators, for any choice of i and j, but also their operator products, are mappings into the space of continuous functions on the last surface, and are compact unless a diagonal element of $\underset{\sim}{T}$ is involved. Unfortunately, more detailed and weaker results will be necessary later for a complete scattering theory and are to be used in the final paper. We have called S_{ij}, etc, "operator elements" and we have implicitly defined their "operator product", e.g. $S_{ki}T_{ij}$, by

$$(S_{ki}T_{ij}\ f)(x) = \int_{S_i} \Phi(x,t)\ ds(t)\ \frac{\partial}{\partial\nu_t}\int_{S_j} \frac{\partial\Phi(t,y)}{\partial\nu_y}\ f(y)\ ds(y)$$
$$(\forall\ f \in E(S_j)), \quad (x \in S_k)\ . \tag{35a}$$

Now, an "operator element" like S_{ij} applies to any f defined on S_j - say $f_j(x)$. We define a function f(x) on S by the equalities :

$$f(x) = f_j(x) \quad (x \in S_j),\ j = 0,1,\dots,N \tag{36a}$$

If each $f_j \in E(S_j)$, $f \in E(S)$. We define a "surface operator" $\underset{\sim}{S}$, or $\underset{\sim}{K}$, etc, as a mapping from E(S) to F(S) such that if $f_j(x)$ is the restriction of f to $x \in S_j$,

$$(\underset{\sim}{S}\ f)(x) = \sum_j \int_{S_j} \Phi(x,t)\ f_j(t)\ ds(t) \qquad (x \in S) \tag{36b}$$

With these definitions, the operator product of two surface operators, say $\underset{\sim}{S}\underset{\sim}{T}$, is the operator such that for any $f \in E(S)$, $x \in S_k$, $k = 0,1,\dots,N$,

$$(\underset{\sim}{S}\ \underset{\sim}{T}\ f)(x) = \left(\sum_i S_{ki}(T\ f)\ \ (x \in S_i)\right)(x) = \left(\sum_i \sum_j S_{ki}\ T_{ij}\ f_j\right)(x) \tag{35b}$$

Since N is finite, $\underset{\sim}{S}$, $\underset{\sim}{K}$, $\underset{\sim}{K}'$ are compact mappings from E(S) to F(S), whereas $\underset{\sim}{T}$ is not, because of its "diagonal" elements.

VI.3 Relations Between The Surface Operators

In the standard case of one surface, there are simple relations between T and the other operators. They are not quite as simple in our case, and the proofs are postponed to the appendix. The results are

$$\underset{\sim}{K}\ \underset{\sim}{K} - \underset{\sim}{S}\ \underset{\sim}{T} = \underset{\sim}{1} + \dot{\underset{\sim}{K}} \tag{37}$$

$$\underset{\sim}{K}'\ \underset{\sim}{K}' - \underset{\sim}{T}\ \underset{\sim}{S} = \underset{\sim}{1} - \dot{\underset{\sim}{K}}' \tag{38}$$

$$\underset{\sim}{T}\ \underset{\sim}{K} - \underset{\sim}{K}'\ \underset{\sim}{T} = \dot{\underset{\sim}{T}} \tag{39}$$

where $\dot{M}$ is obtained from the matrix M made of operator elements M_{ij} by the ansatz

$$\dot{M}_{kj} = \begin{cases} 2(k-j-1)\ M_{kj} & \text{for} \quad j \leq k-1 \\ 0 & \text{for} \quad j = k \\ 2(k+1-j)\ M_{kj} & \text{for} \quad j \geq k+1 \end{cases} \tag{40}$$

The dotted matrices involve only "non diagonal" elements. $\dot{\underset{\sim}{T}}$ therefore is compact on $E(S)$.

VI.4 Potentials

The right hand side of (31), (32), (33), (34), can be continued for any $x \in \mathbb{R}^3 \setminus S$ as simple and double layer potentials and their derivatives. Well known results apply (Ref.5) and readily show that the single layer potential (31 cont.) and the double layer potential (32 cont.) define solutions of the Helmholtz equation in $\mathbb{R}^3 \setminus S$ that satisfy the Sommerfeld condition (21). In addition, it is well known that the single layer potential is continuous throughout $\mathbb{R}^3$ but its derivative has known discontinuities when crossing the surface where it is defined. The reverse is true for the double-layer potential. These results will be used to construct our Green Function.

VI.5 Construction of the Green Function

We can choose between constructing directly a solution of (26,27), by writing down $G(x,y)$ as the sum $\Phi(x,y)$ + a single layer potential + a double layer potential, or using the same ansatz to solving the inhomogeneous equation

$$\Delta\, u(x) + k^2\, u(x) = -\, f(x) \tag{41}$$

completed by the conditions (27) (on $u(x)$). $G(x,y)$ is then identified as the resolvent kernel of this problem. Both approaches are quite similar but the second one avoid writing down "y" several times in each formula and avoid taking care of a singularity at $x = y$. Hence we use this second approach and write down

$$u(x) = F(x) + \sum_{j=o}^{N} \int_{S_j} ds(z)\, \frac{\partial\Phi(x,z)}{\partial \nu_z}\, \psi_j(z)$$

$$+ \sum_{j=o}^{N} \int_{S_j} ds(z)\ \Phi(x,z)\ \varphi_j(z) \tag{42}$$

where

$$F(x) = \int_{\mathbb{R}^3} \Phi(x,y)\ f(y)\ dy \tag{43}$$

The potential representations satisfy the Helmholtz equation and the Sommerfeld condition ; since $F(x)$ satisfies (41) and the Sommerfeld condition, so does also $u(x)$. Now the continuity conditions through S_i imply that for $x \in S_i$, the following equalities must be satisfied, where the index i reminds that $x \in S_i$, $i = 0,1,2,\ldots,N$ and the notations introduced above are used :

$$\left[\psi_i(x)\ (\alpha^-(x) + \alpha^+(x))\right]_{x\in S_i} =$$

$$= \left\{(\alpha^+(x) - \alpha^-(x))\ [2F(x) + (\underset{\sim}{K}\ \psi)(x) + (\underset{\sim}{S}\ \varphi)(x)]\right\}_{x\in S_i} \tag{44}$$

$$\left[\varphi_i(x)\ (\alpha^-(x) + \alpha^+(x)) + \psi_i(x)\ (\alpha'^-(x) + \alpha'^+(x))\right]_{x\in S_i}$$

$$= \Big\{ (\alpha^+(x) - \alpha^-(x))\ [2F'(x) + (\underset{\sim}{T}\psi)(x) + (\underset{\sim}{K}'\varphi)(x)]$$

$$- (\alpha'^+(x) - \alpha'^-(x))[2F(x) + (\underset{\sim}{K}\psi)(x) + (\underset{\sim}{S}\varphi)(x)]\Big\}_{x\in S_i} \tag{45}$$

In these formulas, the prime used with α or F stands for the normal derivative at $x \in S_i$, ie $\alpha'^-(x)$ means $\frac{\partial\alpha}{\partial\nu_x}\left(x\in S_i^-\right)$. φ_i and ψ_i are respectively the restrictions of φ and ψ to $x\in S_i$. The equations (44) and (45) must be satisfied for all i's. It is convenient to rewrite them by introducing the functions on S, or "multiplicative operators" β, β', and τ, hereafter defined

$$\beta(x) = \frac{\alpha^+(x) - \alpha^-(x)}{\alpha^+(x) + \alpha^-(x)} \qquad \beta'(x) = \frac{\alpha'^+(x) - \alpha'^-(x)}{\alpha^+(x) + \alpha^-(x)}$$

$$\gamma(x) = \frac{\alpha'^+(x) + \alpha'^-(x)}{\alpha^+(x) + \alpha^-(x)} \tag{46}$$

Sufficient assumptions on β, β', γ is that they belong to $C^{1,\alpha}(S)$. Hence, with these condensed notations, the continuity conditions through S yield the following equations for the functions $\varphi(x\in S)$ and $\psi(x\in S)$:

$$\psi = 2\,\beta\,F + \beta\,\underset{\sim}{S}\,\varphi + \beta\,\underset{\sim}{K}\,\psi \tag{47}$$

$$\varphi + \gamma\psi = 2\,\beta\,F' - 2\,\beta'\,F + \beta\,\underset{\sim}{T}\,\psi - \beta'\,\underset{\sim}{K}\,\psi + \beta\,\underset{\sim}{K}'\,\varphi - \beta'\,\underset{\sim}{S}\,\varphi \tag{48}$$

If the system obtained from (47-48) by setting $F = F' = 0$ has a solution, ψ^o, ψ^o, inserting it into (42) yields a solution $\Psi^o(x)$ of the homogeneous chain (18)-(21) with $a(x) = \alpha(x)$, $b(x) = \dfrac{\partial\alpha(x)}{\partial\nu_x}$, $\lambda_i = \mu_i = 1$.

We know from section V that Ψ^o necessarily vanishes, and using jump relations across layers shows that so do ψ^o and φ^o. Hence, if the system (47-48) has a solution, it is unique.

So as to get the solution, we may try solving (47) in terms of ψ and insert the result into (48). However $(\underset{\sim}{1}-\beta\,\underset{\sim}{K})^{-1}$ exists as a bounded operator if and only if 1 is not an eigenvalue of $\beta\underset{\sim}{K}$. Since $\underset{\sim}{K}$ is bounded there is at least a range of values of β, small enough (say, $\|\beta\| < \|\underset{\sim}{K}\|^{-1}$) where 1 cannot be an eigenvalue. We shall call it the Neuman range. On the other hand, if 1 is an eigenvalue of $\beta\underset{\sim}{K}$, the equation (47) yields ψ if and only if $\beta(2F + \underset{\sim}{S}\varphi)$ is orthogonal to the n-dimensional null-space of the adjoint of $(\underset{\sim}{1}-\beta\underset{\sim}{K})$. Then ψ contains itself n parameters corresponding to the null-space $N(\underset{\sim}{1}-\beta\underset{\sim}{K})$, inserting it into (48) can give φ up to n parameters, which are eventually determined by the condition on $\beta(2F + \underset{\sim}{S}\,\varphi)$. This process works butis complicated. For the sake of simplicity, we shall assume in this preliminary paper that

$$N(\underset{\sim}{1} - \beta \underset{\sim}{K}) = 0 \tag{49}$$

Since $\underset{\sim}{K}$ is compact, the Fredholm alternative holds for (47), considered as an equation for ψ, $(\underset{\sim}{1} - \beta \underset{\sim}{K})^{-1}$ is a bounded operator and it enables us to construct ψ from φ. Reinserting the result into (48), we meet the operator

$$\underset{\sim}{U} = \underset{\sim}{T}\ (\underset{\sim}{1} - \beta \underset{\sim}{K})^{-1}\ \beta\ \underset{\sim}{S} \tag{50}$$

which needs a special study, because it involves the product of an unbounded operator and a compact one. From (39) it is easy to get

$$\underset{\sim}{T}(1 - \beta \underset{\sim}{K}) + [\underset{\sim}{T},\beta]\underset{\sim}{K} + \beta\ \dot{\underset{\sim}{T}} = (\underset{\sim}{1} - \beta \underset{\sim}{K}')\underset{\sim}{T} \tag{51}$$

where $[.,.]$ is the usual symbol for a commutator. Hence

$$(\underset{\sim}{1} - \beta \underset{\sim}{K}')^{-1}\underset{\sim}{T} + (\underset{\sim}{1} - \beta \underset{\sim}{K}')^{-1}([\underset{\sim}{T},\beta]\underset{\sim}{K} + \beta\ \dot{\underset{\sim}{T}})(\underset{\sim}{1} - \beta \underset{\sim}{K})^{-1} = \underset{\sim}{T}(1 - \beta \underset{\sim}{K})^{-1} \tag{52}$$

On the other hand

$$\underset{\sim}{T}\ \beta\ \underset{\sim}{S} = [\underset{\sim}{T},\ \beta]\underset{\sim}{S} + \beta\ \underset{\sim}{T}\ \underset{\sim}{S} \tag{53}$$

Inserting (38) into (53), then (52) and (53) into (50) we obtain

$$\underset{\sim}{U} = (\underset{\sim}{1}-\beta\ \underset{\sim}{K}')^{-1}\Bigg\{[\underset{\sim}{T},\beta]\underset{\sim}{S} - \beta\ \Big(1-\underset{\sim}{K}'\underset{\sim}{K}' - \dot{\underset{\sim}{K}}'\Big)$$

$$+ \Big([\underset{\sim}{T},\beta]\underset{\sim}{K} + \beta\ \ \dot{\underset{\sim}{T}}\Big)\ (1 - \beta\ \underset{\sim}{K})^{-1}\beta\underset{\sim}{S}\Bigg\} \tag{54}$$

let us show that $[\underset{\sim}{T},\beta]$ is a bounded operator. The only elements which need a proof are the "diagonal ones", which read for $x \in S_k$ (and applied to $f_k(x)$) :

$$D_k(x) = \frac{\partial}{\partial \nu_x} \int_{S_k} \frac{\partial\varphi(x,t)}{\partial \nu_t}\ \beta(t)\ f_k(t)\ ds(t) - \beta(x) \int_{S_k} \frac{\partial\varphi(x,t)}{\partial \nu_t}\ f_k(t)\ ds(t)$$

$$= \left(\frac{\partial}{\partial\nu_x}\beta(x)\right)\int_{S_k}\frac{\partial\varphi(x,t)}{\partial\nu_t}f_k(t)\,ds(t)$$

$$+\frac{\partial}{\partial\nu_x}\int_{S_k}\frac{\partial\varphi(x,t)}{\partial\nu_t}[\beta(t)-\beta(x)]\,f(t)\,ds(t) \tag{55}$$

The first term in the last right hand side of (55) is the product of $\frac{\partial}{\partial\nu_x}\beta(x)$ by $(K_{kk}f_k)(x)$ and thus defines a bounded operator. The boundedness of the next one follows from the lemma (2.10) of Ref.5 (after a few easy calculations). Q.E.D.

Going back to U in (54), and since the product of a compact operator by a bounded operator is compact, we see that $U + \beta\,1$ is compact :

$$\underset{\sim}{U} = -\beta\,\underset{\sim}{1} + \underset{\sim}{C} \tag{56}$$

where C is complicated, but trivially written down from (54). Inserting the solution ψ of (47) into (48) and taking into account (56) yields

$$(1+\beta^2)\varphi = \underset{\sim}{A}\,F + \underset{\sim}{B}\,\varphi \tag{57}$$

where

$$\underset{\sim}{A}F = 2\beta F' + [2\beta\underset{\sim}{T}\,(\underset{\sim}{1}-\beta\underset{\sim}{K})^{-1}\beta - 2\beta' - 2\beta'\underset{\sim}{K}\,(\underset{\sim}{1}-\beta\underset{\sim}{K})^{-1}\beta - 2\gamma\,(\underset{\sim}{1}-\beta\underset{\sim}{K})^{-1}\beta]F \tag{58}$$

$$\underset{\sim}{B} = -\gamma\,(\underset{\sim}{1}-\beta\underset{\sim}{K})^{-1}\beta\,\underset{\sim}{S} + \beta\,\underset{\sim}{C} - \beta'\underset{\sim}{K}\,(\underset{\sim}{1}-\beta\underset{\sim}{K})^{-1}\beta\,\underset{\sim}{S} + \beta\underset{\sim}{K}' - \beta'\underset{\sim}{S} \tag{59}$$

$(1+\beta^2)^{-1}\,\underset{\sim}{B}$ is compact, the Fredholm alternative holds, the homogeneous solution yields a solution of (18)-(21) and therefore vanishes. Hence φ can be constructed as

$$\varphi = \left[(1+\beta^2)\underset{\sim}{1} - \underset{\sim}{B}\right]^{-1}\underset{\sim}{A}\,F \tag{60}$$

Suppose now the Neuman series converges. If β, β', γ are small (ie if there is a finite number of small reflectors only), the linear term (simple Born approximation) or the next one (second order Born approximation) may be of interest. They are studied in a lecture prepared for another workshop (Ref.6).

Constructing the Green function is very much simplified in two cases.

Assumption A $\alpha(x)$ is continuous throughout $\mathbb{R}^3$.

Then one can use (42) with $\psi = 0$, and β vanishes identically. The remaining equation (48) reduces to

$$\varphi = - \beta'(2\, F + \underset{\sim}{S}\, \varphi) \tag{61}$$

With our assumptions on α, $\beta'\underset{\sim}{S}$ is compact, the homogeneous equation has only the zero solution because of section V results, and

$$\varphi = - 2(\underset{\sim}{1}+\beta'\underset{\sim}{S})^{-1}\beta' F = : - 2(\underset{\sim}{1}-\beta'\underset{\sim}{R})\ \beta' F \tag{62}$$

It follows from (42) and (43) that the Green's function is then :

$$G(x,y) = \Phi(x,y) - 2\sum_{k=o}^{N} \int_{S_k} ds(z)\ \Phi(x,z)\ \beta'(z)\ \Phi(z,y)$$

$$+ 2 \sum_{k=o}^{N} \int_{S_k} ds(z)\ \Phi(x,z)\ \beta'(z) \sum_{j=o}^{N} \int_{S_j} ds(t)\ R(z,t)\ \beta'(t)\ \Phi(t,y) \tag{63}$$

where $R(z,t)$ is the kernel of $\underset{\sim}{R}$.

It is very remarkable that problems where $\alpha(x)$ satisfy a much weaker assumption than A can be reduced to a related construction of their Green function :

Assumption B. For each surface, the "relative discontinuity" $\alpha^+(x)/\alpha^-(x)$ does not depend on the position of x on the surface.

In this case, it is possible to construct, like in the one dimensional case, a piecewise constant "singular data" function say $\sigma(x)$ equal to $\sigma_{N+1}=1$ for $x \in \Omega_{N+1}$, and

$$\sigma_i = \sigma_{i+1}\alpha\left(x \in S_i^-\right)\Big\backslash\alpha\left(x \in S_i^+\right) \qquad (64)$$

in each domain Ω_i ($i = N, N-1,\dots,0$). Now, if we want to study the mixed equation (1) we can set $\tilde{\varphi} = \alpha\varphi/\sigma$. It follows from the theorem I that $\tilde{\varphi}(k,x)$ satisfies the equation (15) for any $x \in \mathbb{R}^3\setminus S$, is continuous throughout $\mathbb{R}^3$, and satisfies through each S_i continuity conditions on $\tilde{\varphi}$ and $\sigma^2\left(\frac{\partial\tilde{\varphi}}{\partial\nu} - \frac{\tilde{\varphi}}{\alpha}\frac{\partial\alpha}{\partial\nu}\right)$. By the same token, we can solve the problem made of Equation (41) completed by the Sommerfeld condition and the impedance continuity conditions (26)-(27) by setting $v = u/\sigma$, $g = f/\sigma$ and solving rather the problem

$$\Delta v(x) + k^2 v(x) = -g(x) \qquad x \in \mathbb{R}^3\setminus S \qquad (65a)$$

$$v(x) \text{ and } \sigma^2\left[\frac{\partial v}{\partial\nu} - \frac{v}{\alpha}\frac{\partial\alpha}{\partial\nu}\right] \text{ continuous} \qquad (65b)$$

Sommerfeld condition

The analysis of §V applies to the homogeneous form of (65) and guarantees the uniqueness of the solution, that can be constructed by making use of (42), with $\psi = 0$, solving the Fredholm equation

$$\varphi = 2\,\beta\,G' - 2\,\beta' G + \beta\,\underset{\sim}{K}'\varphi - \beta'\underline{S}\,\varphi \qquad (66)$$

where $G(x) = \int_{\mathbb{R}^3} \Phi(x,y)\, g(y)\, dy$ and

$$\beta = \frac{(\sigma^+)^2 - (\sigma^-)^2}{(\sigma^+)^2 + (\sigma^-)^2}\,, \quad \beta' = \frac{(\sigma^+)^2\,\alpha'^+/\alpha^+ - (\sigma^-)^2\,\alpha'^-/\alpha^-}{(\sigma^+)^2 + (\sigma^-)^2} \qquad (67)$$

The Fredholm alternative guarantees that $(1 + \beta'\underline{S} - \beta\,\underset{\sim}{K}')^{-1}$ always exists as a bounded operator. It yields $v(x) = \int_{\mathbb{R}^3} g(x,y)\, g(y)\, dy$.

This resolvent $g(x,y)$ (which is of course not symmetric), readily yields the Green function of the original problem as the product $G(x,y) = \sigma(x)\ g(x,y)/\sigma(y)$. Thus, we see that our problem is much more simple than the general three-dimensional problem. Yet, this case is probably very appealing for physicists. From the mathematical point of view, it is the direct generalization of the one-dimensional case.

VII. THE THREE-DIMENSIONAL IMPEDANCE SCATTERING

VII.1 The Fundamental Integral Equation

We come back to Equation (1), with $\alpha(x) \in C^2$ on the Ω_i's and jumping together with its normal derivative through the surfaces S_i, which are defined like in Section V. We are looking for $\varphi(k,x)$ satisfying

$$\begin{aligned} &(\alpha^{-2}\ \mathrm{div}\ \alpha^2\ \mathrm{grad} + k^2 - V)\ \varphi(k,x) = 0 \quad (x \in \mathbb{R}^3) \\ &\alpha\varphi - \exp[ik.x] \qquad \text{outgoing as } |k| \to \infty \end{aligned} \tag{68}$$

The "outgoing condition" is that precisely settled by the Sommerfeld condition (21). Now, let $\psi = \alpha\varphi$. As it is noticed at the end of Section IV, ψ satisfies :

$$\begin{aligned} &\Delta\psi + k^2\psi = (V + \alpha^{-1}\Delta\alpha)\psi = 0 && (x \in \mathbb{R}^3\backslash S) \\ &(\alpha^+)^{-1}\ \psi^+ = (\alpha^-)^{-1}\ \psi^- && (x \in S) \\ &\alpha^+\ \frac{\partial\psi^+}{\partial\nu} - \psi^+\ \frac{\partial\alpha^+}{\partial\nu} = \alpha^-\ \frac{\partial\psi^-}{\partial\nu} - \psi^-\ \frac{\partial\alpha^-}{\partial\nu} && (x \in S) \\ &\psi(k,x) - \exp[ik.x] \qquad \text{outgoing} \quad (|x| \to \infty) \end{aligned} \tag{69}$$

So as to construct ψ, we shall use the Green function $G(x,y)$ defined by (25), (26), (27). Let R be large enough so that $\bar{D}_N \subset B(0,R) = \{x : |x| < R\}$. Applying the Green's theorem to ψ and G in $\omega(R) = B(0,R)\backslash \bar{D}_N$, we get

$$\int_{\omega(R)} [\Delta_y\psi(k,y)\ G(x,y) - \psi(k,y)\ \Delta_y\ G(k,y)]\ dy$$

$$= \int_{\omega_{(R)}} G(x,y)\ [V(y) + \alpha^{-1}(y)\ \Delta\alpha(y)]\ \phi(k,y)\ dy + \begin{cases}\phi(x) & \text{if } x \in \omega(R) \\ 0 & \text{otherwise}\end{cases}$$

$$= \int_{|y|=R} \left(\frac{\partial\phi}{\partial\nu_y} G - \phi \frac{\partial G}{\partial\nu_y}\right) ds(y) - \int_{S_N^+} \left(\frac{\partial\phi}{\partial\nu_y} G - \phi \frac{\partial G}{\partial\nu_y}\right) ds(y) \qquad (70)$$

The integral on $|y| = R$, say $I(R)$, can be divided into two parts

$$I(R) = I_1(R) + I_2(R) = \int_{|y|=R} \left[\frac{\partial(\phi - e^{ik.y})}{\partial\nu_y} G - (\phi - e^{ik.y}) \frac{\partial G}{\partial\nu_y}\right] ds(y)$$

$$+ \int_{|y|=R} \left[\frac{\partial e^{ik.y}}{\partial\nu_y} G(x,y) - e^{ik.y} \frac{\partial G(x,y)}{\partial\nu_y}\right] ds(y) \qquad (71)$$

If we apply the Green theorem to $\exp[ik.x]$ and $G(x,y)$ in $\omega(R)$ we obtain

$$I_2(R) = \int_{S_N^+} \left[\frac{\partial e^{ik.y}}{\partial\nu_y} G(x,y) - e^{ik.y} \frac{\partial G}{\partial\nu_y}(x,y)\right] + \begin{cases}e^{ik.x} & \text{if } x\in\omega(R) \\ 0 & \text{otherwise}\end{cases} \qquad (71)$$

Since $\phi(k,y) - \exp[ik.y]$ and $G(x,y)$ satisfy the Sommerfeld condition, $I_1(R)$ goes to zero as $R \to \infty$. Hence, letting $R \to \infty$ in (70) yields for any $x \in \Omega_{N+1}$:

$$\phi(k,x) = \phi_{in}(k,x) - \int_{\Omega_{N+1}} G(x,y) \left[V(y) + \frac{\Delta\alpha(y)}{\alpha(y)}\right] \phi(k,y)\ dy$$

$$- \int_{S_N^+} \left(\frac{\partial\phi(k,y)}{\partial\nu_y} G(x,y) - \phi(k,y) \frac{\partial G(x,y)}{\partial\nu_y}\right) ds(y) \qquad (72)$$

where $\phi_{in}(k,x)$ is defined by the formula :

$$\psi_{in}(k,x) = \int_{S_N^+} \left[\frac{\partial e^{ik.y}}{\partial \nu_y} G(x,y) - e^{ik.y} \frac{\partial G(x,y)}{\partial \nu_y} \right] ds(y) + \begin{cases} e^{ik.x} & \text{if } x \in \Omega_{N+1} \\ 0 & \text{otherwise} \end{cases} \quad (73)$$

Now, in each domain Ω_i, the Green theorem yields

$$- \int_{\Omega_i} G(x,y) \left[V(y) + \frac{\Delta\alpha(y)}{\alpha(y)} \right] \psi(k,y)\, dy$$

$$+ \int_{S_{i+1}^-} \left[\frac{\partial \psi(k,y)}{\partial \nu_y} G(x,y) - \psi(k,y) \frac{\partial G(k,y)}{\partial \nu_y} \right] ds(y)$$

$$- \int_{S_i^+} \left[\frac{\partial \psi(k,y)}{\partial \nu_y} G(x,y) - \psi(k,y) \frac{\partial G(k,y)}{\partial \nu_y} \right] ds(y)$$

$$= \begin{cases} \psi(k,x) & \text{if} \quad x \in \Omega_i \\ 0 & \text{otherwise} \end{cases} \quad (74)$$

and the continuity conditions (27) imply that $\frac{\partial \psi}{\partial \nu} G - \psi \frac{\partial G}{\partial \nu}$ has the same value on S_i^+ and S_i^-. Adding (72) and the formulas (74) written for $i = (N-1), (N-2), \ldots, 0$, we obtain the scattering integral equation :

$$\boxed{\psi(k,x) = \psi_{in}(k,x) - \int_{\mathbb{R}^2 \setminus S} G(x,y) \left[V(y) + \frac{\Delta\alpha}{\alpha} \right] \psi(k,y)\, dy \qquad x \in \mathbb{R}^2 \setminus S \quad (75)}$$

It generalizes to our problem the usual scattering integral equation, and reduces to it when there are no discontinuities : according to (73), ψ_{in} reduces then to $\exp[ik.x]$, and $G(x,y)$ reduces to $\Phi(x,y)$ given by (28)). Hence there are two steps in our way of processing, and they have a different physical meaning :

<u>1st step.</u> Construct ψ_{in} and $G(x,y)$ from a knowledge of the singular

surfaces and the singular data only - what can be called the step of reflectors (or hard scatterers).

<u>2nd step.</u> Construct ψ by solving the integral equation (75), where the "diffuse scatterers" represented by $V(x)$ and $\alpha^{-1}\Delta\alpha$ appear for the first time - we call it the step of diffuse (or soft) scatterers.

VII.2 Properties of the Incident Wave Function

The argument used above also shows that ψ_{in} is the solution of

$$\left.\begin{aligned} &(\Delta + k^2)\,\psi_{in} = 0 \qquad x \in \mathbb{R}^3\backslash S \\ &\psi_{in}/\alpha \text{ and } \alpha\,\frac{\partial\psi_{in}}{\partial\nu} - \psi_{in}\,\frac{\partial\alpha}{\partial\nu} \text{ continuous}/S \\ &\psi_{in} - \exp[ik.x] \text{ outgoing at } \infty \end{aligned}\right\} \tag{76}$$

Hence ψ_{in} is the wave function that corresponds to the physical problem of impedance scattering by an impedance factor $\bar{\sigma}(x)$ which would be the (piecewise harmonic) solution of (76) with $k = 0$. To find the behaviour of ψ_{in} as $|x| \to \infty$, we apply the Green's theorem to ψ_{in} and $\Phi(x,y)$ in $\omega(R)$. For $x \in \omega(R)$, we readily obtain

$$\psi_{in}(x) = \int_{|y|=R}\left[\frac{\partial\psi_{in}(k,y)}{\partial\nu_y}\,\Phi(x,y) - \psi_{in}(k,y)\,\frac{\partial}{\partial\nu_y}\,\Phi(x,y)\right]ds(y)$$

$$- \int_{S_N^+}\left[\frac{\partial\psi_{in}(k,y)}{\partial\nu_y}\,\Phi(x,y) - \psi_{in}(k,y)\,\frac{\partial}{\partial\nu_y}\,\Phi(x,y)\right]ds(y) \tag{77}$$

Since $\psi_{in} - e^{ik.x}$ is outgoing, as R goes to infinity, the integral over the sphere tends to e^{ikx}, giving the representation :

$$\psi_{in}(k,x) = \exp[ik.x] - \int_{S_N^+}\left[\frac{\partial\psi_{in}(k,y)}{\partial\nu_y}\,\Phi(x,y)\right.$$

$$- \psi_{in}(k,y) \left.\frac{\partial\Phi}{\partial\nu}\right] ds(y) \qquad (x \in \Omega_{N+1}) \tag{78}$$

Using now the asymptotic behaviour $(4\pi)^{-1} \exp\left[i|k| \, (|x| - \hat{x}.y)\right]$ of $\Phi(x,y)$ (see (28)) we obtain for $|x| \to \infty$

$$\varphi_{in}(k,x) = \exp[ik.x] - (4\pi)^{-1} \frac{e^{i|k||x|}}{|x|} A_o\,(|k|\,\hat{x},\ k) + o(|x|^{-1}) \tag{79}$$

with $\hat{x} = x/|x|$, and

$$A_o(\xi,k) = \int_{S_N^+} e^{-ix.\xi}\left[\frac{\partial\psi_{in}}{\partial\nu}(k,x) + i\xi.\nu(x)\ \psi_{in}(k,x)\right] ds(x) \tag{80}$$

This "first step" scattering amplitude gives information not only on S_N but also, (by means of ψ_{in}), on all the reflectors

It is interesting to see that $\psi_{in}(k,x)$ is also related to the asymptotic behaviour of $G(x,y)$. So as to prove it, we apply Green's theorem to $G(x,y)$ and $\Phi(x,z)$ in $z \in \omega(R)$, with x,y fixed $(x \neq y)$, x in $\omega(R)$. We get

$$\int_{\omega(R)} [\Phi(x,z)\ \Delta_z G(z,y) - G(z,y)\ \Delta\Phi(x,z)]\ dz = \begin{cases} G(x,y) - \Phi(x,y) & \text{if } y \in \overline{D}_N \\ G(x,y) & \text{if } y \in D_N \end{cases}$$

$$= \left(\int_{|z|=R} - \int_{S_N^+}\right) \left[G_o(x,z) \frac{\partial G}{\partial\nu_z}(z,y) - \frac{\partial G_o}{\partial\nu_o}(x,z)\ G(z,y)\right] ds(z) \tag{81}$$

Letting $R \to \infty$, and since G and Φ are outgoing, we obtain for any $x \in \Omega_{N+1}$:

$$G(x,y) = - \int_{S_N^+} \left[\Phi(x,z) \frac{\partial G}{\partial\nu_z}(z,y) - \frac{\partial\Phi(x,z)}{\partial\nu_z} G(z,y)\right] ds(z)$$

$$+\begin{cases} \Phi(x,y) & \text{if} \quad y \in \bar{D}_N \\ 0 & \text{if} \quad y \in D_N \end{cases} \tag{82}$$

As $|x| \to \infty$, this yeields

$$G(x,y) \sim \frac{1}{4\pi} \frac{e^{i|k||x|}}{|x|} \psi_{in}(-|k|\,\hat{x},y) \quad (|x| \to \infty) \tag{84}$$

VII.3 The Full Scattering Amplitude

Substituting (84) in (75) yields

$$\psi(k,x) \sim \psi_{in}(k,x) - \frac{1}{4\pi} \frac{e^{i|k||x|}}{|x|} A_1(|k|\hat{x},\ k) + o(|x|^{-1}) \text{ as } |x| \to \infty \tag{85}$$

where

$$A_1(\xi,k) = \int_{R^3 \setminus S} \psi_{in}(-\xi,x) \left[V(x) + \frac{\Delta\alpha}{\alpha}(x)\right] \psi(k,x)\, dx \tag{86}$$

Combining (79) and (85) finally yields the asymptotic representation of $\psi(k,x)$:

$$\psi(k,x) = e^{ik\cdot x} - \frac{1}{4\pi} \frac{e^{i|k|\ |x|}}{|x|} A(|k|\hat{x},k)) + o(|x|^{-1}) \tag{87}$$

where

$$A(\xi,k) = A_0(\xi,k) + A_1(\xi,k) \tag{88}$$

A_0 is the scattering amplitude which would be due to reflectors only. A_1 is the scattering amplitude due to diffuse scattering *the reflectors being present*. A is the full scattering amplitude, the quantity to be measured. It is easy to see now that these results trivially reduce to the usual ones for the Schrödinger problem when no reflector is present. Notice also that the generalizations to R_n with $n \geqslant 3$ is trivial.

VII.4 On the Class of Equivalent Impedance Factors

The "standard equivalence" (Ref 1) between impedance factors α which lead to the same function $\Delta\alpha/\alpha$ inside the domains Ω_1 and the same singular data on the S_1 leads us to the question : how can a "representative" of this class can be constructed, which in addition goes to 1 as $|x| \to \infty$. Thus we have to construct α^0 such that

$$\Delta\alpha^0 - \alpha_0(\alpha^{-1}\,\Delta\alpha) = 0 \tag{89a}$$

$$\alpha^0/\alpha \text{ and } \alpha^0\,\frac{\partial\alpha}{\partial\nu} - \alpha\,\frac{\partial\alpha^0}{\partial\nu} \quad \text{continuous through } S \tag{89b}$$

$$\alpha\text{-}1 \quad \text{going} \tag{89c}$$

It is a limit case of the problem (69) (for $k = 0$) and can be solved in a similar way.

Going further into the "standard equivalence", we can see that the Green's function should not be modified if α is replaced by $\tilde{\alpha}$ which satisfies with α the condition (89b). This invariance is related to that of the chains when continuity conditions are mixed or when, equivalently, a part of Equation (47) is added to Equation (48). It follows that β' can be replaced by $\beta' + q\beta/\alpha^+\alpha^-$, γ by $\gamma - q/\alpha^-\alpha^+$, where $q(x)$ is any function continuous through S. A clever choice yields the "standard" coefficients, completely expressed in terms of coefficients (3) :

$$\bar{\beta} = \frac{1-t+r}{1+t+r} \qquad \bar{\beta} = -s/t \qquad \bar{\gamma} = -\,\bar{\beta}\,\bar{\beta}' \tag{90}$$

where we used the $-$ to denote the standard coefficients, instead of o, to mean that these two reductions can be used quite independently.

VIII. FURTHER GENERALIZATIONS

Extensions to vectorial and tensorial problems related to electro-magnetic or elastic wave theory are now in progress but almost no result has been published as yet. Such results are particularly important both for theoretical studies and in view of applications to practical problems like identification problems from the results of radar or seismic soundings. Some of them will be published soon (Born approximations etc). Generalizations of perturbation theory, studies of bound states and exceptional points are also in progress. Generalizations of the three dimensional results to n dimensional cases ($n \geq 3$)

are straightforward. Generalizations to "plane-like" or "cylindrical-like" geometries are also in progress.

ACKNOWLEDGEMENTS

Several parts of the one-dimensional studies have been the object of fruitful discussions or more active collaboration with A. Degasperis and B. Dolveck-Guilpart. The one-dimensional study is now almost completely published (see in particular Ref.2 and 3). Parts of the three-dimensional study have been the object of fruitful discussions with B. de Facio. Other ones result from an active collaboration with A. Nachman (in particular §VI and VII) ; they will be the object of more complete publications later.

Appendix : Proofs of (37)-(38)-(39).

The Proofs of Formulas (37)-(38)-(39) are quite similar with each other and it is enough to reproduce here one of them - say (39). Thus, we are led to calculate, for $f_j \in E(S_j)$

$$Q_{kij} = 4 \frac{\partial}{\partial\nu_x} \int_{S_i} \left[\frac{\partial\Phi(x,t)}{\partial\nu_t} g_j(t) - \Phi(x,t) \frac{\partial g_j(t)}{\partial\nu_t}\right] ds(t) \qquad (90)$$

where $x \in S_k$ and

$$g_j(t) = \int_{S_j} \frac{\partial\Phi(t,y)}{\partial\nu_y} f_j(y)\, ds(y) \qquad (91)$$

g_j is a continuous solution of Helmholtz equation across any element of S except S_j, across which

$$g_j(t^+) - g_j(t) = g_j(t) - g_j(t^-) = \frac{1}{2} f_j(t) \qquad (t \in S_j) . \qquad (92)$$

The result applies of course to $\int_{S_i} \frac{\partial\Phi(x,t)}{\partial\nu_t} h(t)\, ds(t)$, but its normal derivative is continuous. Now, so as to proceed, either we see a direction (i.e inwards to Ω_o or outwards to Ω_{N+1}) where Φ and g_i are continuous solutions of the Helmholtz equation up to and including S_i, or there is not and we must repeatedly apply the Green's theorem to go from S_i^- to S_{i-1}^+ and continuity relations to go from S_{i-1}^+ to S_{i-1}^-, etc, or from S_i to S_{i+1}, etc, up to a surface after which Φ and g_i remain continuous. Of course, jump relations must be applied when necessary. With these remarks, the following calculations are easy to understand :

(1) $\boxed{i < k}$;

(a) $\underline{i < j}$; $Q_{kij} = 0$

(b) $\underline{i = j}$;
$$Q_{kij} = 4 \frac{\partial}{\partial\nu_x} \int_{S_j} \left[\frac{\partial\Phi(x,t)}{\partial\nu_t} g_j(t^-) - \Phi(x,t) \frac{\partial}{\partial\nu_t} g_j(t^-) \right] ds(t)$$
$$+ 2 \frac{\partial}{\partial\nu_x} \int_{S_j} \frac{\partial\Phi(x,t)}{\partial\nu_t} f_j(t)\, ds(t) = 2 \frac{\partial}{\partial\nu_x} g_j(x)$$

(c) $\underline{i > j}$;
$$Q_{kij} = 4 \frac{\partial}{\partial\nu_x} \int_{S_i} \left[\frac{\partial\Phi(x,t)}{\partial\nu_t} g_i(t) - \Phi(x,t) \frac{\partial}{\partial\nu_t} g_j(t) \right] ds(t)$$
$$= \dots = 4 \frac{\partial}{\partial\nu_x} \int_{S_j} \left[\frac{\partial\Phi(x,t)}{\partial\nu_t} g_j(t^-) - \Phi(x,t) \frac{\partial}{\partial\nu_t} g_i(t^-) \right] ds(t) +$$
$$+ 4 \frac{\partial}{\partial\nu_x} \int_{S_j} \frac{\partial\Phi(x,t)}{\partial\nu_t} f_j(t)\, ds(t) = 4 \frac{\partial}{\partial\nu_x} g_j(x)$$

(2) $\boxed{i = k}$

(a) $\underline{k < j}$;
$$Q_{kkj} = 4 \frac{\partial}{\partial\nu_x} \int_{S_k} \left[\frac{\partial\Phi(x^-,t)}{\partial\nu_t} g_j(t) - \Phi(x^-,t) \frac{\partial}{\partial\nu_t} g_i(t) \right] ds(t)$$
$$+ 2 \frac{\partial g_j(x)}{\partial\nu_x} = 2 \frac{\partial g_j(x)}{\partial\nu_x}$$

$$(b)\ \underline{k = i}\ ;\ Q_{kkk} = 4\ \frac{\partial}{\partial \nu_x} \int_{S_k} \left[\frac{\partial \Phi(x^-,t)}{\partial \nu_t}\ g_k(t) - \Phi(x^-,t)\ \frac{\partial}{\partial \nu_t}\ g_k(t) \right] ds(t)$$

$$+\ 2\ \frac{\partial g_k(x)}{\partial \nu_x} = 4\ \frac{\partial}{\partial \nu_x} \int_{S_k} \left[\frac{\partial \Phi(x^-,t^+)}{\partial \nu_t^+}\ g_k(t^+) - \Phi(x^-,x^+)\ \frac{\partial}{\partial \nu_t^+}\ g_k(t^+) \right] ds(t)$$

$$+\ 2\ \frac{\partial g_k(x)}{\partial \nu_x} - 2\ \frac{\partial g_k(x)}{\partial \nu_x} = 0$$

$$(c)\ \underline{k > i}\ ;\ Q_{kkj} = 4\ \frac{\partial}{\partial \nu_x} \int_{S_k} \left[\frac{\partial \Phi(x^+,t)}{\partial \nu_t}\ g_j(t) - \Phi(x^+,t)\ \frac{\partial}{\partial \nu_t}\ g_j(j) \right] ds(t)$$

$$-\ 2\ \frac{\partial g_j(x)}{\partial \nu_x} = -\ 2\ \frac{\partial g_j(x)}{\partial \nu_x}$$

By the same token we easily derive for $i < k$ the cases

$$(i>j)\ Q_{kij} = 0\ ;\ (i=j)\ Q_{kij} = -\ 2\ \frac{\partial g_j(x)}{\partial \nu_x}\ ;\ (i<j)\ Q_{kij} = -\ 4\ \frac{\partial g_j(x)}{\partial \nu_x}.$$

Then a simple summation over the index i yields the result (39).

REFERENCES

[1] Sabatier P.C., "Remark on the three-dimensional mixed impedance potential equation" Inverse Problems 3, L83-L86 (1987) + Corrigendum in Inverse Problems 4, L1 (1988).

[2] Sabatier P.C., "For an impedance scattering theory" p.723-745 in "Non linear Evolutions" J. Léon Ed. World Scientific Publ.

Singapore 1988.

[3] Sabatier P.C. & Dolveck-Guilpart B., "On modeling discontinuous data" "One dimensional approximations" J. Math. Phys. (1988).

[4] Chadan K. & Sabatier P.C., "Inverse Problems in Quantum Scattering Theory" 2nd Edition revised and expanded. Springer Verlag. (in press)

[5] Colton D. & Kress R., "Integral Equation Methods in Scattering Theory" J. Wiley. New York 1986.

[6] to be published in the Proceedings of a Workshop on Direct & Inverse methods in Radar Polarimetry. W. Boerner Ed.

SURVEY OF OPTIMIZATION ALGORITHMS APPLIED TO INVERSE PROBLEMS

A. ROGER
Société Elf Aquitaine (Production)
Méthodes et Simulations Mathématiques
64018 PAU CEDEX

and

Laboratoire d'Optique Electromagnétique
Faculté Saint Jérôme
13397 MARSEILLE CEDEX 13

ABSTRACT

Optimization algorithms are rapidly surveyed and the specific properties of each algorithm are described. Then two inverse problems are tackled with the Optimization approach in order to point out some difficulties often encountered in practical inversion procedures.

INTRODUCTION

Inverse problems are of strategical importance in Industry. For instance medical imaging and soil probing offer major challenges which could lead to great achievements. However they are dependent upon phenomenons of various physical nature and often great complexity. Moreover in the industrial context an existence theorem or even an abstract constructive algorithm is not sufficient: the actual values of the searched-for quantities in practical cases are demanded, thus making numerical computations necessary. So what is needed is a versatile, powerful and numerically practicable tool for solving inverse problems. Nowadays the most widely used technique is the **Optimization approach**.

OPTIMIZATION ALGORITHMS

Definitions

The Optimization algorithms deal with the general minimization problem (P):

$$\text{(P)} \quad \min_{x \in \mathbb{R}^N} f(x) : \mathbb{R}^N \rightarrow \mathbb{R} \tag{1}$$

Throughout this paper f is assumed to be sufficiently differentiable. For the sake of simplicity this paper will not deal with constraints.

The vector x contains the N parameters to be retrieved. In inverse problems the "cost function" f is chosen to measure the gap between real data and simulated data. The user of the algorithm is requested to yield a formula or a computer code which enables one to compute simulated data and hence the value f(x) for any set of parameters x. The computation of the **gradient** $G \in \mathbb{R}^N$ of f is also required by most algorithms; this is a requisite for the numerical efficiency. Of course G can be calculated by finite differences formulas; but no general rule is available to estimate a correct stepsize, which must be small enough to make second order terms negligible and large enough to avoid rounding errors in differences. Moreover the evaluation of the N components of G by finite differences requires at least N + 1 cost computations: this is not feasible in most industrial problems, because N can be very large (greater than 500) and a cost computation very expensive. Fortunately it is often possible to compute **directly** the gradient by a technique called "adjoint state"[1], which requires only **two** cost computations whatever N. This point will be enlightened by two examples in the following.

The second order term in the Taylor expansion of f implies the so-called "Hessian" H, i.e. the matrix of second order derivatives. The simplest case is that of **quadratic** cost functions, which have a constant Hessian. Usually the Hessian is not available by direct means and is approximated by a "Quasi Hessian". The values of the gradient and of the Quasi Hessian at a point x_k will be denoted by G_k and H_k respectively.

Finally the cost function f is approximated locally by a quadratic form $q_k(x)$:

$$f(x_k + s) - f(x_k) \cong q_k(s) = G_k^t s + \frac{1}{2} s^t H_k s \tag{2}$$

A **local** optimization algorithm constructs a sequence x_k, which under certain conditions, converges towards a **local minimum** of f. To this aim it makes use of the local model (2) to deduce x_{k+1} from x_k .

Note that if H_k is positive definite, the quadratic form $q_k(s)$ actually has a minimum at $-H_k^{-1}G_k$ which can be a candidate for the next iterate. Setting $x_{k+1} = x_k - H_k^{-1}G_k$ gives the well-known Newton algorithm. The case of an indefinite Hessian requires special care.

Only local optimization algorithms will be described in this paper. Indeed the very few methods available for finding a global minimum cannot be employed in most practical cases because they generally demand very large numbers of cost and gradient computations.

An algorithm is said to be **locally** convergent iff the convergence is ensured only if the starting point is in a neighborhood of the solution. On the opposite an algorithm is said to be **globally** convergent iff the convergence holds in the whole space .i.e. whatever the starting point. As is well known the Newton Algorithm is locally but not globally convergent [2]. Note that a local optimization algorithm may nevertheless be globally convergent: examples will be given below.

The Case Of Least Squares Algorithms

This case is encountered when the cost function is a sum of M squares, i.e. may be written as:

$$f(x) = \frac{1}{2}F^t(x)\,F(x)\ ,\ F \in \mathbb{R}^M \tag{3}$$

The **Jacobian** is defined as the $M \times N$ matrix J , the coefficient J_i^j of which is the partial derivative $\frac{\partial F_i}{\partial x_j}$ of the i-th component of F.

The gradient and the Hessian are then given by:

$$G = J^tF \tag{4}$$

$$H = J^tJ + \frac{\partial^2 F}{\partial x^2}F \tag{5}$$

Quasi Newton Approximations

The principle of Quasi Newton methods is to approximate the Hessian at the step k with a Quasi Hessian H_k calculated by an additive correction formula. The simplest correction is given by the **rank one** formula:

$$H_{k+1} = H_k + a_k u_k . u_k^t \quad a_k \in \mathbb{R}\ ,\ u_k \in \mathbb{R}^N$$

where a_k, u_k, v_k are given by simple formulae [2]. Rank two corrections are also well known:

$$H_{k+1} = H_k + a_k u_k . u_k^t + b_k v_k . v_k^t \qquad a_k, b_k \in \mathbb{R} \qquad u_k, v_k \in \mathbb{R}^N$$

Among others the most famous formulae are the DFP formula [3] and the BFGS formula [4]. Rank two formulae are often preferred because under certain conditions they yield a sequence of **positive definite** Quasi Hessians.

These correction formulae are constructed in such a way that they yield the exact Hessian in N iterations when applied to quadratic cost functions. Note that there exist also formulae which directly yield a sequence $\{ H_k^{-1} \}$ of inverses of Quasi Hessians.

Gauss Newton Approximation

This approximation is used in the least squares case. The Hessian is simply approximated by:

$$H_k = J_k^t J_k \tag{6}$$

which is easily computed since the Jacobian, which contains only first order derivatives, is known in most cases. Note that this Quasi Hessian is a semi positive definite matrix.

The Line Search Strategy

This is the most classical strategy. The iterative process is as follows:

1. Set $k = 0$, choose a starting point x_0
2. Compute $f(x_k)$, G_k, H_k
3. Compute a descent direction $D_k \in \mathbb{R}^N$
4. Minimize $f(x_k + \theta D_k)$, $\theta \in \mathbb{R}$ (line search)
5. End if the stopping criterion is satisfied
6. Increment k and go to 2

The crucial point for the practical implementation is the line search: the minimization must be only approximate to avoid too many cost evaluations and yet must be precise enough to ensure a quick convergence.

Many different directions can be used. The steepest descent is given by:

$$D_k = - G_k$$

The resulting algorithm is the **steepest descent** algorithm. It is **globally convergent** but its asymptotic speed of convergence is rather low.

Quasi-Newton algorithms use the Newton direction defined by:

$$D_k = -H_k^{-1}G_k$$

This direction is significant only when the Newton step $-H_k^{-1}G_k$ actually corresponds to a minimum of the local quadratic model, i.e. when H_k is positive definite. This algorithm has a good asymptotic speed of convergence ("superlinear") and is also **globally convergent**. However it implies storages and manipulations of Quasi-Hessians.

The **conjugate gradient** algorithms make use of descent directions defined by:

$$D_{k+1} = a_k D_k + B_k \quad , a_k \in \mathbb{R} \, , B_k \in \mathbb{R}^N$$

The most famous conjugate gradient formulae are that of Fletcher-Reeves [5] and that of Polack-Ribière [6] . These algorithms are also **globally convergent** and superlinearly convergent. In addition they imply no Hessian handling or storage; thus they are very interesting for large optimization problems ($N > 500$). However they generally lead to more cost and gradient computations than Quasi Newton algorithms.

The Trust Region Strategy

This strategy has been developed in the last ten years [7] The corresponding algorithms may be sketched as:

1. Set $k = 0$, choose a starting point x_0 and a starting radius r_0 and compute $f(x_0)$
2. Compute G_k, H_k ($f(x_k)$ was calculated in 1 or 4)
3. Minimize the local model $q_k(s)$ inside the sphere $|s| \leq r_k$ (trust region) in order to obtain the local solution s_k
4. Compute the "quality coefficient" $\rho_k = \dfrac{f(x_k + s_k) - f(x_k)}{q_k(s_k)}$
5. End if the stopping criterion is satisfied
6. If ρ_k is satisfactory, set $x_{k+1} = x_k + s_k$, increase r_k, increment k and go to 2
7. If ρ_k is not satisfactory, set $x_{k+1} = x_k$, decrease r_k, increment k and go to 3

The rules for updating r_k and checking ρ_k are precisely fixed by the theory so as to ensure good properties of the algorithm.

The Trust Region approach makes full use of the local model and takes into account its local validity. Very strong properties have been demonstrated for these algorithms. They exhibit both global convergence properties and super linear asymptotic speed of convergence. In addition, they generally require much less cost and gradient computations than any line search algorithm, bnt more computations have to be performed on the local model.

AN INVERSE PROBLEM IN OPTICS

An example of optical inverse problem is given by the **diffraction gratings**. A diffraction grating is a periodically modulated surface (Figure 1). Illuminated by a plane electromagnetic wave such as a laser beam, it generates a diffracted field which at infinity is composed of plane waves called **orders**. The direct problem is an exterior problem governed by the Helmholtz equation. It has been solved in many different ways in the last twenty years [8].

The Adjoint State

The inverse problem consists in retrieving the profile of the surface modulation from measures on the diffracted field at infinity and was solved only recently [9].

The algorithms which solve the direct problem enable one to calculate the diffracted field and thus the values of any cost function associated with it.

The computation of the gradient is not an obvious matter. The figure 1 shows the definition of the "adjoint state" for the computation of the gradient of the complex amplitude B of the field in a given direction. As an example let us consider the TE case of polarization, which corresponds to a Dirichlet boundary condition on the grating surface. In this case the following formula holds [10]:

$$\delta B = K \int_{\text{period}} \frac{\partial u}{\partial n} \frac{\partial v}{\partial n} \, \delta h(x) \, dx$$

where K is a constant, $\delta h(x)$ is the variation of the surface profile $y = h(x)$ and δB the corresponding variation of B. So whenever the initial and the adjoint direct problems are solved, the quantities $\frac{\partial u}{\partial n}$ and $\frac{\partial v}{\partial n}$ are known and so is the gradient of B or any related quantity.

It is worth noting that the adjoint state is precisely the direct problem involved by the well known reciprocity theorem. This remarkable property was found to be always verified in Electromagnetics [11]: different polarization cases lead to different boundary conditions and different gradient formulae, but the adjoint state remains the same.

The optimization algorithms have been successfully applied to this inverse problem and grating profiles were accurately reconstructed from intensity measurements at infinity [12].

Invariances And Saddle Points

By applying successively the reciprocity theorem and a symmetry with respect to the Oy axis, it is easily shown that two grating profiles symmetrical to each other give rise to the same diffracted

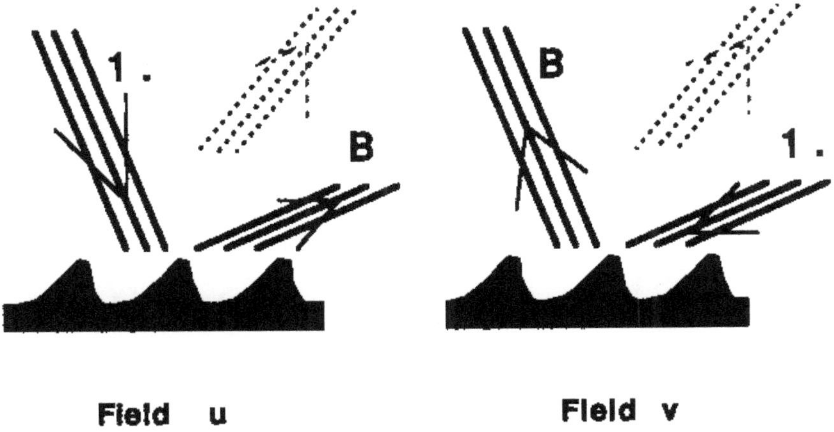

Figure 1. **The adjoint state for a diffraction grating:** At left hand the initial direct problem with field u. At right hand the adjoint direct problem with field v. The adjoint state is that implied by the reciprocity theorem.

intensity in the zero order (plane wave in the direction of specular reflection) (figure 2). This property has an important consequence: if the profile is an even function the variation of the intensity generated by any odd perturbation is null at the first order; this means that the gradient has an infinity of zero components. As a consequence, starting from an even profile, one gets a sequence of even profiles and the limiting point of the algorithm is a **saddle point** with zero gradient and **not** a local minimum.

The solution is easily found: it is enough to start from a non symmetrical profile in this case. But a more general rule can be remembered: any invariance theorem corresponds to zero components of the gradient which in turn can prevent the algorithm from finding actual minima. So it is necessary, before any numerical computation, to check all the invariance and symmetry properties of the problem. This removes many numerical difficulties; in addition it often suggests a way to improve the numerical implementation.

AN INVERSE PROBLEM IN SEISMICS

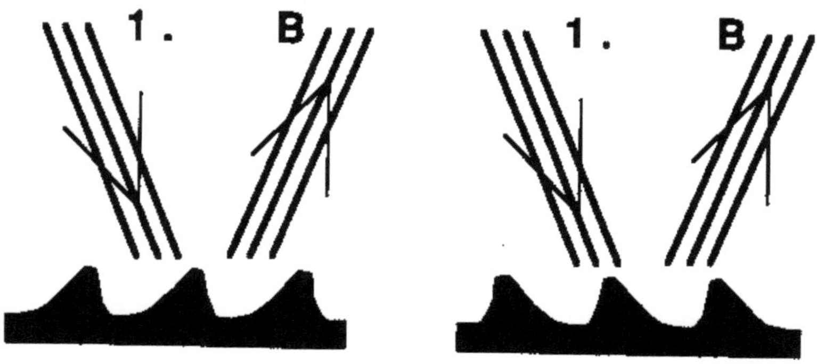

Figure 2. The invariance property: The reciprocity theorem shows that two profiles symmetrical of each other with respect to the Oy axis yield the same diffracted intensity in the direction of specular direction.

Geophysical sounding is of the highest interest in Oil industry. Sources are placed at the soil surface. They generate acoustic waves which are diffracted by the underground structures and are measured by sonic surface receivers. The direct problem is governed by the acoustic wave equation. It is a time dependent exterior problem with **initial** conditions in time and boundary conditions on the surface.

The Adjoint State

The seismic inverse problem consists in retrieving the acoustic parameter distribution of the soil. The adjoint state is governed by the same wave equation and the same boundary conditions as the initial direct problem; but the sources are now the differences between measured and simulated

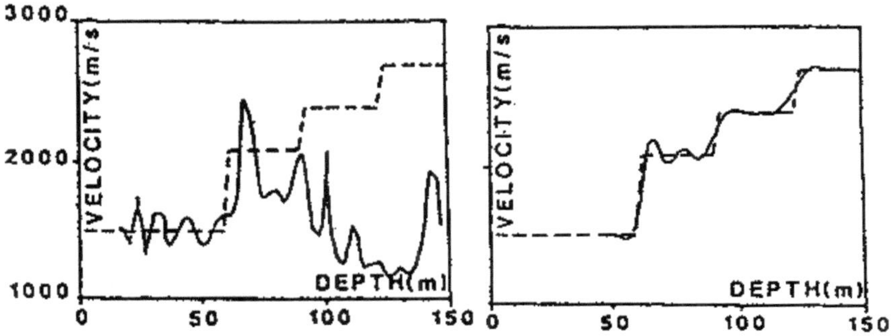

Figure 3. **Reconstructed velocity profiles.:** The model to be retrieved is a four layer profile, in dotted line on the two figures. The reconstructed profile is in full line: result obtained with an initial homogeneous medium (left) and result obtained with a ramp as initial profile (right).

pressures and the initial conditions in time are replaced by final conditions. One often says that the adjoint state corresponds to the "backpropagation of the residuals".

Strategies For Local Minima

The main difficulty encountered in this problem is that most of the time the algorithm converges towards a local minimum which may be very different from the true sought-for global minimum. Figure 3 presents a four layer example [13] where the reconstruction is unsuccessful. Figure 4 presents the corresponding seismograms of the real profile and the reconstructed profile [13] . Though the profiles are very different, the seismograms are very similar; this is an indication that additional a-priori information is needed.

Fortunately many techniques allow one to overcome this difficulty. Figure 3 presents also the reconstructed profile when starting with a better initial guess: this alternate profile is quite satisfactory. Among other techniques for avoiding local minima, one can mention variation constraints on the solution and progressive downward identification.

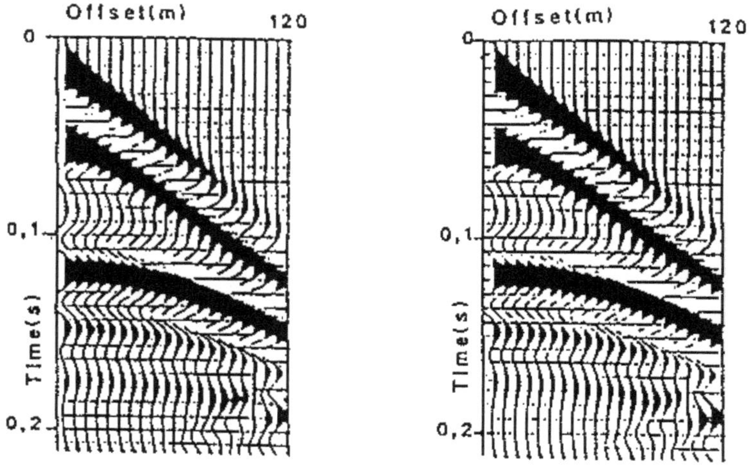

Figure 4. Four layer seismograms: Seismograms of the exact four layer medium (left) and of the reconstructed medium of figure 3 (right).

CONCLUSION

The optimization approach of inverse problems is for sure widely used, especially for industrial purposes, because it is a very versatile and easy to use method. It enables one to deal with highly complex problems of very large sizes and are well adapted for numerical implementations.

However the moral of the story is the following: when the problem is complex, a naïve implementation often leads to numerical difficulties. It is necessary to examine the physics, which is somewhat neglected in the definition of the general optimization algorithms, and to take it into account. This can be achieved by various means such as refining the modelization, changing the cost function, choosing better starting points, putting constraints on the parameters. Moreover the physics often suggests specific means which are even more efficient.

With this in mind, the optimization approach is a powerful tool to solve inverse problems numerically, and often the only available approach when the problem is complex.

ACKNOWLEDMENT

The author is indebted to P. Terpolilli for many aspects of this paper, especially the convergence properties of algorithms.

REFERENCES

1. J.L.Lions, "Contrôle optimal des systèmes gouvernés par des équations aux dérivées partielles", Dunod, Gauthier Villars, Paris (1968)

2. M.Minoux, "Programmation Mathématique. Théorie et Algorithmes" Dunod, Bordas, Paris (1983)

3. W.C. Davidon, "Variable metric methods for minimization", Argonne National Labs Report ANL-5990 (1959)

4. C.G. Broyden, "The convergence of a class of double-rank minimization algorithms"J.I.M.A.,76-90,222-236 (1970)

5. R. Fletcher, C.M. Reeves, "Function minimization by conjngate gradients", Computer Journal 7, pp149-154 (1964)

6. E. Polak, "Computational methods in Optimization, a unified approach", Academic Press, New York (1971)

7. J.J. Moré, "Recent developments in Algorithms and software for Trust Region methods", Mathematical Programming: the state of the art, pp 258-287, Springer-Verlag (1982)

8. R. Petit (editor), "Electromagnetic Theory of Gratings", Springer-Verlag (1980)

9. A.Roger, D. Maystre, "Inverse scattering metbod in electromagnetic Optics: application to the diffraction grating", J.O.S.A., 70, 12, pp 1483-1495 (1980)

10. A.Roger, "Generalised reciprocity relations for perfectly conducting gratings", Optica Acta, 30, pp 575-585 (1982)

11. A.Roger, "Reciprocity theorem applied to the computation of functional derivatives", Electromagnetics, 2, pp 69-83 (1982)

12. A. Roger, M. Breidne, "Grating profile reconstruction by an inverse scattering method", Optics Commun., 35, pp 299-302 (1980)

13. P.Kolb, F. Collino, P. Lailly, "Pre-stack Inversion of a 1-D medium", Proc IEEE, 74, 3, pp 498-508 (1986)

A LINEARISED INVERSE PROBLEM: ACOUSTIC IMPEDANCE TOMOGRAPHY OF BIOLOGICAL MEDIA

J.P. Lefebvre, CNRS-Laboratoire de Mécanique et d'Acoustique
31 Ch. Joseph-Aiguier, 13402 Marseille cédex 9, France.

ABSTRACT

Biological materials behave the peculiarity to be nearly completely transparent to ultrasonic waves: they only give rise to a very weak scattering, sufficient however to allow processing with adapted electronics. That is the reason of the succes of such an imaging technique as echography.

From the physical and the mathematical point-of-view, a linear approximation of scattering, as Born's one, is well suited for biological tissues, since they consist largely of water. That leads to a linear approximation of the inverse scattering problem known as the Inverse Born Approximation.

But the first problem which arises is acoustic modelling, i.e. the choice of a simple but realistic equation governing wave propagation and scattering in such inhomogeneous media. We have chosen a partial differential equation with two space-varying parameters, one of which, the impedance fluctuation, plays a dominating rôle in back scattering, and the other, the velocity fluctuation, the symmetrical one in forward scattering. That is shown through a Born approximation of the direct problem.

The related inverse scattering problem is then solved by inversion of the direct Born approximation (Inverse Born Approximation), within near back scattering, where only impedance fluctuations play a significant rôle. Two pecular scannings are proposed for the tomographic reconstruction of the impedance profile, i.e. for Acoustic Impedance Tomography. They lead to simple one dimensional inverse Fourier transform, easy to perform on a micro-computer.

We then briefly describe the experimental apparatus especially designed in order to test a large variety of reconstruction methods for ultrasonic tomography, with emphasis one our ones. We expose the present status of our project, and show some results given by the first step, i.e. Reflectivity Tomography, that is still a qualitative and uncoherent imaging method.

INTRODUCTION

An important trend in current ultrasonic imaging is that of image quantifying. Pulse-echo imaging, on the other hand, is a qualitative method since the grey scale or colour scale used has no physical interpretation and is completely arbitrary. The reflectivity of the medium is known to be the significant parameter, i.e. the ability to produce an echo. This depends not only on the physical characteristics of the medium, but also on the viewing angle, the waveform used, etc. If we wish to proceed further and attribute to each image point or pixel a specific value for a specific physical parameter of the medium (as is done in X-ray tomography), we must define the problem in physical terms, i.e. in terms of the inverse acoustic propagation problem. This brings into focus the importance of the acoustic modelling problem. In essence we need a simple model, one which is realistic but at the same time reduced to the least number of parameters so as to permit an effective inversion.

We have succeeded in reducing the number of parameters to unity in the situation where we confine our attentiou to reflexion only, ignoring transmission. In this case we show that a single parameter, acoustic impedance, plays the only significant rôle. Solving the inverse propagation problem then leads to impedance imaging. In particular, we solve only the linearised inverse problem, using tomographic reconstruction algorithms derived from those of X-ray tomodensitometry. Heuce we call our method ultrasonic tomo-impediometry, or Acoustic Impedance Tomography (AIT).

1. MODELLING

At first we omit absorption-dispersion phenomena due to dissipative processes to focus on the single phenomenon of scattering from inhomogeneities in the medium. Through a one-dimensional study, Leeman has shown [1] that for biological media the effect on wave propagation of variations in the absorption coefficient is negligible in the frequency range 1 to 10 MHz. The error in neglecting absorption is in the order of $\alpha\lambda$ (absorption coefficient α times wavelength λ) i.e. less than 1 percent.

Hence we will characterise our acoustically inhomogeneous medium by its fluctuations in density ρ and velocity c only . The equation for acoustic pressure P propagation is thus written [2]:

$$- \frac{1}{c^2} \frac{\partial^2 P}{\partial t^2} + \rho \, div \left(\frac{1}{\rho} \, grad \, P \right) = 0$$

In practice, the ultrasound is transmitted by a probe which contacts the skin via a gel film (in vivo testing) or which is submerged in a water tank to ensure coupling (in vitro testing). For convenience, we take the coupling medium as a homogeneous reference of known characteristics ρ_0, c_0.

The wave propagation equation is then written:

$$-\frac{1}{c_0^2}\frac{\partial^2 P}{\partial t^2} + \Delta P = \left(\frac{1}{c^2} - \frac{1}{c_0^2}\right)\frac{\partial^2 P}{\partial t^2} + \frac{1}{\rho}\,\mathrm{grad}\rho . \mathrm{grad}P$$

We substitute $\alpha = (c^2 - c_0^2) / 2c^2$ and $\zeta = \mathrm{Log}(Z/Z_0)$ as characterising quadratic fluctuations in wave velocity c and logarithmic variations in impedance $Z = \rho c$, respectively, in the above equation, obtaining:

$$-\frac{1}{c_0^2}\frac{\partial^2 P}{\partial t^2} + \Delta P = -\frac{2\alpha}{c_0^2}\frac{\partial^2 P}{\partial t^2} + \frac{1}{2}\,\mathrm{grad}(\mathrm{Log}(1-2\alpha)).\mathrm{grad}P + \mathrm{grad}\zeta.\mathrm{grad}P$$

Since biological tissue is characterised by its weak inhomogeneity ($\alpha \approx \zeta \approx 10^{-2}$), we can write to first order in α and ζ that:

$$-\frac{1}{c_0^2}\frac{\partial^2 P}{\partial t^2} + \Delta P = -\frac{2\alpha}{c_0^2}\frac{\partial^2 P}{\partial t^2} - \mathrm{grad}\alpha.\mathrm{grad}P + \mathrm{grad}\zeta.\mathrm{grad}P$$

We have chosen this model to characterise biological media for acoustic purposes. Next, we propose to solve the inverse problem, i.e. to find α and ζ from measurements of the wave propagation and scattering. Having posed the hypothesis that these parameters are small for biological media, we may safely solve the linearised inverse problem by assuming simple monoscattering in the direct problem which we will then invert. The monoscattering approximation belongs to the class of Born approximations, and the related inverse is often referred as the Inverse Born Approximation (IBA).

2. THE DIRECT PROBLEM

We consider a propagating plane wave in the coupling fluid with direction $\overline{n_0}$ (unit vector) and angular frequency ω:

$$P_0 = \mathcal{P}_0\, e^{i\omega t} = e^{i(\omega t - k\overline{n}_0.\overline{x})}$$

where $k = \omega / c_0$ is the wavenumber in the homogeneous medium.

We look for a solution which is also periodic, $P = p\, e^{i\omega t}$, where p satisfies the partial differential equation:

$$\Delta p + k^2 p = 2k^2\alpha p - \mathrm{grad}\alpha.\mathrm{grad}p + \mathrm{grad}\zeta.\mathrm{grad}p$$

or the resulting equivalent integral equation:

$$p(\bar{x},\omega) = p_0(\bar{x},\omega) + \int_{R^3} g(\bar{x},\bar{x}';\omega)\{2k^2\alpha p - \text{grad}\alpha.\text{grad}p + \text{grad}\zeta.\text{grad}p\}_{\bar{x}'} d^3x'$$

where

$$g(\bar{x},\bar{x}',\omega) = -\frac{e^{-ik|\bar{x}-\bar{x}'|}}{4\pi|\bar{x}-\bar{x}'|}$$

g is the Green's function for the Helmholtz operator for propagation in an infinite medium:

$$\Delta g + k^2 g = \delta(\bar{x}-\bar{x}')$$

Since the parameters of inhomogeneity, α, ζ, are small ($\approx 10^{-2}$) for biological media (which consist largely of water) we may limit ourselves to the simple monoscattering assumption, the Born approximation, i.e. the first iterate of the integral equation solution. To first order, then, the diffracted wave field may be written:

$$p_d^{(1)}(\bar{x},\omega) = p^{(1)}(\bar{x},\omega) - p_0(\bar{x},\omega)$$
$$= \int_{R^3} g(\bar{x},\bar{x}';\omega)\{2k^2\alpha p_0 - \text{grad}\alpha.\text{grad}p_0 + \text{grad}\zeta.\text{grad}p_0\}_{\bar{x}'} d^3x'$$

Finally, to further simplify the calculations, we work far from the transducers so as to avoid the rather complicated near-field structure. This mode of operation poses no special problem in practice, neither in the in vitro case (large enough tank) nor in the in vivo case (flexible transducer mounts for pulse-echo).

We suppose, then, that the equipment dimensions are large compared to the inhomogeneous diffracting medium dimensions:

$$\frac{|\bar{x}|}{|\bar{x}'|} \gg 1 \quad , \quad \forall\, \bar{x}' \in \text{Supp}(\alpha,\zeta)$$

Hence:

$$g(\bar{x},\bar{x}';\omega) \# g_\infty(\bar{x},\bar{x}';\omega) = -\frac{e^{-ik|\bar{x}|}}{4\pi|\bar{x}|}\, e^{ik\bar{n}\cdot\bar{x}'}$$

where $\bar{n} = \bar{x}/|\bar{x}|$ is the unit vector in the receiver direction.

In that case, the asymptotic wave field may be written:

$$p_{d\infty}^{(1)}(\bar{x},\omega) = \frac{e^{-ik|\bar{x}|}}{4\pi|\bar{x}|}\, h^{(1)}(\bar{n}_0,\bar{n},\omega)$$

where

$$h^{(1)}(\bar{n}_0,\bar{n},\omega) = -k^2\left[(1+\bar{n}_0.\bar{n})\hat{\alpha}(\bar{K}) + (1-\bar{n}_0.\bar{n})\hat{\zeta}(\bar{K})\right]_{\bar{K}=k(\bar{n}_0-\bar{n})}$$

and $\hat{\alpha}(\bar{K})$, $\hat{\zeta}(\bar{K})$ are spatial Fourier transform of $\alpha(\bar{x})$ and $\zeta(\bar{x})$ which characterise the inhomogeneity of the medium and which we designate as "object", following optics terminology:

$$\hat{\alpha}(\bar{K}) = \int_{\mathbb{R}^3} \alpha(\bar{x}) e^{-i\bar{K}.\bar{x}} d^3x \quad ; \quad \hat{\zeta}(\bar{K}) = \int_{\mathbb{R}^3} \zeta(\bar{x}) e^{-i\bar{K}.\bar{x}} d^3x$$

h is the asymptotic wave scattering frequency response of the medium for wave incidence direction $\bar{n}_0$ and observation direction $\bar{n}$.

The above expressions show that the fluctuations in velocity and impedance uniquely determine the wave scattering frequency response. On the other hand, the inverse problem is not a priori unique.

However, a study of the scattering plots for variations in velocity α and impedance ζ (figure 1) show that the former predominates in forward scattering (i.e. is maximum for $\bar{n} = \bar{n}_0$) and the latter in back scattering (i.e. is maximum for $\bar{n} = -\bar{n}_0$). Both plots are cardioids, one a symmetrical reflection of the other. This highly directive property results in the fact that impedance alone plays a significant rôle in reflection while only velocity is important in transmission. Hence we can devide the problem and adopt a single parameter model: impedance for reflection, velocity for transmission.

In reflection we have:

$$h^{(1)}(\bar{n}_0,\bar{n},\omega) \# -k^2(1-\bar{n}_0.\bar{n})\,\hat{\zeta}(k(\bar{n}_0-\bar{n}))$$

and in transmission:

$$h^{(1)}(\bar{n}_0,\bar{n},\omega) \# -k^2(1+\bar{n}_0.\bar{n})\,\hat{\alpha}(k(\bar{n}_0-\bar{n}))$$

One can see that this single parameter model (impedance in reflexion, velocity in transmission) only produces an error on the order of 1 percent in the scattered field for changes of $\pm$ 12° about pure reflection ($\bar{n} = -\bar{n}_0$) or pure transmission ($\bar{n} = \bar{n}_0$), on the order of 10% for changes of $\pm$ 37°.

In what follows we will be concerned only with the reflection zone, which corresponds to the pulse-echo model. Hence we will be concerned with the single parameter $\zeta = \mathrm{Log}(Z/Z_0)$ which characterises impedance fluctuations:

$$h^{(1)}(\bar{n}_0,\bar{n},\omega) = -k^2(1-\bar{n}_0.\bar{n})\,\hat{\zeta}(k(\bar{n}_0-\bar{n}))$$

Reducing the inverse problem to a single parameter ensures uniqueness for the linearised case that is the Inverse Born Apptoximation. Solving the problem is simple: Fourier-invert judiciously collected wave field data and thus obtain the object $\zeta(\bar{x})$ which is the pattern of the logarithmic impedance fluctuations. This constitutes Acoustic Impedance Tomography.

A further important point is that due to the symmetry of the scattering patterns for acoustic impedance variation and for wave velocity variation. What we develop for the first one in reflection will apply simply to the other one in transmission. We note this, since other writers, in contrast, have taken approaches where either they could not take advantage of symmetry due to their formulation [2,3,4] or they have totally neglected one parameter (such as the variation in ρ which is actually important for medical imaging) [5,6].

The essential merit of our formulation is, as we have already mentionned, that it is exact (within the Born approximation) and reduced to only a single parameter over a large sector of the plane about the line of either pure reflection or pure transmission.

In what follows we limit ourselves to the reflection zone, of high practical interest since it directy concerns medical echography; and adopt the single parameter approximation, i.e.the logarithmic impedance ζ modelling. So we propose reconstructing the object $\zeta(\bar{x}) = \mathrm{Log}(Z(\bar{x})/Z_0)$, the logarithmic impedance profile, from the scattering measurements h $(\bar{n}_0, \bar{n}, \omega)$.

3. THE INVERSE PROBLEM

The relation established between the wave scattering frequency response of the medium $h(\bar{n}_0, \bar{n}, \omega)$ and the spatial Fourier transform of the object $\zeta(\bar{x})$ shows that, with appropriate scattering measures, we can fill the Fourier space of the object $\zeta(\bar{x})$. It would be possible to fill it at random by action on the three parameters $\bar{n}_0, \bar{n}, \omega$, then make an interpollation in the Fourier space in order to have a rectangular grid, and finally proceed a multidimensional inverse Fourier transform to reconstruct the object ζ [7]. That procedure needs a large memory capacity computer. We prefer to perform adequate space and frequency scannings in order to have only one dimensional inverse Fourier transforms in the reconstruction algorithm. We propose two possible seannings and their associated reconstruction algorithms.

3.1. Space-Frequency Scanning

Let the diffraction angle $\theta = (\vec{n}_0,\vec{n})$ be fixed (not too far from back scattering $\vec{n} = -\vec{n}_0$, i.e. $\theta = 180°$, in order to have a good one parameter ζ approximation) and perform a double scanning: the first in frequency between ω_{min} and ω_{Max}, the second in space on the incidence angle $\varphi = (\vec{Ox}, \vec{n}_0)$ between $-\pi$ and $+\pi$ (we consider for more simplicity a two dimensional problem).

One can show (figure 2) how that action fills the Fourier plane of the object ζ by successive straight line segments over a ring of center O and extreme radii $2k_{min} \sin \theta/2$ and $2k_{Max} \sin\theta/2$.

That produces a band-pass filtering with band $B = [\, 2k_{min} \sin \theta/2 ,\; 2k_{Max} \sin\theta/2]$.

Let ζ^B be the filtered object, and $h\varphi\ (\theta, \omega)$ the measures (θ is kept fixed). Introducing:

$$\tilde{P}^B_\varphi(\theta,\chi) = \hat{\zeta}^B_\varphi(\theta,k) = -h_\varphi(\theta,\omega)/[k^2(1-\cos\theta)]$$

with $\chi = 2k \sin \theta/2$;

and using the frame change $[x,y] \to [\xi,\eta]$ which results from the rotation of the $[Ox, Oy]$ frame of an angle $\varphi + \theta/2 - \pi/2$, one can show that :

$$\zeta^B(x,y) = \frac{1}{2\pi}\int_{-\pi}^{+\pi} \Pi_\varphi(\xi,\eta)\, d\varphi$$

with:

$$\Pi_\varphi(\xi,\eta) = P^B_\varphi(\theta,\xi) \underset{\xi}{*} H(\xi)$$

and

$$\begin{cases} P^B_\varphi(\theta,\xi) = \dfrac{1}{2\pi}\displaystyle\int_{\mathbb{R}} \tilde{P}^B_\varphi(\theta,\chi)\, e^{i\chi\xi}\, d\chi \\ H(\xi) = \dfrac{1}{2\pi}\displaystyle\int_{\mathbb{R}} \chi\, Y(\chi)\, e^{i\chi\xi}\, d\chi = -P_f(1/\xi^2) \end{cases}$$

That formulation is the classical Tomographic Reconstruction formula by summation of the filtered back-projections $\Pi\varphi$ encountered in X-ray tomography [8].

That algorithm is easy to perform on a micro-computer because of its iterative nature.

The maximum spatial frequency band is covered when $\theta = \pi$, i.e. in back-scattering. That position leads to the best image resolution (the maximum spatial frequency is $1/2k_{Max}$), but also to the strongest low frequency filtering: there is a big hollow of radius $1/2k_{min}$ at the center of the Fourier plane. That might bring errors in quantitative reconstruction. So the first thing to do will be to deconvoluate the pulse-echo data in order to cover a maximum frequency band with a minimum central hollow.

3.2. Double Spatial Scanning

If we let the frequency ω fixed and perform a double spatial scanning with varying the diffraction angle $\theta = (\vec{n}_0, \vec{n})$ between θ_{min} and θ_{Max} (values between 0 and π) and the incidence angle between $-\pi$ and $+\pi$, one can show (figure 3) how that action fills the Fourier plane of the object, by successive arcs of a circle of center kn_0 and radius k, over a ring of center O and extreme radii $2k \sin \theta_{min}/2$ and $2k \sin\theta_{Max}/2$. That produces a band-pass spatial filtering of band $[2k \sin \theta_{min}/2 \,,\, 2k \sin\theta_{Max}/2]$.

Let ζ^B be the filtered object, and $h\varphi(\theta, \omega)$ the measures (ω is kept fixed). Introducing:

$$\tilde{\Gamma}^B_\varphi(\chi, k) = \frac{1}{\cos\theta} \hat{\zeta}^B_\varphi(\theta, k) = -\frac{1}{k^2 \cos\theta(1-\cos\theta)} h_\varphi(\theta, \omega)$$

with $\chi = k \sin \theta$;

and using the frame change $[x,y] \to [\xi,\eta]$ which results from the rotation of the $[Ox, Oy]$ frame of an angle $\varphi - \pi/2$, one can show that :

$$\zeta^B(x,y) = \frac{1}{2\pi} \int_{-\pi}^{+\pi} \Pi_\varphi(\xi, \eta)\, d\varphi$$

with:

$$\Pi_\varphi(\xi,\eta) = \Gamma^B_\varphi(\xi, k) \underset{\xi}{*} H(\xi) \underset{\xi}{*} G(\xi, \eta)$$

and:

$$\begin{cases} \Gamma^B_\varphi(\xi, k) = \frac{1}{2\pi} \int_{\mathbb{R}} \tilde{\Gamma}^B_\varphi(\chi, k)\, e^{i\chi\xi}\, d\chi \\ H(\xi) = \frac{1}{2\pi} \int_{\mathbb{R}} \chi\, Y(\chi)\, e^{i\chi\xi}\, d\chi = -\mathrm{Pf}(1/\xi^2) \\ G(\xi, \eta) = \frac{1}{2\pi} \int_{\mathbb{R}} e^{i\{\chi\xi + [k - (k^2 - \chi^2)^{1/2}]\eta\}}\, d\chi \end{cases}$$

By analogy with the previous formulation and following Devaney's terminology [5], one can say that the problem is solved by summation of filtered back-propagations. Our formulation is the analogue of Devaney's one for forward scattering.

4. PRESENT STATUS

Until now our attention has been focalized on the first procedure, i.e. Acoustic Impedance Tomography by a joint space and frequency scanning, since frequency scanning is easily performed by fast-Fourier analysis of broad-band pulses (like those delivered by standard echographs) captured by fast analogue-to-digital converters (as usual digital storage oscilloscopes). In order to use only a single transducer for emission and reception and also to have the better image-resolution, we confine to pure back-scattering.

The complete procedure to recover the logarithmic acoustic impedance profile $\zeta(\vec{x})$, that is for Acoustic Impedance Tomography, would so be:

- 1. Deconvolution of emission-reception pulse-echo data in order to obtain the impulse response in pure reflection of the medium $H\varphi$ $(\theta=\pi, \omega)$, and fast-Fourier transform of that impulse response in order to get the frequency response in pure reflection of the medium $h\varphi$ $(\theta=\pi, \omega)$ on the largest possible frequency band,

- 2. Filtering and back-projecting that frequency response $h\varphi$ in order to construct the filtered back-projection $\Pi\varphi$ of the logarithmic impedance profile ζ (two-dimensional image),

- 3. Summation of that filtered back-projection $\Pi\varphi$ with the image obtained at the previous iterate, i.e. at the previous angle φ,

- 4. Iteration of the process, by angular scanning over 360° with angle φ, the result being a slice of the medium, in words of logarithmic acoustic impedance ζ.

So Acoustic Impedance Tomography is based on measures of the impulse response in back-scattering of the medium. Quantitative imaging in reflection mode needs use of very wide-band transducers and efficient deconvolution procedures. For that second point our team relies on G.Demoment's one at "Laboratoire Signaux et Systèmes" (L2S), Paris, which has already found a solution adapted to echography [9] in order to improve image resolution. But no deconvolution procedure will ever be able to restore the very low frequencies, lost by any ultrasonic transducer. That lack in low frequencies, as we have shown above, causes in reflection imaging a central hole in the spatial frequencies plane. So impedance images will ever be not only low-pass filtered as in echography (resolution limited by the upper limit of the transducer's frequency-band) but also high-pass filtered. Particularly, except in the case of a-priori information, there is no hope of absolute measure of impedance by this way.

So we decided to go step by step, and begin by process only the signal's amplitude, like in echography. That is what is named Reflectivity Tomography, which is also an incoherent imaging technique, less sensitive to speckel noise than usual coherent techniques. In a second step, we shall enhance image resolution by adapting the deconvolution algorithms settled by L2S for echography. Finally we shall deal with the quantitative problem, that is Acoustic Impedance Tomography, by mainly studying the low frequencies restoration problem.

So we begin by Reflectivity Tomography, and substitute to the first operation of the general procedure a simple amplitude-demodulation operation based on a Hilbert transform.

4.1. The Experimental Apparatus

We realized a system especially designed in order to implement as well the general procedure of Acoustic Impedance Tomography as other more classical procedures, as transmission tomography or more simplely echography. Consequently we shall be able to compare all these techniques. So we decided to build a robot designed to handle the transducers with high versatility and high precision, that precision being typically tenth a wave-length of the acoustic waves, that is 1.5 hundredth a millimeter at ten megahertz in water (figure 4).

From the mechanical point of view, the general architecture of the system is a copy of X-ray scanners of the first generation: a main symmetrical arm holds two transverse arms allowing the parallel translation of two transducers. Angular scanning is allowed by the rotation of either the main arm or the object holder. The transducers also may be pointed at distance with high precision. All motions are insured by six stepping motors driven in sequence by only one programmable translator-indexer device fitted with a home-made power multiplexer. Translator-indexer device and power multiplexer are integrated in a control rack which includes also other remote controls, as the distance adjustement of the transverse arms, or the system out of water setting automatism. Increments are multiples of 0.75 hundredth a millimeter for translations and of 1 hundredth degree for rotations.

From the electroacoustic point of view, the system is composed of an echographic device and a commercial digitizer. The echographic device is made of a pulser-receiver (60 Volts level, 50-100-250 nanoseconds duration pulses; 70 dB receiver gain) manufactured by the Institut de Dynamique Cardiovasculaire Broussais Hospital (U.256 INSERM), of a PANAMETRIX stepless gate, of PANAMETRIX wide-band transducers (3.5 and 5 MHz at now). The data collection is carried out by a digital storage oscilloscope, model 1602 from GOULD (maximum 20 MHz sample rate, 10 K-word memory, 8 bit resolution).

The whole system is architectured around a compatible IBM-PC micro-computer (M-24-SP model from Olivetti) of 640 K-bytes memory with a 20 M-bytes hard-disk and a high definition (640X400) color screen. That micro-computer controls all operations:

robot command, trigger action of the echographic device, of the data acquisitions, data processing, image construction and processing.

4.2. Present state of the project

The first stage of the project is at now performed: reflectivity imaging with standard definition (no deconvolution) is efficient, at least for faisibility studies in the Biomedical Engineering field (previous prime objective). With a 5 MHz transducer and a 2° angular scanning rate, one can build 400 X 400 pixels images with a 0.3mm X 0.3mm pixel size (correlated with the wavelength). The thickness of the slice, with a 12.7mm diameter transducer focalized at 117mm, is 3mm (focal spot width).

The tests were carried out in immersion mode, with various targets, one of which is a standard test object made of three brass rods of 3, 6, 12 mm diameter (figure 5). That standard test object is used by all the teams which work on Reflectivity Tomography [10,11], since it is a perfect example of specular object, hardly imaged wity using classical echography.

5. FUTURE

The applied field preferred up to now was the Biomedical Engineering one. We are now interessed by Non Destructive Evaluation of materials, that is, for the scientific point-of-view, the case where media are characterized by a larger number of parameters (anisotropic media) and where the amplitude changes of those parameters may be important, leading to a true non linear inverse problem. But the first thing to do is to correct the rough errors due to the refraction and the acceleration of waves at the crossing of the water-material interface. It is what we are doing now.

We also try to enhance calculus speed (trials with simplified reconstruction filters) and the image quality (settlement of specific image processing algorithms).

So the system, which allows only faisibility studies in the Biomedical Engineering field ("in vitro" imagery, too much important consuming time for data acquisition and processing), could be, in the Non Destructive Evaluation field, a true imaging system prototype, adapted to immersion inspecion of compact objects.

REFERENCES

[1] Leeman, S., "Impediography revisited", Acoustical Imaging 9, 513-520, K.Y.Wang ed., Plenum Press, New-York (1980).

[2] Morse, P.M., Ingard, K.U., "Theoretical Acoustics", Mc.Graw-Hill, New-York (1968).

[3] Norton, S.J., "Generation of separate density and compressibility images of tissue", Ultrasonic Imaging, 5, (3), 240-252 (1983).

[4] Chernov, L.A., "Wave propagation in random medium", Dover Publications Inc., New-York (1967).

[5] Devaney, A.J., "A filtered back-propagation algorithm for diffraction tomography", Ultrasonic Imaging, 4, 336-350 (1982).

[6] Gray, S.H., Bleistein, N., "One dimensional velocity inversion for acoustic waves, numerical results", J. Acoust. Soc. Amer., 67, 1141-1144 (1981).

[7] Stark, H., Paul, I., "An invesrigation of computerized tomography by direct Fourier inversion and optimum interpolation", IEEE Trans. Biomed. Eng., BME-28, 177-185 (1981).

[8] Mersereau, R.M., Oppenheim A.V.,"Digital reconstruction of multidimensional signals from their projections", Proceedings of the IEEE, 62, (10), 1319-1337 (1974).

[9] Herment, A., Demoment, G.,"Improvements in resolution of echographs through 1D on-line deconvolution", Ultrasonic Tissue Characterization, 4, 67-86, J.M. Thijssen & M. Smith Ed. (1984).

[10] Moshfeghi, M., "Ultrasound Reflection-Mode Tomography Using Fan-Shaped-Beam Insonification", IEEE. Trans. Ultra. Ferro. Freq. Cont., UFFC-33, (3), 299-314 (1986).

[11] Wade, G., Elliot, S., Kogger, I., Flesher, G., "Acoustic echo computer tomography", Acoustical Imaging, 8, 565-576, A.F. Metherell Ed., Plenum Press, N.Y. (1979).

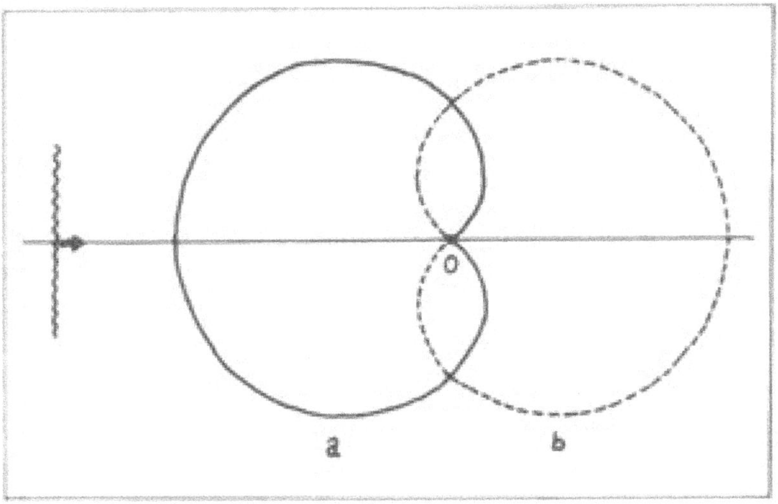

Figure 1: Scattering diagram for Impedance fluctuations (a) and for velocity fluctuations (b).

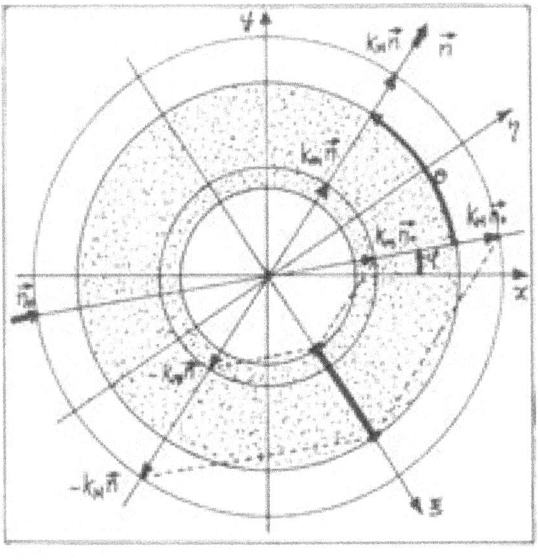

Figure 2: Spectral cover for space-frequency scanning.

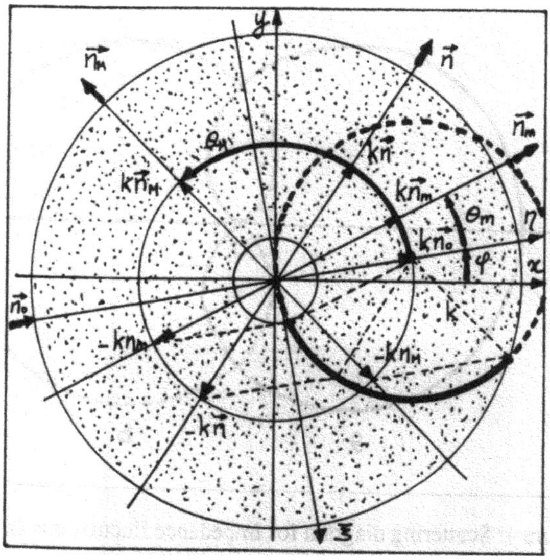

Figure 3: Spectral cover for double spatial scanning.

Figure 4: General view of the system.

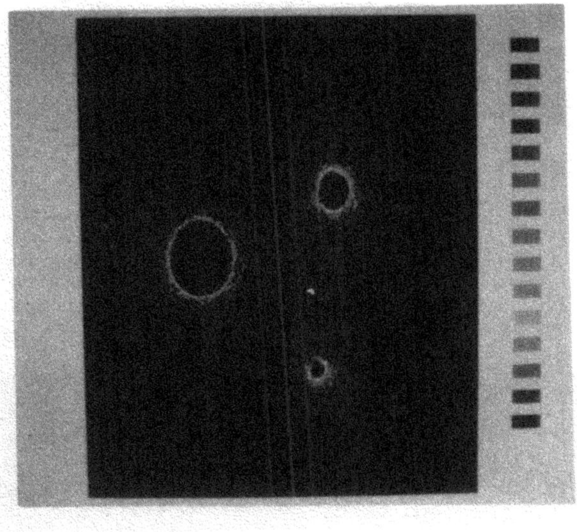

Figure 5: Tomographic image of the standard test object.

PROBING ONE-DIMENSIONAL INHOMOGENEOUS MEDIA: HOW CAN IT BE DONE ?

D. LESSELIER*, W. TABBARA*,**

*Laboratoire des Signaux et Systèmes
C.N.R.S. - E.S.E.
** Laboratoire d'Electronique Générale
Université Pierre et Marie Curie

E.S.E., Plateau de Moulon
91192 Gif-sur-Yvette Cedex, France.

Abstract: Main solutions of the one-dimensional inverse scattering problem in electromagnetics and linear acoustics are discussed in this paper. We are interested in characterizing an inhomogeneous slab (or half-space) whose parameters only vary along one direction from a generally incomplete knowledge of the reflected field for a given excitation or set of excitations. Indeed, as we recently approached questions related to the acoustical probing of the sea floor, we faced up to a large body of literature but a limited number of original solutions, with only a few confronted to real data. Then, we attempted to classify most of the known inversion procedures and show how they perform. A detailed review can be found in ref. 1. Here, we attempt to draw simple guidelines for a prospective user; we only choose a handful of references without seeking exhaustivity to illustrate our conclusions and we also don't go into complicated mathematics. We successively consider: exact solutions of the continuous inverse problem derived from quantum mechanics, deductive solutions developed in time-domain from a discrete model of the medium and which are closely related to the previous ones, iterative solutions based on progressive adjustement of some key parameters of the unknown medium, approximate solutions which are perturbations of a reference solution, and solutions based on analyzing the resonances of the scattered field. Seismic solutions should require a separate treatment, even though they often relate to one of the previous solutions, and are not analyzed here.

I. MODEL

This study is restricted to a simple model of the physical reality. We consider an unknown or partly unknown medium whose electric or acoustic parameters only vary along one direction (depth) in a three-dimensional space, and which occupies either a limited portion of that space or a whole half-space (i.e., a plane slab of finite or infinite thickness d). We assume isotropy and linearity, and only consider deterministic situations. Most references available in the literature which deal with the probing of such a one-dimensional object discuss time-invariant, locally reacting media, with -save exception- instantaneous reaction, i.e., lossy dielectric media in electromagnetics characterized by a relative permittivity and a conductivity independent of frequency, non-viscous elastic media in acoustics. Also, in electromagnetics, non-magnetic media are almost always the only ones investigated. In acoustics, emphasis is put on lossless fluid media where only compressional (P) waves are propagated and which can be characterized by their acoustic impedance as a function of the travel-time. In both domains, continuous or stratified media are investigated. In the first case, most references limit their field of investigation to slabs whose parameters vary smoothly enough with depth in order to be represented by (sometimes twice-) continuously differentiable functions. In the second case, parameters are taken as piecewise constants, the slab generally consisting of layers of equal travel-time.

Given sources are located above the slab. In most cases, a plane wave of normal or oblique incidence, or a set of such plane waves, is impinging upon the upper interface. In some cases, in acoustics essentially, sources are at finite distance from the slab and are of limited extent (line source or point-source); this enables to take into account the presence of the air-water interface (assumed to be impenetrable) in underwater probing situations. The incident waves then interact with the slab. We generally have access to the reflected field (in the case of a slab of finite thickness, we might also be able to measure the transmitted field). We either operate in the frequency domain, i.e., we assume purely harmonic time-dependence of the wavefield, or in the time-domain, i.e., impulse excitation is considered, often restricted to a ideal δ-function. The inverse problem to be solved is "quantitatively estimate all the parameters of the slab, or some of them, from measurements (or simulations) of the field outside the slab". In the case of a plane illumination, we assume we know the reflection coefficient R at a given (generally normal) incidence, either as a function of the wavenumber k in the embedding medium (where the incident field is generated) on the whole line $[0,+\infty[$ or a part of it, in a continuous manner or at discrete values of k, or as a function of time t on $[0,+\infty[$ or a more limited window; sometimes, multiincidence data are taken into account at a given frequency instead of (or in addition to) multifrequency data. When line- or point-sources are used, to model situations of seismic sounding, we generally assume that we know the field on a line above and parallel to the upper interface of the unknown slab, the receivers being placed at discrete locations and increasing distance from the source. (In addition, source and receivers can be simultaneously translated in the transverse direction.)

II. CLASSIFICATION:

The classification problem has neither an unique nor an obvious solution[2]. Here, we would like to draw simple guidelines useful to the "inverse scattering community", and to do it almost without mathematics and with reference to only a few dozen publications -we don't pretend to exhaustivity-; the reader still willing to learn more on the 1-D inverse problem could find a fairly complete review of the literature in ref. 1.

We first choose to separate exact solutions[3] from approximate ones[4]. Once the source and the medium have been modelized and exact wave-equations relating the field to the slab parameters written, even if this model can be oversimplified, exact approaches are supposed not to use any approximations to retrieve the unknown parameters of the slab from the field data, whereas approximate ones do use them. We further separate exact solutions of the continuous inverse problem (we assume continuous variations of the parameters) from those of the discrete one (we assume that the slab consists of layers with generally known -and equal- thickness). The first ones are derived from quantum mechanics[5] and use transformation of the wave-equations into Schrödinger equations solved by Gelfand-Levitan type techniques[6]. The second ones have been introduced in the 60's in geophysics, aim at a direct deductive reconstruction[7] of the unknown stratified medium from transient data (by "peeling" the slab), and are in fact closely related to the first ones as already shown quite early[8]. A third type of exact solutions, of more recent introduction but of increasing importance and large field of applications, deserves attention and is considered next: iterative approaches in time- or frequency-domain. They are based on the building of an optimal solution[9], in some sense, to the inverse scattering problem by progressive (step-by-step) adjustement of a set of parameters describing the slab. At each step, the adjustement itself can be approximate, using some kind of linearisation of the wave-equations, but the sequence of iterates should converge to the exact parameters. Then, we consider the approximate methods. We refer to them as perturbation methods because they provide us with a perturbation of an exact reference solution by linearisation of the inverse problem. Note that most of them are either low-frequency approximations (e.g., Born approximation) or high-frequency ones (e.g., WKB, geometrical optics), and that they fail when the size of the inhomogeneities in the slab is of the order of the wave-length. We finally consider a last class of solutions, those based on analyzing the resonances[10] of the slab response, even though they are of limited application in the probing of penetrable media. Seismic solutions[11] whose particular scope -and language- requires separate treatment but which borrow much of their foundations from the previous classes are not dealt with herein.

Some general introductory remarks are of interest. First, references on the 1-D inverse scattering problem in electromagnetics and acoustics already appear in the early 50's (with specific attention devoted to solving the Schrödinger equation), but we essentially took a close look at more recent and often more complete works (those carried on since 1970 or so). Even though the literature is now quite replete and papers are still coming on the subject, we only discovered a few principal approaches, which we focused on, and many variations around them, which we set aside in a first stage of the analysis. Second, from a practical point-of-view and though there seems to be a renewed interest in real-world 1-D inverse problems these last few years with direct applications to biomedical and civil engineering, generally missing are the technical performances of the existing or needed experimental devices, the accuracy of the probing itself which we would like to reach, especially when taking into account the simplifications of the model, and the validation of the method with respect to real band-limited sparse noisy data. Evaluation of the applicability of a solution method to a particular realistic problem is then challenging... Third, many models seem to be oversimplified, which at least would require to study the direct scattering problem using a more accurate model as a first (mandatory) step to solving the inverse one using a simplified model.

III. EXACT GELFAND-LEVITAN TYPE METHODS:

In short, these methods[3,6] aim at constructing the scattering potential q of a Schrödinger equation satisfied by the field V, for example in the time-domain

$$V''_{xx}(x,t) - V''_{tt}(x,t) = q(x)\, V(x,t), \tag{1}$$

letting x be the travel-time deduced from depth by Liouville transformation. The Schrödinger equation is obtained through various transformations of the wave-equations. q is a more or less complicated function of the electric or acoustic parameters of the slab and of their derivatives with respect to x, which is null at $x<0$, sufficiently regular at $x>0$

and has adequate behavior at infinity. In traditional situations of the electromagnetic and acoustic sounding, there are no eigenmodes as is henceforth assumed. Classical techniques enable us to write down the Marchenko equation

$$R(x+t)+K(x,t)+\int_{-t}^{x} K(x,y)\, R(y+t)\, dy = 0 \qquad -x \leq t \leq x, \tag{2}$$

$$K(x,-x)=0, \qquad 2\, K'_x(x,x) = q(x),$$

which we have to solve for K knowing the impulse reflection coefficient R. (Note that this equation is obtained for a real-valued spectral coefficient R(k); losses are neglected.)

Many sophisticated analyses of the above problem, or others related to more complicated formulations, exist, which we won't dwell upon here. However, we would like to note a few points as follows. First, various methods of perturbation can be readily derived under some more or less stringent conditions from the exact Gelfand-Levitan type formulations; for example, developing R and K into series in equ. (2) -renormalization techniques can also be used[12]- leads to the second-order approximation $q(x)=-2\, R'(2x)+4\, R^2(2x)$, which at the first-order reduces to the well-known impediography[13-14] equation $(\mathrm{Log}\ Z(x))'=4\, R(2x)$ in a lossless fluid medium with twice-differentiable impedance Z. Second, if the theoretical aspects of the analysis are well developed by many authors, numerical applications are only a few whereas exceptional remain experimental ones, in particular due to the fact that numerically deriving the slab parameters, e.g., the impedance profile, from the potential is a delicate matter[15]. In electromagnetics, reconstructing the permittivity of a purely dielectric space has been numerically investigated by some authors[15-18] and a recent paper[19] shows experimental results obtained on the probing of a low-contrast plexiglas slab from band-limited microwave data (in fact, a deductive discrete procedure -see next section- is used for that). Also, for example when dealing with a isotropic plasma, modelizing the reflection coefficient by rational functions allows to successfully compute[12, 20] the permittivity from analytical formulations of the potential, on condition that the rational model suitably fits the experimental data. In acoustics, leaving aside deductive resolutions we will deal with later, most methods have only been illustrated by canonical examples, save exception (e.g., ref. 21). Third, as we need to know the reflection coefficient on the entire frequency spectrum, in particular its behavior at $k=0$ and $k\to\infty$, careful care must be taken when modeling the absorption. Consistent formulations[22] are derived in electromagnetics, and some numerical results are available in the case of a lossy dielectric slab, where the conductivity or the permittivity profile can be determined. On the contrary, correctly accounting for the absorption in acoustical inverse scattering is not easy, since the usually assumed linear dependence of the absorption on the frequency is inconsistent with causality. (Note the precursor work[23]) where the relaxation modulus of a homogeneous medium is reconstructed.) Fourth, investigations almost always assume plane wave illuminations, except in acoustics the involved Coen's approaches[24]) which are not illustrated by numerical results and ref. 21. Finally, and this is not to be overlooked, suitable low-frequency content of the scattered field is a prerequisite to a good probing, e.g., in the case of high-pass filtered data, the trend of the reconstructed profiles is absent. (In any case, smooth variations of the slab parameters are needed to mathematically formulate the inverse potential problem.)

IV. DEDUCTIVE METHODS:

Such procedures have been introduced in the beginning of the 60's in order to retrieve the acoustical impedance profiles of a lossless fluid stratified medium from a measure of its interaction with a plane wave of given transient dependence. The impulse reflection coefficient (spike seismogram) could be calculated step-by-step using a Goupillaud[25]) model (constant travel-time layers) of the impedance and then the question was how can a similar procedure be employed to calculate the impedance step-by-step from the seismogram. One of the most classical procedures is based on Claerbout's paper[26]), where it is shown that the inversion is straightforward for it amounts to solving a Toeplitz system of linear equations (for which fast solution algorithms are available). Many other procedures have been proposed since then[7]). For example, on condition that the incident wave contains a δ-function, some of them construct (Choleski's algorithm[27])) the causal solution of a system of first-order differential equations obtained after splitting the compressional wavefield into a elementary wave P^+ propagating downwards and an other one P^- propagating upwards:

$$P^{+\prime}_t + P^{+\prime}_x = -A\ P^- \quad \text{and} \quad P^{-\prime}_t - P^{-\prime}_x = +A\ P^+ \tag{3}$$

where $A(x)=2\,P^-(x, t=x^+)$ is proportional to the logarithmic derivative of the impedance. If we assume (P^+, P^-) determined at x, the reflectivity A(x) -constant in interval [x, x+Δ[- can be computed; solving the system above provides (P^+, P^-) at x+Δ and so on.

In short there is identification of the layer [x, x+Δ [and downward continuation of the wavefield with elimination of the previously identified layer[28]. A good summary of the various methods developed these last twenty years and a careful numerical evaluation of their performances with respect to filtered and/or noisy synthetic data is given in ref. 29. Numerous numerical comparisons in realistic situations are also found in refs. 30-31.

We draw several conclusions from the above studies. First, these methods are quite easy to use for they take full advantage of the causality[3]. Fast algorithms can be obtained. Second, they more or less have the same mathematical properties (existence of a solution, unicity, stability) and require piecewise constant (at the most smoothly varying) impedance profiles, with an adequate choice of the sampling of the reflected field and of the profile with respect to their variations. Third, as the reconstruction is a deductive one, errors accumulate from one iterate to the next ones and the procedure tends to breakdown at high depths; even a limited amount of noise can be dangerous. (Note that simple iterative procedures can sometimes be incorporated in the algorithm to limit the divergence.) Other common sources of errors are a wrong timing of the source wave or of the seismogram, whose time origin must be correctly identified, and the presence of absorption when the model, as is usual, does not take this absorption into account. Fourth, as it was already emphasized in the preceding chapter, a small error on the low-frequency spectrum of the reflection coefficient may have dramatic consequences (the inverse problem is ill-posed in that respect).

Similar studies have been conducted in electromagnetics on lossy dielectric media. Exact second kind Fredholm integral formulations of the transient electric field solved by marching-on in time procedures are used[32] by one of the authors (DL) to numerically retrieve the conductivity or the permittivity of an inhomogeneous lossy dielectric half-space. Numerous simulations clearly show the need of a space-time sampling such that fields and slab profiles are suitably described and of an accurate measurement of the reflected field. (They also emphasize shadow effects of highly conducting or dielectric layers.) In the case of a lossless dielectric medium, deductive resolution of the Marchenko equation after convenient space-time discretization is also carried on in ref. 15, the derivative of the potential being calculated by finite-difference. In ref. 18, the procedure is about the same but using an other integral formulation avoids the numerical derivation of the potential. Note that, in short, they both present a clever numerical solution algorithm for a Gelfand-Levitan type integral equation, with divergence of the reconstruction overcome by iterative correction, and are already refered to in the previous section. (As aforementioned, experimental results have been obtained using the second technique.)

IV. ITERATIVE METHODS:

Such methods can be considered from an optimisation theory[9] point-of-view as follows. We know the response Π of the slab to a given excitation and the relationship f between the field F and the slab profile P of a space-variable x in a given space Ω (we are able to exactly compute F knowing P). We are looking for a profile P* in Ω such that $f(P^*)$ is close enough to Π in some sense. To get this optimal solution P*, we minimize a real-valued positive cost function J(P) which is generally but not necessarily written as $\|\Pi - f(P)\|^2$ + some regularizing term (constraints on P,...), letting $\| \ \|$ be an appropriate norm. To do so, we construct a sequence of profiles converging to P*. (In case of non-convexity, we generally obtain a local minimum.) Steepest descent, conjugate-gradient, or quasi-Newton procedures can be employed, which in particular requires the calculation of the gradient of J with respect to P in each iteration. It can be shown from functional analysis that the gradient is easily obtained once we obtain the solution A of an adjoint problem where the source term is the difference between Π and the response of a slab with profile P. The main advantage to that approach is avoiding time-consuming and often inaccurate computations of the gradient by finite-difference. Mathematically setting the adjoint problem itself is often a little complicated, but the numerical burden of solving it exactly is about the same as the direct problem's one. (As a general rule in the time-domain we solve the adjoint problem by marching-on at decreasing time[33].)

Early papers -where the adjoint formulation is not employed- are refs. 34-35. Since then, this approach has been employed in a systematic way to probe lossy dielectric or fluid slabs from transient data using second-kind Fredholm's integral formulation[32] and in a little different setting in optics[36] (regularized solution of first-kind integral equations, as is described in ref. 3), with experimental results available in both cases, whereas a very complete theoretical and numerical investigation of iterative solutions has recently been published[37]. Recent references are, among others, refs. 38-40. Note that a lot of attention is now paid to iterative distorted-Born procedures for which at each iteration approximate linear integral equations are solved, i.e., letting $\Delta F_n = \Pi - F_n$ associated to the nth. iterate P_n, and L the linear operator (tangent to f) which links small variations of the profile to small variations of the field, obtain the next iterate $P_{n+1} = P_n + \Delta P_n$ by directly computing the solution of $\Delta F = L \Delta P_n$ (this is a Newton-Kantorovitch algorithm). Intense investigations[41-43] are also going on about acoustic sounding of (modelized as) fluid media from pre-stack noisy (generally simulated) data.

Such an approach is obviously quite general. But its effectiveness strongly depends on the performances of the solutions of the direct and adjoint problems in each iteration. Furthemore, when conjugate-gradient algorithms are used, once the conjugate direction along which we move in the space of profiles is determined, finding the minimum of J on this line requires several values of J, then solving several times the direct problem. The numerical burden is reduced when employing Born-type procedures but convergence may not be observed and solving first-kind integral equations as is often needed remains challenging. Obviously, the choice of the iterative procedure itself is critical, especially when a huge amount of data is handled as in seismic sounding. Also, an optimal solution with respect to a given cost is not necessarily unique, or physically sound, and it is very important to be able to incorporate constraints on the profiles which are either related to physical criteria or to some *a priori* knowledge of the solution. In addition, we must carefully select the first iterate. In seismic sounding, starting from a low-frequency estimate of the slab profiles looks like a prerequisite considering that low-frequencies are generally absent from the scattered field data. In electromagnetic cases, Born approximated profiles, or better, distorted-Born ones, often provide convenient initial guesses. Finally, we may note that the simultaneous probing of two different sets of parameters, say, the permittivity and the conductivity of a dielectric slab, or the sound speed and the density of a fluid one, is not easy (most references are interested in determining one set or the other) and in any case requires two independent sets of data.

Iterative solutions of the Riccati equation are also worth mentioning[44] (a cost function is not explicitly introduced). For example, for a fluid slab with impedance Z function of the travel-time x and known thickness d, this equation is

$$r' = 2j\omega r - C(1-r^2) \qquad C(x) = 0.5\, Z'/Z \tag{4}$$

where r is the local reflectivity (ratio between the upward and downward propagating wave) with $R(\omega)=r(0)$ the reflection coefficient, and $r(d)=0$; its integral formulation is

$$r(x) = e^{2j\omega x} \int_x^d P(y)\, e^{-2j\omega y}\, (1 - r^2(y))\, dy. \tag{5}$$

Neglecting r^2, the initial (Born-approximate) iterate of the sequence of profiles $\{P_n\}$ is the Fourier transform of R. Next iterates are deduced straightforwardly. Performances of such a method, which can take into account absorption losses but requires the knowledge of r(d), are better when variations of the impedance (or permittivity) are of low amplitude. In that case, it can effectively handle a limited amount of noise. However, generalizations to dispersive or elastic media do not seem available yet.

V. PERTURBATION METHODS:

We are now looking for an approximate solution of the inverse problem. Typical solutions are low-frequency ones based on the Born approximation, which assimilates the field within the unknown slab to the incident field. For example, the propagation constant k(z) of a fluid half-space with same density than the embedding and illuminated under normal incidence (time-dependence $\exp(+j\omega t)$, wave-number k_0), with reflection coefficient $R(k_0)$, is given by

$$q(z)=(k(z)/k_0)^2-1=(2\pi)^{-1}\int_{-\infty}^{+\infty}(4\,j/\beta)\,R_B(\beta)\,e^{j\beta z}\,d\beta\,,\qquad \beta=2k_0, \tag{6}$$

where $R_B(\beta)=R(\beta)$ when Born's approximation holds. Similar results are obtained in electromagnetics. In both cases, such a relationship does not provide acceptable results when q is larger than a few percents. For higher contrasts, we obtain the distorted-Born approximation[45-46]

$$q(z)=4\,(k(x)-k_0)/(k(x)+k_0)\approx(2\pi)^{-1}\int_{-\infty}^{+\infty}(4\,j/\beta)\,R(\beta)\,e^{j\beta z}\,d\beta\,,\quad \beta=2k_0, \tag{7}$$

for a slab of thickness d sandwiched between two identical media with propagation constant k_0, letting x be the travel-time to depth z. The relationship is exact for a homogeneous slab and neglects multiple reflections for an inhomogeneous one. It can be extended to non-identical media and oblique incidence. One of the authors (WT) led many numerical simulations which show that this approach gives good results up to ratios k/k_0 of about 10, and confirmed this validity by experimental results in time-domain reflectometry[45]. Various versions of the above relationship and extensions to other sources than plane waves have been proposed, e.g., refs. 47-50, whereas a systematic investigation[37] of this approximation has been conducted recently for a lossy dielectric slab. Changing directly z to x in the Helmholtz equation satisfied by the field and applying Born's approximation leads[47] to the impediography relationship

$$\mathrm{dLog}\, Z(x)/dx = (2/\pi)\int_{-\infty}^{+\infty} R(\beta)\, e^{2j\beta x}\, d\beta = 4\, R(2x), \tag{8}$$

which has been initially derived directly in the time-domain by neglecting multiply reflected pulses in a Goupillaud medium whose layers'thickness is made closer and closer to zero[13,14]. Note that neglecting multiply reflected signals is of current use in seismic sounding (e.g., refs. 7, 29), and has been also applied to other areas[51]. Among other approximate solutions, a few are based on the WKB approximation, whose efficiency is well-known when dealing with the direct scattering problem for slowly-varying profiles, but which seems to be much more difficult to employ for solving the inverse problem (however, ref. 52 presents an interesting study of such a WKB solution). Others -with quite particular scope- are based on asymptotic high-frequency solutions, as in ref. 53. In shallow-water ocean acoustics, there also exist works[54] based on expansion of the pressure in the far-field into normal modes thar are related to the properties of the bottom.

The approximate solutions above generally have a limited range of application. In short, each approximate solution available in the literature seems to be tailored to its own domain of application and it is very difficult for a prospective user to know if this solution can be applied to other domains, or worst, how it compares to other approximate solutions. However, a few remarks seem to be pertinent. As a general rule, approximate solutions deal with non-dispersive media, and experience some difficulties when losses cannot be neglected. Extensions to elastic media in acoustics are *a priori* very complicated and almost absent. Very restrictive also is the fact that we need to measure the reflection coefficient on a large frequency band, which poses practical problems but also theoretical ones, for most of the relationships we use to get the slab profiles from this reflection coefficient are based on low-frequency expressions of the fields, e.g., Born, Born-distorted approximations. (Other approximations should fail, too, e.g., WKB ones when total reflection occurs within the slab.)

VI. <u>RESONANCE METHODS</u>:

The basis of these Singularity Expansion Methods is given by ref. 10. Grossly speaking, the impulse reflection coefficient or its Laplace transform can be expanded into series

$$R(t) = \Sigma_n R_n{}^{(t)} \exp(s_n{}^{(t)} t) \text{ and } R(p) = \Sigma_m R_m{}^{(p)} \exp(s_m{}^{(p)} p), \quad (9)$$

where $s_n{}^{(t)}$ and $s_m{}^{(p)}$ are poles and $R_n{}^{(t)}$ and $R_m{}^{(p)}$ residues. Resonances or singularities (branch points may also exist) contain information on the geometrical, acoustical or electric parameters of the slab, and can be used in order to characterize them. One of the first studies led on that problem is presented in ref. 55, where it is shown that for a three-layer lossy dielectric medium knowledge of the three first poles suffices to completely probe the structure. The most complete one on the same kind of medium, is ref. 37, where a detailed study is carried on the continuously varying case. In that case, singularities can only be found using a numerical procedure (if the permittivity slowly varies, WKB type methods give acceptable estimates of the poles and residues). The classical Prony method of research is revisited and is shown to provide nice results on noisy transient data on condition that the incident pulse is correctly sampled and does not excite poles of high order. In acoustics, thorough analyses of the resonances with respect to frequency and/or angle of incidence can be found in refs. 56-58, with applications to fluid or elastic media. Possibilities of applying such a method to acoustic sensing of a sediment-covered sea floor are especially emphasized in ref. 56; in the same situation using finite-length sinusoidal trains also appears to be an effective tool [59]). However, even though the above references present interesting results, such an approach seems still quite in infancy, at least for penetrable one-dimensional structures like ours. Most people deal with stratified media with a small number of layers and rare are the investigation of continuously varying ones. Also emphasis is put on the direct scattering problem. Experimental measurements have been made in the laboratory, but are restricted to validating the resonances formalism, and we do not know of any solution of the inverse problem using data obtained in a real environment.

REFERENCES:

1) Lesselier D. and Tabbara W., J. Acoust., to appear, 1988.
2) Sabatier P. C., Inverse Prob., 1, 1, Introd., 1985.
3) Sabatier P. C., Radio Sci., 18, 1-18, 1983.
4) Bolomey J. Ch. *et al.*, Radio Sci., 17, 6, 1567-1578, 1982.
5) Chadan K. and Sabatier P. C., *Inverse Problems in Quantum Scattering Theory*, Springer-Verlag, New York, 1977.
6) Newton R. G., Geophys. J. R. Astr. Soc., 65, 191-215, 1981.

7) Mendel J. M., IEEE ASSP Magazine, 4-17, 1986.
8) Ware J. A. and Aki K., J. Acoust. Soc. Am., 45, 4, 911-921, 1968.
9) Cea J., *Lectures on Optimization. Theory and Algorithms*, Springer-Verlag, New York, 1978.
10) Baum C. E., in *Transient Electromagnetics Fields*, R. Mittra Ed., Springer-Verlag, New York, 169-179, 1976.
11) Berkhout A. J., *Seismic Migration. Imaging of Acoustic Energy by Wave Field Extrapolation. Theoretical Aspects*, Elsevier, New York, 1985.
12) Ladouceur H. D. and Jordan A. K., J. Opt. Soc. Am. A, 2, 11, 1916-1921, 1985.
13) Wright H., J. Acoust. Soc. Am., 53, 5, 1356-1359, 1973.
14) Kak A. C. *et al.*, in *Ultrasound: its Applications in Medicine and Biology*, Fry F. J. Ed., Elsevier, Amsterdam, 495-536, 1978.
15) Kritikos H. N. *et al.*, Proc. IEEE, 70, 3, 295-297, 1982.
16) Schaubert D. H. and Mittra R., IEEE Trans. Antennas Propagat., AP-25, 2, 261-265, 1977.
17) Coen S., IEEE Trans. Antennas Propagat., AP-29, 5, 726-732, 1981.
18) Jaggard D. L. and Olson K.E., J. Opt. Soc. Am. A, 2, 11, 1931-1936, 1985.
19) Jaggard D. L. *et al.*, J. Opt. Soc. Am. A, 4, 7, 1228-1232, 1987.
20) Jordan A. K. and Lakshmanasamy S., in *Inverse Problems: an Interdisciplinary Study*, Sabatier P. C. Ed., Academic Press, London, 78-88, 1987.
21) Stickler D. C., J. Acoust. Soc. Am., 74, 3, 994-1005, 1983.
22) Corones J. P. *et al.*, in *Inverse Methods in Electromagnetic Imaging*, Boerner W.M. *et al.* Eds., Reidel Publ. Comp., Dordrecht, Part I, 121-130, 1985.
23) Ammicht E. *et al.*, J. Acoust. Soc. Am., 81, 4, 827-834, 1987.
24) Coen S., J. Acoust. Soc. Am., 70, 1, 172-175 and 5, 1473-1479, 1981.
25) Goupillaud P., Geophys., 26, 754-760, 1961.
26) Claerbout J. F., Geophys., 33, 2, 264-269, 1968.
27) Habashy T. M.and Mittra R. , J. Electromagn. Waves Appl., 1, 1, 25-58, 1987.
28) Bruckstein A. M *et al.*, SIAM J. Appl. Math., 45, 2, 312-335, 1985.
29) Ursin B. and Berteussen K. A., Proc. IEEE, 74, 3, 389-400, 1986.
30) Gray S. and Symes W., Geophys. J. R. Astr. Soc., 80, 149-163, 1985.
31) Carrion P., Inverse Prob., 2, 1-22, 1986.
32) Lesselier D., *Diagnostic Optimal de la Lame Inhomogène en Régime Temporel. Applications à l'Electromognétisme et à l'Acoustique.*, Thèse d'Etat, Univ. P.et M. Curie, Paris, 1982.

33) Crosta G., same ref. as 22), 111-120.

34) Mostafavi M. and Mittra R., Radio Sci., 7, 12, 1105-1111, 1972.

35) Mittra R. and Itoh T. C, IEEE Trans. Microwave Theory Techn., MTT-23, 1, 176-177, 1975

36) Roger A., *Problèmes Inverses de Diffraction en Electromagnétisme. Théorie, Traitement Numérique et Applications à l'Optique*. Thèse d'Etat, Univ. Aix-Marseille III, Marseille, 1981.

37) Tijhuis A. G., *Electromagnetic Inverse Profiling. Theory and Numerical Implementation.*, VNU Science Press, Utrecht, 1987.

38) Coen S. *et al.*, IEEE Trans. Antennas Propagat., AP-29, 2, 298-306, 1981.

39) Roger A., same ref. as 22), 111-120.

40) Uno T. and Adachi S., IEEE Trans. Antennas Propagat., AP-35, 12, 1456-1466, 1987.

41) Bamberger A. *et al.*, Geophys., 47, 5,757-770, 1982.

42) Kolb P. *et al.*, Proc. IEEE, 74, 3,499-508, 1986.

43) Santosa F. *et al.*, Inverse Prob., 3, 3, 477-499, 1987.

44) Gjevik B. *et al.*, Geophys. Prospect., 24, 492-505, 1976.

45) Tabbara W., *Etudes de Problèmes de Diffraction Inverse Electromagnétique par Analyse Multi-fréquentielle des Champs Diffractés*, Thèse d'Etat, Univ. P. et M. Curie, Paris, 1976.

46) Cohen J. K. and Bleistein N., SIAM J. Appl. Math., 32, 4, 784-799, 1977.

47) Raz S., Radio Sci., 16, 3, 347-353, 1981.

48) Gray S. and Hagin F., SIAM J. Appl. Math., 42, 2, 346-355, 1982.

49) Jaggard D. L. and Kim Y. , J. Opt. Soc. Am., 2, 11, 1922-1930, 1985.

50) Tobocman W., J. Comput. Physics, 64, 230-245, 1986.

51) Gazanhes C. *et al.*, J. Acoust. Soc. Am., 69, 3, 720-727, 1981.

52) Smith P. R. *et al.*, Inverse Prob., 4, 1, 235-248, 1988.

53) Ladouceur H. D. and Huston F. L., same ref. as 20), 89-96.

54) Rajan S. D. *et al.*, J. Acoust. Soc. Am., 82, 3, 998-1017, 1987.

55) Lytle R. J. and Lager D. L., Radio Sci., 11, 3, 199-209, 1976.

56) Nagl A. *et al.*, IEEE Trans. Geosci. Remote Sens., GE-20, 3, 332-337, 1982.

57) Maze G. *et al.*, J. Acoust. Soc. Am., 80, 1, 295-301, 1986.

58) Jackins P. D. et Gaunaurd G. C., J. Acoust. Soc. Am., 80, 6, 1762-1776, 1986.

59) Nagl A. *et al.*, Inverse Prob., 1, 1, 99-110, 1985.

CARACTERISATION OF OCEAN FLOORS BY AN INVERSE METHOD

Luc LEVIANDIER
THOMSON-SINTRA
1 Av. Aristide Briand
94117 ARCUEIL CEDEX

No written contribution received.

PROBLEME DE DIFFRACTION INVERSE EN IMAGERIE MICROONDE. EFFETS DE L'ANALYSE VECTORIELLE.

M. Baribaud, R. Salazar Cardozo

ENSERG-LEMO 23 Avenue des Martyrs 38016 GRENOBLE Cédex FRANCE
Tél. 76.87.69.76.

ABSTRACT :

A scattering inverse problem synthesis has been carried out. A more general approach taking account of microwave scattering vectoriel character has shown that the usual assumptions come from the equivalent current density divergence the magnitude of which is taken as zero. This theoretical approach has permited to establish a general relation between three fundamental terms : the equivalent current density (unknown), the measured scattered field and the supplementary information which had to be added to obtain the inversion process. An interative method is proposed which use an a priori information. It will permit to benefit from the scattered microwave field information.

I. CHAMP DIFFRACTE PAR UN OBJET

Considérons la propagation des ondes électromagnétiques dans un milieu homogène caractérisé par sa perméabilité magnétique μ_1 , sa conductivité électrique σ_1 et sa permittivité électrique ϵ_1 . A l'intérieur de ce milieu se trouve un volume (v) de substance caractérisée par une perméabilité magnétique μ_2, une conductivité σ_2 et une permittivité ϵ_2.

Dans un système tridimensionel de coordonnées cartésiennes, soit $\vec{r}$ un vecteur de position repérant un point P à l'extérieur du volume (v), dans le milieu ($\mu_1,\epsilon_1,\sigma_1$), et soit $\vec{\rho}$ le vecteur repérant un point P' à l'intérieur du volume (v), comme le montre la figure 1.

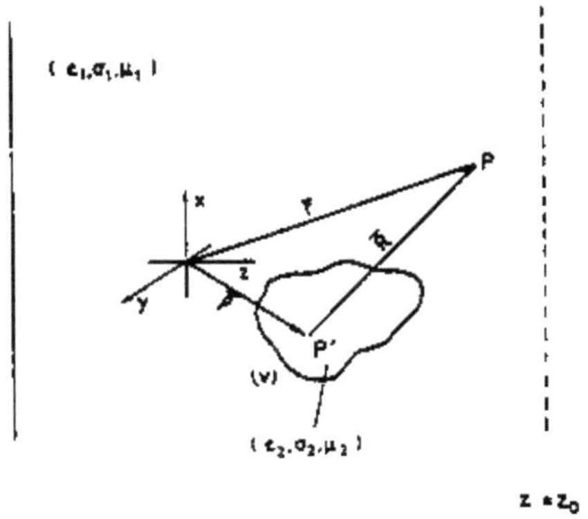

Figure 1.
Schéma de principe pour l'étude théorique.

Par définition le champ électrique diffracté $\vec{E}_s(\vec{r})$ est la différence entre le champ mesuré $\vec{E}_2(\vec{r})$ en présence du volume (v) et le champ $\vec{E}_1(\vec{r})$ qui existerait en l'absence du volume (v) [1], [2]. De la même façon le champ magnétique diffracté est la différence entre le champ $\vec{H}_2(\vec{r})$ mesuré en présence du volume (v) et le champ $\vec{H}_1(\vec{r})$ qui existerait au même endroit en l'absence de ce volume. On a :

$$\vec{E}_s(\vec{r}) = \vec{E}_2(\vec{r}) - \vec{E}_1(\vec{r}) \tag{1}$$

$$\vec{H}_s(\vec{r}) = \vec{H}_2(\vec{r}) - \vec{H}_1(\vec{r}) \tag{2}$$

Dans l'hypothèse d'un milieu non-magnétique, c'est-à-dire $\mu_1 = \mu_2 = \mu_0$ l'écriture des équations de Maxwell conduit aux relations suivantes :

$$\text{rot } \vec{H}_s = \vec{J}_s + \sigma_1 \vec{E}_s + \epsilon_1 \frac{\partial \vec{E}_s}{\partial t} \tag{3}$$

$$\text{rot } \vec{E}_s = -\mu_0 \frac{\partial \vec{H}_s}{\partial t} \tag{4}$$

Dans ces relations $\vec{J}_s$ est le vecteur donné par l'expression :

$$\vec{J}_s = (\sigma_2-\sigma_1)\,\vec{E}_z + (\epsilon_2 - \epsilon_1)\,\frac{\partial \vec{E}_z}{\partial t} \tag{5}$$

Cette relation met en évidence le rôle de la présence du volume (v). En effet, en regardant l'équation (3) on peut constater que le volume (v) peut être remplacé par une source équivalente de courant ayant la valeur donnée par l'expression (5), pour obtenir les mêmes champs $\vec{E}_s$ et $\vec{H}_s$ que ceux produits par la présence du volume (v). Ceci est d'autant plus valable que la source équivalente de l'expression (5) n'existe qu'à l'intérieur du volume (v), puisque à l'extérieur σ_2 devient σ_1, ϵ_2 devient ϵ_1 et en conséquence $\vec{J}_s$ est nul.

Le calcul du champ électrique à l'extérieur du volume (v) dû à la présence d'une source de champ à l'intérieur de ce volume est fait à l'aide des potentiels scalaire V et vecteur $\vec{A}$ [3], [4] :

$$\vec{E} = -\,\mathrm{grad}\ V - \frac{\partial \vec{A}}{\partial t} \tag{6}$$

Dans notre cas le potentiel vecteur $\vec{A}$ est calculé directement à partir de la densité de courant équivalente $\vec{J}_s$ tandis que le potentiel scalaire V est calculé à partir des deosités volumiques et surfaciques de charge équivalentes. On obtient :

$$\vec{E}_s(\vec{r}) = -i\omega\mu_0 \int_{(v)} \vec{\vec{G}}(\vec{r},\vec{\rho}) \cdot \vec{J}_s(\vec{\rho})\, dv \tag{7}$$

$\vec{\vec{G}}(\vec{r},\vec{\rho})$ est la dyade liée à la fonction de Green scalaire par la relation :

$$\vec{\vec{G}}(\vec{r},\vec{\rho}) = \left[\vec{\vec{I}} - \frac{1}{k^2}\,\mathrm{grad}_{\vec{r}}\left(\mathrm{grad}_{\vec{\rho}}\right)\right] \Phi\,(\vec{r},\vec{\rho}) \tag{8}$$

où $\overleftrightarrow{I}$ représente la dyade unité.

L'expression (7) montre la relation directe existant entre le champ diffracté $\vec{E}_s$ (effet) et la densité de courant équivalente $\vec{J}_s$ (cause). La connaissance de $\vec{J}_s$ dans le volume (v) permet à travers (7) le calcul du champ électrique diffracté. On note que lors de la déduction de cette relation aucune hypothèse n'a été émise sur la nature de $\vec{J}_s$.

II. PROBLEME DE DIFFRACTION INVERSE

Dans le paragraphe précédent on a déterminé le champ électrique diffracté en fonction des sources équivalentes de champ se trouvant dans le volume (v). Dans la pratique on peut envisager la mesure de $\vec{E}_s$ et on veut savoir si à partir de cette mesure il est possible de déterminer la source qui l'a produit. On va aborder cette analyse inverse à partir de la représentation du volume (v) par une source équivalente : la densité de courant équivalente $\vec{J}_s$. Même si la connaissance de $\vec{J}_s$ ne permet pas de passer directement aux inhomogénéités ϵ_2 et σ_2, s'il est possible de connaitre $\vec{J}_s$ alors on pourra délimiter sans ambiguïté le volume (v) parce que $\vec{J}_s$ n'est différente de zéro qu'à l'intérieur de (v), en conformité avec la relation (5).

La relation (7) représente une convolution dans le domaine spatial entre la dyade $\overleftrightarrow{G}$ et les sources de champ $\vec{J}_s$. On sait que dans le domaine de Fourier les convolutions se transforment en produits algébriques, alors on a intérêt à faire une représentation spectrale de la relation (7) ; on doit faire apparaître la transformée de Fourier de $\vec{J}_s$.

Reprenons l'expression de la dyade $\overleftrightarrow{G}$:

$$\overleftrightarrow{G} = \left(\overleftrightarrow{I} - \frac{1}{k^2} \nabla\nabla'\right) \Phi(\vec{r},\vec{\rho})$$

Les coordonnées du vecteur $\vec{r}$ dans le système cartésien de la figure 1 sont (x,y,z) tandis que celles du vecteur $\vec{\rho}$ on les notera (ξ,η,ζ).

La fonction de Green scalaire admet la représentation spectrale suivante [5] :

$$\Phi(\vec{r},\vec{\rho}) = \frac{i}{8\pi^2}\int_{-\infty}^{+\infty}\int_{-\infty}^{+\infty}\frac{1}{\gamma}\, e^{i[\alpha(x-\xi) + \beta(y-\eta) + \gamma|z-\zeta|]}\, . \, d\alpha d\beta \qquad (9)$$

Dans cette relation α,β,γ sont les variables dans le domaine de Fourier correspondantes aux trois directions spatialles x,y,z respectivement. Elles sont liées dans la relation (9) par la relation

$$\alpha^2 + \beta^2 + \gamma^2 = k^2 \qquad (10)$$

L'introduction de la relation (14) dans l'expression de la dyade $\bar{\bar{G}}$ conduit à [6] :

$$\vec{E}_s(\vec{r}) = \frac{\omega\mu_0}{8\pi^2}\int_{-\infty}^{+\infty}\int_{-\infty}^{+\infty}\frac{1}{\gamma}\vec{U}\wedge\left(\vec{U}\wedge\vec{\mathcal{J}}_s\right)e^{ik\vec{r}\cdot\vec{U}}. d\alpha d\beta \qquad (11)$$

où le vecteur $\vec{U}$ de coordonnées (u,v,w) est définit par les relations

$$u=\frac{\alpha}{k}\ ,\quad v=\frac{\beta}{k}\ \text{ et }\ w=\frac{\gamma}{k} \qquad (12)$$

et $\vec{\mathcal{J}}_s$ est la transformée de Fourier tridimensionnelle de $\vec{J}_s$.

Afin de mettre en évidence une relation dans le domaine de Fourier entre le champ $\vec{E}_s(\vec{r})$ mesuré à l'extérieur du volume (v) et la densité de courant équivalente $\vec{\mathcal{J}}_s$ à l'intérieur de ce volume, il convient de modifier la relation (16) et l'écrire sous la forme :

$$\vec{E}_s(\vec{r}) = \frac{1}{4\pi^2}\int_{-\infty}^{+\infty}\int_{-\infty}^{+\infty}\left[-\frac{\omega\mu_0}{2\gamma}e^{i\gamma z}\vec{U}\wedge\left(\vec{U}\wedge\vec{\mathcal{J}}_s\right)\right]e^{i(\alpha x+\beta y)}d\alpha d\beta \qquad (13)$$

Considérons maintenant un plan dans le milieu de propagation caractérisé par $z=z_0$; dans ce plan on va mesurer le champ électrique diffracté ; on calcule la transformée de Fourier bidimensionnelle du champ mesuré, c'est-à-dire $\vec{E}_s(x,y,z_0)$: d'après la définition d'une transformée de Fourier bidimensionnelle des coordonnées x et y on obtient :

$$\vec{\mathcal{E}}_s(\alpha,\beta,z_0) = \int_{-\infty}^{+\infty}\int_{-\infty}^{+\infty} \vec{E}_s(x,y,z_0)e^{-i(\alpha x+\beta y)}dxdy \qquad (14)$$

La transformation de Fourier inverse permet d'exprimer $\vec{E}_s(x,y,z_0)$ en fonction de sa transformée de Fourier par la relation :

$$\vec{E}_s(x,y,z_0) = \frac{1}{4\pi^2}\int_{-\infty}^{+\infty}\int_{-\infty}^{+\infty} \vec{\mathcal{E}}_s(\alpha,\beta,z_0)e^{i(\alpha x+\beta y)}d\alpha d\beta \qquad (15)$$

La comparaison de la relation (15) et la relation (13) permet alors d'écrire la relation cherchée :

$$\vec{\mathcal{E}}_s(\alpha,\beta,z_0) = -\frac{\omega\mu_0}{2\gamma}e^{i\gamma z_0}\vec{u} \wedge \left[\vec{u} \wedge \vec{\mathcal{J}}_s(\alpha,\beta,\gamma)\right] \qquad (16)$$

Cette importante relation va permettre de mener la discussion sur la possibilité de trouver une solution au problème inverse : est-il possible de dégager, à partir de la relation (16), la valeur de $\vec{\mathcal{J}}_s(\alpha,\beta,\gamma)$ en fonction de $\vec{\mathcal{E}}_s(\alpha,\beta,z_0)$? Si la réponse à cette question est affirmative alors on pourrait revenir sur la densité de courant équivalente dans le volume (v) par une transformation de Fourier inverse.

La figure 2 montre une représentation graphique de la relation (16) : dans l'espace (α,β,γ) il est toujours possible de trouver un plan contenant les vecteurs $\vec{u}$ et $\vec{\mathcal{J}}_s$. A partir de ces deux vecteurs on peut déterminer graphiquement $\vec{\mathcal{E}}_s(\alpha,\beta,z_0)$, tenant compte de la relation (16) et du fait que le vecteur $\vec{u}$ est unitaire.

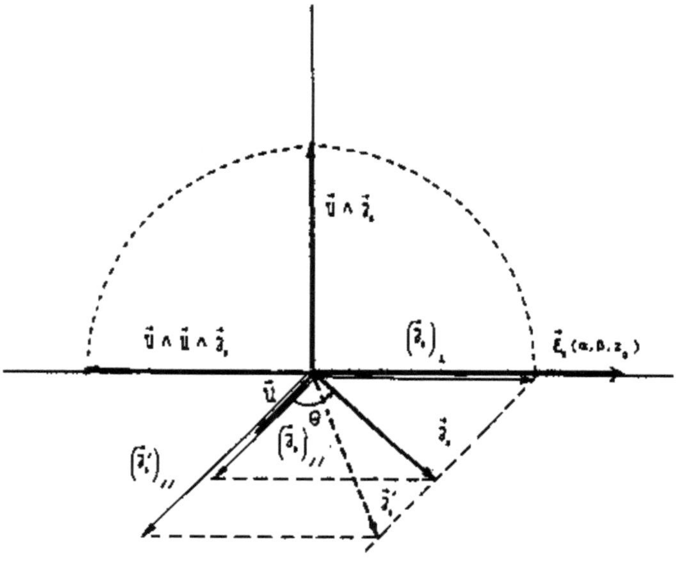

Figure 2

Détermination par voie graphique de $\vec{\mathcal{E}}_s(\alpha,\beta,z_0)$.

Sur la figure 2 on a dessiné en pointillé un deuxième vecteur $\vec{J}_s$ qui a la même composante perpendiculaire à $\vec{U}$ que $\vec{J}_s$, c'est-à-dire $(\vec{J}'_s)_\perp = (\vec{J}_s)_\perp$. D'après la construction graphique, le champ $\vec{\mathcal{E}}_s$ produit par $\vec{J}'_s$ sera le même que celui produit par $\vec{J}_s$. On conclut alors qu'il y a une infinité de distributions $\vec{J}_s$ permettant d'obtenir le même $\vec{\mathcal{E}}_s$. On retrouve donc, à partir de la relation vectorielle (16), la même non-unicité du problème de source inverse qu'avait déjà donné lieu à plusieurs points de vue, parfois contradictoires, dans le cas scalaire [7...10] : la relation (16) avec sa représentation de la figure 2 constituent une généralisation de la non-unicité du problème de source inverse au cas vectoriel.

La différence entre toutes les densités de courant produisant le même champ diffracté se trouve au niveau de la composante spectrale parallèle à la direction du vecteur $\vec{U}$. On peut constater que pour les deux vecteurs $\vec{J}_s$ et $\vec{J}'_s$ on peut écrire :

$$\left(\vec{U}\cdot\vec{J}_s\right) = \left(\vec{J}_s\right)_{//} \neq \left(\vec{U}\cdot\vec{J}'_s\right) = \left(\vec{J}'_s\right)_{//} \tag{17}$$

Il est facile de montrer que le produit scalaire $\vec{U}\cdot\vec{J}_s$ n'est autre chose que la transformée de Fourier de la divergence de $\vec{J}_s$.

A partir de la relation de définition de $\vec{J}_s$ (5), on obtient :

$$\text{div}\ \vec{J}_s = \text{div}\left\{[(\sigma_2-\sigma_1) + i\omega(\epsilon_2-\epsilon_1)]\vec{E}_2\right\} \tag{18}$$

En utilisant les relations (1) et (3) on aboutit à [7] :

$$\text{div}\ \vec{J}_s = \frac{\sigma_1+i\omega\epsilon_1}{\sigma_2+i\omega\epsilon_2}\ \overrightarrow{\text{grad}}\ (\sigma_2+i\omega\epsilon_2)\cdot\vec{E}_2 \tag{19}$$

On peut constater que la divergence de la densité de courant équivalente porte une information sur les variations de conductivité et permittivité dans le volume (v). A l'extérieur du volume σ_2 et ϵ_2 deviennent des constantes, leur gradient est nul et la divergence sera nulle aussi.

La relation qui relie la densité de courant équivalente à l'intérieur du volume (v) et le champ diffracté créé par celle-ci dans un plan à l'extérieur de ce volume, n'est pas biunivoque ; plusieurs densités de courant (plusieurs objets) peuvent créer le même champ diffracté, même s'il y a quelque chose de commun entre elles (la composante spectrale perpendiculaire au vecteur $\vec{U}$).

Cette ambiguité a conduit à des approximations de l'équation de propagation cherchant une relation biunivoque entre le champ mesuré et la source que l'a produit. Parmi les plus répandues on peut citer

- Approximation de non dépolarisation du champ incident
- Approximation de Born
- Approximation de Rytov
- Approximation d'une densité de courant équivalente constante.

On peut ajouter aussi la limitation la plus utilisée pour le test des différentes méthode et algorithmes de reconstituation d'image.

- Approximation du volume (v) rempli de substance homogène.

En premier lieu il faut souligner que les deux dernières approximations ($\vec{J}_s$ constant et ϵ^* constant) bien que différentes dans leur nature sont en général interprétées comme si elles étaient équivalentes. Au contraire, les trois premières peuvent être discutées à la lumière de la premiére parce que, dans les problèmes de diffraction microondes, avant de pouvoir utiliser les approximations de Born ou de Rytov, il faut que l'équation de propagation ait déjà subi l'approximation de non dépolarisation du champ incident [11].

Nous avons montré [6] que les approximations utilisées pour simplifier l'équation de propagation visent le même terme de cette équation, même si elles sont présentées différamment. Elles conduisent toutes à écrire :

$$\text{div}\ \vec{J}_s = 0 \qquad (20)$$

Or, compte tenu de la relation (19), cette condition conduit à négliger les variations des caractéristiques ϵ_2 et σ_2 dans le volume (v), réduisant ainsi le domaine d'application de la méthode au cas de très faible diffraction.

IV. VERS UNE RECONSTITUTION VECTORIELLE

Toute hypothèse qui puisse conduire à l'inversion de la relation (16) doit mettre en évidence quelque chose sur la composante spectrale $\left(\vec{\mathcal{J}}_s\right)_{//}$ donc sur la valeur de div $\vec{J}_s$.

Il est impératif d'ajouter une information supplémentaire. Cette information "a priori" doit pouvoir se mettre sous forme analytique. Il convient d'insister sur le rôle de cette information supplémentaire qui n'est pas de simplifier la mathématique du problème mais de permettre d'aboutir à la détermination des propriétés du volume (v). Nous avons déduit une relation qui explicite le rôle des trois variables mises en jeu dans la reconstitution de (v) : le champ diffracté mesuré, la densité de courant équivalente à déterminer et l'information supplémentaire à ajouter, soit [6].

$$\vec{\mathcal{J}}_s(\alpha,\beta,\gamma_0) = \frac{\vec{\mathcal{E}}_s(\alpha,\beta,z_0)}{f(\gamma)} + M(\alpha,\beta,\gamma)\ \vec{U} \tag{21}$$

ou $$\vec{U}\cdot\vec{\mathcal{J}}_s(\alpha,\beta,\gamma) = M(\alpha,\beta,\gamma) \quad \text{et} \quad f(\gamma) = \frac{\omega\mu_0}{2\gamma}\, e^{i\gamma z_0} \tag{22}$$

La relation (21) met en évidence les deux composantes nécessaires à la détermination de la densité de courant équivalente : le premier terme correspond à l'information obtenue après la mesure du champ diffracté à l'extérieur du volume (v), tandis que le deuxième contient l'information à priori sur ce volume.

Une fois $\vec{\mathcal{J}}_s$ connu, l'algorithme de détermination des propriétés du volume (v) suit le schéma classique proposé dans la littérature, c'est-à-dire l'obtention de coupes tomographiques de la densité de courant équivalente par transformation de Fourier bidimensionnelle inverse de (21). La valeur de M doit être compatible avec le fait que tous les volumes donnant lieu au même champ diffracté doivent présenter, dans le domaine spectral, la même composante de $\vec{\mathcal{J}}_s$ normal au vecteur $\vec{U}$. On peut envisager un processus d'optimisation des valeurs de $\vec{\mathcal{J}}(\alpha,\beta,\gamma)$ à partir des valeurs de $M(\alpha,\beta,\gamma)$; le champ diffracté (et sa transformée de Fourier) est supposé connu ; on considère une valeur initiale du paramètree $M(\alpha,\beta,\gamma)$ et, avec la relation (21), on calcule une première valeur de $\vec{\mathcal{J}}_s(\alpha,\beta,\gamma)$.

Le processus d'optimisation peut se résumer schématiquement sous la forme suivante :

<u>Données d'entrée</u> :

* le champ diffracté $\vec{E}_s(x,y,z_0)$ dans sa représentation spectrale $\vec{\mathcal{E}}_s(\alpha,\beta,z_0)$;

* la valeur initiale de la divergence de $\vec{J}_s$, aussi dans sa forme spectrale $M_0(\alpha,\beta,\gamma)$.

<u>Traitements</u> :

* Calcul de $\vec{J}_s$ avec les valeurs de $\vec{\xi}_s$ et M_0 par la relation (21).

* Calcul de M'_0 avec la relation $M'_0 = \vec{u} \cdot \vec{J}_s$.

* Calcul de la composante de $\vec{J}_s$ normale au vecteur $\vec{u}$, notée $\left(\vec{J}_s\right)'_{\perp}$. Ici on doit utiliser la valeur de $\vec{J}_s$ trouvée lors du premier traitement.

* La différence entre M_0 et M'_0 d'une part, entre $\left(\vec{J}_s\right)_{\perp}$ et $\left(\vec{J}_s\right)'_{\perp}$ d'autre part, serviront à déterminer une nouvelle valeur de M par le choix d'une fonction d'optimisation convenable.

* On arrête le processus quand les différences évoquées sont inférieures à un epsilon donné.

Mais la détermination de $\vec{J}_s$ peut se faire aussi à partir du champ mesuré, l'information supplémentaire et l'utilisation des méthodes variationelles d'estimation comme la Méthode du Maximum d'Entropie [12],[13], utilisée déjà avec un relatif succès dans des problèmes semblables, où la caractéristique essentielle est le manque d'information.

Il convient de préciser que les méthodes utilisées jusqu'ici prennent la plus simple valeur pour l'information supplémentaire :

$$M(\alpha,\beta,\gamma) = 0 \qquad (23)$$

avec les limitations importantes déjà citées. De ce point de vue l'algorithme décrit ci-dessus est une approche beaucoup plus générale, car il n'impose en principe aucune restriction ni sur la taille du volume diffractant ni sur sa composition interne. La diffraction microondes sera ainsi abordée sous un angle vectoriel.

Cette approche du problème inverse est en parfaite concordance avec les différentes méthodes utilisées actuellement pour des problèmes de même nature, mais qui admettent une approximation scalaire comme par exemple l'exploration des fonds sous-marins pour déterminer ses propriétés physiques ou la détermination de la composition des couches terrestres, que ce soit avec des excitations acoustiques ou électromagnétiques. Ici l'information supplémentaire joue un rôle très important et conditionne la reconstruction des profils cherchés ; dans ces applications l'information à priori est un profil initial.

CONCLUSION

Une synthèse de l'aspect théorique du phénomène de diffraction inverse a été faite. Par une approche plus générale, capable de tenir compte du caractère vectoriel de la diffraction microondes, nous avons montré que les différentes hypothèses utilisées pour chercher l'unicité du problème inverse de diffraction, font appel à la divergence de la densité de courant équivalente qui caractérise mathématiquement l'objet physique. Nous avons montré qu'elle est toujours considérée comme étant nulle. Cette même approche nous a permis d'établir une relation générale entre les trois termes qui interviennent dans l'étude du problème inverse : la densité équivalente de courant (inconnue), le champ diffracté à l'extérieur de l'objet sur une surface plane (mesuré) et l'information supplémentaire qu'il est nécessaire d'ajouter pour que le processus d'inversion puisse avoir lieu. Les méthodes utilisées jusqu'ici ont donné à cette information à priori sa valeur la plus simple (zéro), mais les limitations sont alors trop importantes. Nous envisageons une méthode itérative beaucoup moins restrictive, qui devra permettre de profiter au mieux de l'information contenue dans le champ microondes diffracté, mais dont les résultats obtenus seront étroitement dépendants de l'information à priori utilisée.

BIBLIOGRAPHIE

[1] M. Kom,
"Imagerie microondes en milieu biologiques. Application au diagnostic médical"
Thèse de doctorat de troisième cycle, INPG, Novembre 1981

[2] S. Wang,
"Reconstitution de l'image d'un objet microonde en milieu biologique à partir de la mesure du champ diffracté"
Thèse de Docteur Ingénieur, INPG, 1983

[3] C.I. Mocanu,
"Teoria cimpului electromagnetic"
Editura didactica si pedagogica. Bucuresti, 1981.

[4] E. Nicolau,
"Sinteza elementelor radiante"

Editura academiei R.S.Romania. Bucuresti, 1971.

[5] **M. Baribaud,**
"Cours d'imagerie microondes"
ENSERG

[6] **R. Salazar Cardozo,**
"Multiplexeur planaire pour la mesure du champ diffracté en imagerie microondes"
Thèse de docteur INPG, 1988. A paraître.

[7] **A.J. Devaney, G.C. Sherman,**
"Non uniqueness in inverse source and scattering problems"
IEEE Trans. on Antennas and Propagation, Vol. AP-30, 1982, pp. 1034-1037

[8] **N.N. Bojarski,**
Comments on "nonuniqueness in inverse source and scattering problems"
IEEE Trans. on Antennas and Propagation, Vol. AP-30, n°5, pp. 1037-1038 (1982)

[9] **W.R. Stone,**
Comments on "nonuniqueness in inverse source and scattering problemes"
IEEE Trans. on Antennas and Propagation, Vol. AP-30, n°5, pp. 1039-1041, (1982)

[10] **M. Fischer, K.J. Langerberg,**
"Limitations and defects of certain inverse scattering theories"
IEEE Trans. on Antennas and Propagation, Vol. AP-32, n°10, pp. 1080-1087 (1984)

[11] **M. Slaney, A. Kak, L. Larsen,**
"Limitations of imaging whith first order diffraction tomography"
IEEE Trans. on Microwave Theory and Techniques, Vol. MTT-32, n°8, 1984, pp. 860-873

[12] **R.M. Bevensee,**
"Solution of underdetermined electromagnetic and seismic problems by the Maximum of Entropy Method"
IEEE Trans. on Antennas and Propagation, Vol. AP-29, n°2, pp. (1981)

[13] **A.M. Djafari, G. Demonen,**
"Tomographie de diffraction et synthèse de Fourier à Maximum d'Entropie"
Rev. Phys. Appl. 22, Février 1987, pp. 153-157

Inverse Spectral Theory for a Singular Sturm-Liouville Operator on [0,1]

J.C. GUILLOT
Ecole Polytechnique
and
Université Paris-Nord

J.V. RALSTON
Department of Mathematics
University of California,
Los Angeles, CA 90024

Abstract

The inverse spectral theory for a singular sturm-Liouville operator on [0,1] is solved by using spectral data to coordinatize $L^2_{\mathbb{R}}([0,1])$.

Consider the Schrödinger operator

$$\mathring{H} = -\Delta + q(|\vec{x}|) \ , \qquad \vec{x} \in R^3,$$

acting on functions in $C^\infty(|\vec{x}| \leq 1)$ vanishing on $|\vec{x}| = 1$. We set $x = |\vec{x}|$ and assume $q(x)$ is a real-valued function in $L^2([0,1])$. The closure H of $\mathring{H}$ is a self-adjoint operator with compact resolvent. Hence H has pure point spectrum without finite accumulation points. Moreover, using the mapping $u(\vec{x}) \to xu(\vec{x})$ and expansion in spherical harmonics, one shows that H is unitarily equivalent to a direct sum of Sturm-Liouville operators acting on $L^2[0,1]$, cf. [1]. The first operator in this sum is

$$A_0(q) = -\frac{d^2}{dx^2} + q(x)$$

with domain

$$D_o = \{u \in H^2[0,1] : u(0) = u(1) = 0\},$$

and the second is

$$A_1(q) = -\frac{d^2}{dx} + \frac{2}{x^2} + q(x)$$

with domain

$D_1 = \{u \in L^2[0,1] : u, \frac{du}{dx}$ are absolutely continuous on

$(0,1]$, $A_1(q)u \in L^2[0,1]$ and $u(1) = 0\}$

The inverse spectral theory for $A_o(q)$, i.e., the description of the possible spectra for $A_o(q)$, $q \in L^2_R[0,1]$, and of the isospectral sets of q, is discussed in detail in [2]. Our objective here is to extend these results to $A_1(q)$. The inverse spectral problem for $A_o(q)$ has been studied extensively. The special feature of [2] is that by using spectral data to coordinatize $L^2_R[0,1]$ the authors give complete analytical descriptions of the spectra and isospectral sets. These coordinates are introduced in the following way. Let $\{\mu_n(q)\}_{n=1}^{\infty}$ be the increasing sequence of eigenvalues of $A_o(q)$ and let $\{\phi_n(x,q)\}_{n=1}^{\infty}$ be the associated sequence of eigenfunctions, normalized by $\lim_{x\downarrow 0} x^{-1}\phi_n(x,q) = 1$. One can then consider the mapping

$$\Psi(q) = (\{\mu_n(q)\}_{n=1}^{\infty}, \{\frac{\partial\phi_n}{\partial x}(1,q)\}_{n=1}^{\infty}).$$

That Ψ is the one to one is simply the Borg-Levinson Theorem.

However, the result of [2] is that the related mapping.

$$\Phi(q) = \left(\int_0^1 q\,dx, \{\mu_n(q) - \pi^2 n^2 - \int_0^1 q\,dx\}_{n=1}^{\infty}, \{n \log(\frac{\partial \phi_n}{\partial x}(1,q)/\frac{\partial \phi_n}{\partial x}(1,0))\}_{n=1}^{\infty}\right)$$

is a real analytic (in q) homeomorphism of $L_R^2[0,1]$ into $R \times \ell_R^2 \times \ell_R^2$. Moreover, the range of Φ consists of all sequences in $R \times \ell_R^2 \times \ell_R^2$ such that the corresponding sequence $\{\mu_n(q)\}_{n=1}^{\infty}$ is strictly increasing. From this one concludes that the set of possible spectra is all increasing sequences $\{\mu_n\}_{n=1}^{\infty}$ such that $\{\mu_n - \pi^2 n^2 - c\}_{n=1}^{\infty} \in \ell^2$ and the isospectral manifolds are analytic homeomorphs of ℓ^2.

For $A_1(q)$ we normalize eigenfunctions by $\lim_{x \downarrow 0} x^{-2} \phi_n(x,q) = 1$, but then define Ψ as before. The asymptotic behavior of the eigenvalues here is

$$\mu_n(q) = (n + \tfrac{1}{2})^2 \pi^2 - 2 + \int_0^1 q(x)\,dx + \tilde{\mu}_n(q) \qquad (I.1)$$

where $\{\tilde{\mu}_n(q)\}_{n=1}^{\infty}$ is in ℓ_R^2. This leads us to define

$$\Phi(q) = \left(\int_0^1 q\,dx, \{\tilde{\mu}_n(q)\}_{n=1}^{\infty}, \{n \log(\frac{\partial \phi_n}{\partial x}(1,q)/\frac{\partial \phi_n}{\partial x}(1,0))\}_{n=1}^{\infty}\right).$$

Then the result of this paper is that Φ is again a real analytic homeomorphism of $L_R^2[0,1]$ into $R \times \ell_R^2 \times \ell_R^2$ with range all sequences such that the corresponding sequence $\mu_n(q)$ in (I.1) is strictly increasing. Interesting problems do arise when one comes to the proof that the Fréchet derivative $\Phi'(q)$ is an isomorphism.

Details will be published in the Journal of Differential Equations.

We wish to thank Eugene Trubowitz for many helpful discussions of these problems.

REFERENCES

1. REED, M. AND SIMON B., Methods of Modern Mathematical Physics. Vol. II, Fourier Analysis, Self-adjointness, Academic Press, (1980).

2. PÖSCHEL, J. AND TRUBOWITZ, E., Inverse Spectral Theory. Academic Press, 1987.

LINEAR INVERSE PROBLEMS IN CONFOCAL LIGHT MICROSCOPY

M.Bertero and P.Boccacci

Dipartimento di Fisica dell'Università di Genova and

Istituto Nazionale di Fisica Nucleare, Sezione di Genova

Via Dodecaneso 33, I-16146 Genova, Italy

ABSTRACT

We review several inverse problems related to data reduction and resolution enhancement in confocal scanning light microscopy (CSLM).

1. INTRODUCTION

In recent years the confocal principle, known from acoustic microscopy[1)], has been applied to optical microscopy and has provided a better resolution and image fidelity than the conventional optical microscopes[2,3)]. This technique is known as confocal scanning light microscopy (CSLM).

The reason of the improvement in resolution obtained by means of CSLM is the following. Let us consider the coherent case and let us denote by $S_1(\vec{\varrho}\,')$ the point spread function of the illuminating lens. If $f(\vec{\varrho}\,')$ is the transparency of the object, placed in the focal plane of this lens (the vector $\vec{\varrho}\,' = \{x', y'\}$ gives the position of a point in the focal plane), then the amplitude of the light field at the exit of the object is $S_1(\vec{\varrho}\,')f(\vec{\varrho}\,')$. Now, if the object is also placed in the focal plane of the imaging lens, as required by the confocal principle, and if $S_2(\vec{\varrho}\,')$ is the point spread function of this lens, the amplitude in the image plane is given by

$$g(\vec{\varrho}) = \int\int S_2(\vec{\varrho} - \vec{\varrho}\,')S_1(\vec{\varrho}\,')f(\vec{\varrho}\,')d\vec{\varrho}\,'. \tag{1}$$

In CSLM, for each scanning position, the image is detected only on the optical axis, i.e. $\vec{\varrho} = 0$. Thus, if we put

$$H(\vec{\varrho}\,') = S_2(-\vec{\varrho}\,')S_1(\vec{\varrho}\,') \tag{2}$$

and if we remark that each step in the scanning procedure corresponds to a translation of the object, i.e. $f(\vec{\varrho}\,')$ is replaced by $f(\vec{\varrho}\,' + \vec{\varrho})$, then we conclude that the amplitude of the image provided by CSLM is

$$g_s(\vec{\varrho}) = \int\int H(\vec{\varrho} - \vec{\varrho}\,')f(\vec{\varrho}\,')d\vec{\varrho}\,'. \tag{3}$$

Eqs. (2) and (3) demonstrate the improvement in resolution provided by CSLM. In fact, if Ω_1 and Ω_2 are the bandwidths of $S_1(\vec{\varrho})$ and $S_2(\vec{\varrho})$ respectively, then the bandwidth of $H(\vec{\varrho})$ is $\Omega_1 + \Omega_2$ and therefore the image $g_s(\vec{\varrho})$ contains contributions of the Fourier components of $f(\vec{\varrho})$ in a band broader than the band of the imaging lens. As it is well known, a broader bandwidth corresponds to a better resolution.

The improvement in resolution, however does not coincide with the improvement in bandwidth, since the Fourier transform of $H(\vec{\varrho})$ is not constant over the band. As an example, let us consider the very simple one-dimensional case. If we assume that the two lenses are perfect low-pass filters and that they have the same bandwidth $\Omega = \Omega_1 = \Omega_2$, we have

$$S_1(x) = S_2(x) = \frac{\Omega}{\pi} sinc\left(\frac{\Omega x}{\pi}\right) \tag{4}$$

where $sinc(x) = sin(\pi x)/(\pi x)$. Then the Fourier transform of $S_1(x)$ is constant over the band $[-\Omega, \Omega]$, i.e. $\hat{S}(\omega) = 1$ when $|\omega| \leq \Omega$ and zero elsewhere, while the Fourier transform of $H(\omega)$ is given by: $\hat{H}(\omega) = (\Omega/\pi)(1 - |\omega|/2\Omega)$ when $|\omega| \leq 2\Omega$

and zero elsewhere. It follows that the bandwidth of $H(x)$ is twice the bandwidth of the lenses but $\hat{H}(\omega)$ tends to zero when the spatial frequency ω tends to the boundary points of the band.

In recent years it has been suggested[4)] that a further improvement in resolution can be obtained by recording the full image $g(\vec{\varrho})$ at each step of the scanning procedure and by solving eq. (1) in order to recover $f(\vec{0})$. In this way one does not increase the band, since the restored object has exactly the bandwidth of $g_s(\vec{\varrho})$, but the Fourier transform of the impulse response function describing this new confocal scanning microscope, equipped with an array of detectors in the image plane, is more well-behaved than the Fourier transform of $H(\vec{\varrho})$ over the band. In fact, it is possible to prove[5)], that this new microscope really produces an improvement of a factor of 2 with respect to the resolution of the conventional microscope.

Equations similar to eqs. (1) and (3) hold true also in the case of fluorescence microscopy. Fluorescence in the object destroys the coherence, since the fluorescence field is proportional to the incident radiation, and in this case one can consider a confocal incoherent microscope[6,7)]. Then eq. (1) is replaced by the following one

$$g(\vec{\varrho}) = \int\int |S_2(\vec{\varrho} - \vec{\varrho}\,')|^2 |S_1(\vec{\varrho}\,')|^2 f(\vec{\varrho}\,') d\vec{\varrho}\,' \tag{5}$$

where $g(\vec{\varrho})$ denotes now the intensity distribution in the image plane and $f(\vec{\varrho}\,')$ is the density of the fluorescent centers in the object plane. Analogously eq. (3) is replaced by

$$g_s(\vec{\varrho}) = \int\int |H(\vec{\varrho} - \vec{\varrho}\,')|^2 f(\vec{\varrho}\,') d\vec{\varrho}\,' \tag{6}$$

where $g_s(\vec{\varrho})$ is the intensity on the optical axis for a scanning position corresponding to a translation $-\vec{\varrho}$ of the object.

Another interesting property of CSLM, which has been exploited very recently in the case of fluorescent objects, is the possibility of obtaining a three-dimensional Fourier transform instead of the two-dimensional one.

By Fourier transforming both sides of the eq. (6) we get

$$\hat{g}_s(\vec{\omega}) = \hat{K}(\vec{\omega})\hat{f}(\vec{\omega}) \tag{10}$$

where $\hat{K}(\vec{\omega})$ is the Fourier transform of $|H(\vec{\varrho})|^2$.

When $\hat{g}_s(\vec{\omega})$ is not affected by noise, then $\hat{g}_s(\vec{\omega}) \neq 0$ only on the band of $|H(\vec{\varrho})|^2$. Then the solution of eq. (10) exists but is not unique since one can add to $\hat{f}(\vec{\omega})$ an arbitrary function which is zero on the band of $|H(\vec{\varrho})|^2$. There exists however a unique generalized solution or solution of minimal norm, $f^{\dagger}(\vec{\varrho}\,')$, which is given by

$$\hat{f}^{\dagger}(\vec{\omega}) = \frac{\hat{g}_s(\vec{\omega})}{\hat{K}(\vec{\omega})}, \tag{11}$$

for values of $\vec{\omega}$ inside the band, and which is zero elsewhere. This generalized solution does not exist when the image is affected by noise. In such a case eq. (10) must be replaced by the following one

$$\hat{g}_s(\vec{\omega}) = \hat{K}(\vec{\omega})\hat{f}(\vec{\omega}) + \hat{n}(\vec{\omega}) \tag{12}$$

where $\hat{n}(\vec{\omega})$ is an unknown function describing the effect of noise, and, as a consequence of the behaviour of $\hat{K}(\vec{\omega})$ at the boundary of the band, it is not possible to divide $\hat{g}_s(\vec{\omega})$ by $\hat{K}(\vec{\omega})$. In order to prevent this difficulty, one must use regularization techniques[11]. The most simple of these techniques is the so-called Tikhonov regularization algorithm which, in the case of the eq. (6) or (10), consists in the computation, for some appropriate value of α, of the following approximate solution

$$f_\alpha(\vec{\varrho}) = \frac{1}{(2\pi)^2}\int\int \frac{\hat{K}^*(\vec{\omega})}{\left|\hat{K}(\vec{\omega})\right|^2 + \alpha}\hat{g}_s(\vec{\omega})e^{i(\vec{\varrho},\vec{\omega})}d\vec{\omega} \tag{13}$$

where the star denotes complex conjugation. The parameter α is called the regularization parameter. In the case of noise free data the regularized solution $f_\alpha(\vec{\varrho})$ tends to the generalized solution $f^{\dagger}(\vec{\varrho})$ when $\alpha \to 0$.

b) solution of the integral equations (1),(5) or (7), related to CSLM in the case of an array of detectors in the image plane.

The main aim is to show that, in the second case, it is possible to obtain a better resolution than in the first one.

2. IMAGE DECONVOLUTION IN THE CASE OF FLUORESCENCE MICROSCOPY

The expression (3) for the image produced by a confocal scanning microscope has been derived in the coherent case. In such a case, however, the output of the detector is proportional to the intensity of the image, i.e. $|g_s(\vec{\varrho})|^2$, and therefore the problem of restoring $f(\vec{\varrho}\,')$ from $|g_s(\vec{\varrho})|^2$ is a rather difficult nonlinear deconvolution problem.

The situation is more promising in the case of fluorescence microscopy, since in eq. (6) $g_s(\vec{\varrho})$ is proportional to the output of the detector and therefore we have a linear deconvolution problem. This two-dimensional approximation applies to objects whose thickness along the optical axis is of the order of the axial resolution, i.e. approximately 1 μm. In the case of thick, weakly fluorescent objects eq. (6) is replaced by eq. (9), which is again a linear integral equation.

Both in the case of eq. (6) and in the case of eq. (9), the problems of restoring the objects $f(\vec{\varrho})$ or $f(\vec{\varrho}, z)$ from the images $g_s(\vec{\varrho})$ or $g_s(\vec{\varrho}, z)$ are classical deconvolution problems. Since the kernels $|H(\vec{\varrho})|^2$ and $|H(\vec{\varrho}, z)|^2$ are band-limited and their Fourier transforms tend to zero when the spatial frequencies tend to points on the boundary of the band, these deconvolution problems are ill-posed: the solution is not unique and it may not exist when the image is affected by noise.

As an example, let us consider the two-dimensional problem (6). The three-dimensional problem (9) can be treated in the same way by means of the three-dimensional image of a three-dimensional object[8,9]. This procedure, also called microtomoscopy[8], is based on the fact that, only the fluorescent centers inside the focal volume contribute to the intensity of the image on the optical axis, for a given scanning position. Thus a true resolution is obtained along the optical axis and a three-dimensional image is obtained by means of a three-dimensional scanning.

In the case of a weakly fluorescent object, the intensity distribution in the image plane is given by

$$g(\vec{\varrho}) = \int\int\int |S_2(\vec{\varrho} - \vec{\varrho}\,', z')|^2 |S_1(\vec{\varrho}\,', z')|^2 f(\vec{\varrho}\,', z') d\vec{\varrho}\,' dz' \qquad (7)$$

where $S_1(\vec{\varrho}\,', z')$ and $S_2(\vec{\varrho}\,', z')$ are the amplitudes of the field distributions in the focal region[10], corresponding respectively to the illuminating and to the imaging lens. In the case of the reflection mode of operation[9], the two field amplitudes are different because the primary and the fluorescent wavelengths are different. This effect, however, is not large (of the order of 10%) and it can be neglected in a first approximation.

In CSLM, where the image is detected only on the optical axis, if we put

$$H(\vec{\varrho}\,', z') = S_2(-\vec{\varrho}\,', z') S_1(\vec{\varrho}\,', z') \qquad (8)$$

then the image provided by a three-dimensional scanning is given by

$$g_s(\vec{\varrho}, z) = \int\int\int |H(\vec{\varrho} - \vec{\varrho}\,', z - z')|^2 f(\vec{\varrho}\,', z') d\vec{\varrho}\,' dz'. \qquad (9)$$

Also in this case one can consider the problem of detecting the full image for each scanning position and of solving eq. (7) in order to recover $f(\vec{0}, 0)$.

In this paper we discuss the following problems:

a) deconvolution of the image provided by CSLM in the case of one detector on the optical axis, i.e. solution of the integral equations (6) or (7);

Notice that in eq. (13) the integral is automatically restricted to the band of $|H(\vec{\varrho})|^2$ and therefore the out-of-band noise in $g_s(\vec{\varrho})$ is automatically suppressed.

As concerns the choise of α, several criteria have been proposed, but the most simple one consists in taking α of the order of the inverse of the square of the signal-to-noise ratio[12)].

As follows from eq. (13), the restored object $f_\alpha(\vec{\varrho})$ has the same bandwidth of the kernel $|H(\vec{\varrho})|^2$ and, therefore, of the original image $g_s(\vec{\varrho})$. However the restored object $f_\alpha(\vec{\varrho})$ can exhibit a better resolution than $g_s(\vec{\varrho})$. In order to quantify this statement we can proceed as follows.

Let us neglect the effect of the noise, i.e. let us assume that $\hat{g}_s(\vec{\omega})$ is related to $\hat{f}(\vec{\omega})$ by eq. (10) and not by eq. (12). Then, if we replace eq. (10) in eq. (13), we get

$$f_\alpha(\vec{\varrho}) = \frac{1}{(2\pi)^2} \int\int \frac{\left|\hat{K}(\vec{\omega})\right|^2}{\left|\hat{K}(\vec{\omega})\right|^2 + \alpha} \hat{f}(\vec{\omega}) e^{i(\vec{\varrho},\vec{\omega})} d\vec{\omega} \tag{14}$$

and this equation can be written in the form of a convolution integral

$$f_\alpha(\vec{\varrho}) = \int\int K_\alpha(\vec{\varrho} - \vec{\varrho}\,') f(\vec{\varrho}\,') d\vec{\varrho}\,' \tag{15}$$

where $K_\alpha(\vec{\varrho} - \vec{\varrho}\,')$ is the inverse Fourier transform of the function

$$\hat{K}_\alpha(\vec{\omega}) = \frac{\left|\hat{K}(\vec{\omega})\right|^2}{\left|\hat{K}(\vec{\omega})\right|^2 + \alpha}. \tag{16}$$

As the function $K(\vec{\varrho}) = |H(\vec{\varrho})|^2$ is the impulse response function of the scanning microscope, thus the function $K_\alpha(\vec{\varrho})$ is the impulse response function of the system which consists of the scanning microscope and of the computer where the algorithm (13) has been implemented.

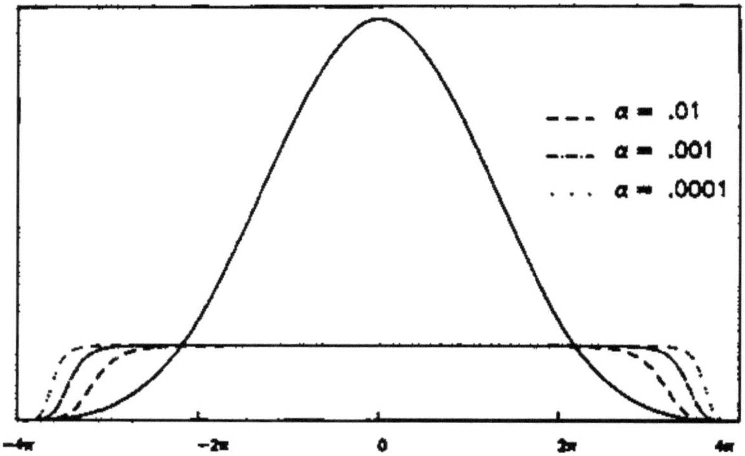

Fig. 1. Comparison of the transfer function (Fourier transform of the impulse response function) of the confocal scanning microscope (full line) with the transfer functions corresponding to various values of the regularization parameter.

In Fig. 1 we compare the functions $\hat{K}$ and $\hat{K}_\alpha$ in the case of the one-dimensional problem (4), with $\Omega = \pi$. Then $K(x) = [sin(\pi x)/\pi x]^4$ and the band of $K(x)$ is $[-4\pi, 4\pi]$. Its Fourier transform, however, is not constant over the band and it tends to zero rather rapidly at the edges of the band. On the other hand the functions $\hat{K}_\alpha$ are pratically constant over a large interval (as one can deduce from (16), $\hat{K}(\vec{\omega}) \simeq 1$ when $\alpha << \left|\hat{K}(\vec{\omega})\right|^2$) and this interval increases as α decreases (remember that α is related to the inverse of the signal- to-noise ratio). This result implies that the use of a regularization algorithm allows an improvement in resolution and that this improvement is related to the value of the signal to noise ratio. We demonstrate this effect in Fig. 2, by means of simulated data.

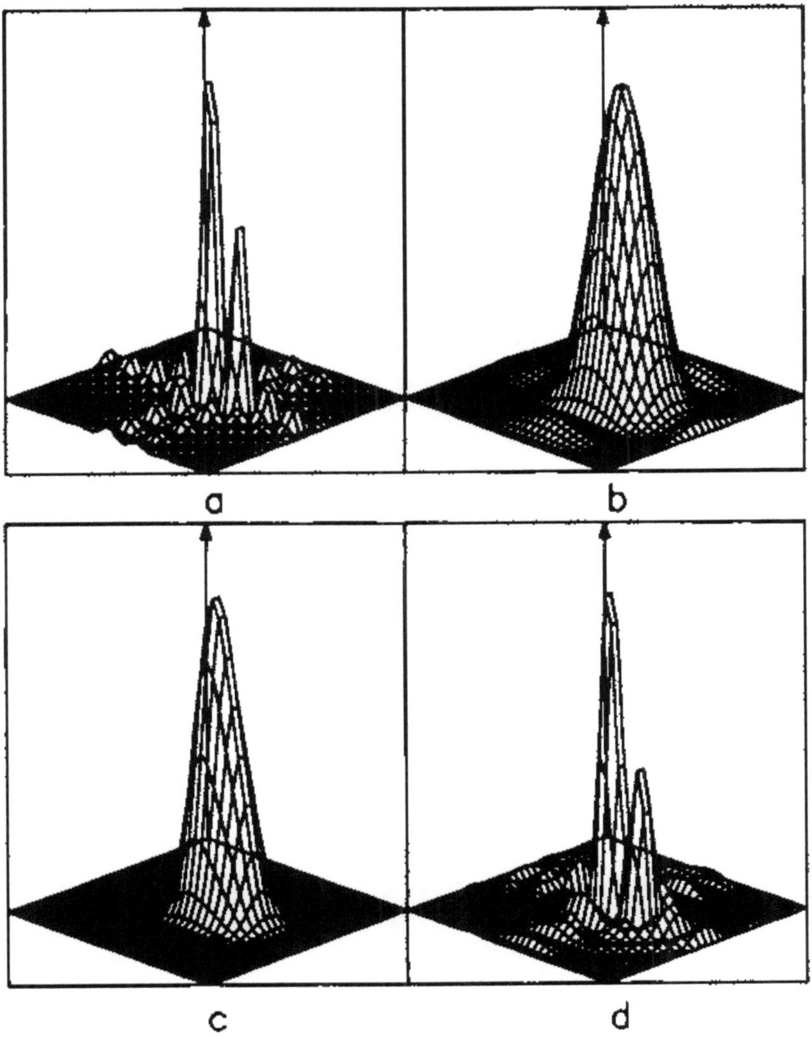

Fig. 2. Comparison of the images of the bandlimited version (bandwidth 4π) of two delta functions. The distance between the two delta functions is $R/2$ (R= Rayleigh distance) and the weights are respectively 1 and 1/2. a) The original object; b) The image provided by a conventional microscope; c) The image provided by the confocal scanning microscope; d) The image obtained by a regularized deconvolution ($\alpha = 10^{-2}$) of the image c).

Work is in progress in order to apply the regularization techniques to the restoration of the three-dimensional images of the scanning microscope developed at the University of Amsterdam[9]. Since it is quite difficult to measure the point spread function of the scanning microscope in the fluorescence mode[13], a theoretical model of the point spread function, which takes into account the main features of the scanning microscope[14], will be used.

3. THE ONE-DIMENSIONAL PROBLEM FOR THE SUPER-RESOLVING MICROSCOPE

The one-dimensional problem is a good mathematical laboratory for investigating the main features of the new scanning microscope equipped with an array of detectors for the measurement of the full image at each step of the scanning procedure.

If we assume that the point spread functions of the lenses are given by the eq. (4) with $\Omega = \pi$, so that the corresponding Rayleigh distance is 1, then the problem of solving eq. (1) implies the investigation of the properties of the integral operator

$$(A_C f)(x) = \int_{-\infty}^{+\infty} sinc(x - x')sinc(x')f(x')dx' \qquad (17)$$

(the subscript C indicates that this operator corresponds to the coherent case) while the problem of solving eq. (6) implies the investigation of the properties of the integral operator

$$(A_I f)(x) = \int_{-\infty}^{+\infty} sinc^2(x - x')sinc^2(x')f(x')dx' \qquad (18)$$

(the subscript I indicates that this operator corresponds to the incoherent case). We consider both A_C and A_I as operators in $L^2(-\infty, +\infty)$.

We first sketch some general features which are common to these two problems. Both the operator A_C and the operator A_I have non-trivial null spaces and this implies that the solution of the corresponding integral equations is not unique. Uniqueness can be restored by considering the generalized solution, i.e. the unique solution which is orthogonal to the null space.

The null spaces of the two operators can be completely characterized[5,15,16]. We do not need, however, a complete formulation of these results. It is sufficient to point out that the orthogonal complement of tbe null space of the operator A_C is a subspace of functions which have bandwidth 2π and zeros at the integer sampling points (except the origin) and that the orthogonal complement of the null space of the operator A_I is a subspace of functions which have bandwidth 4π and also zeros at the integer sampling points (except the origin).

The two integral operators are compact. If we omit the subscripts C and I, this means that for each operator A we can introduce its singular system $\{\alpha_k; u_k, v_k\}_{k=0}^{\infty}$, i.e. the set of solutions of the coupled equations

$$Au_k = \alpha_k v_k \quad , \quad A^* v_k = \alpha_k u_k \tag{19}$$

where A^* is the adjoint of the operator A.

Tben in the case of noise free data, namely data which are in the range of the operator A, the generalized solution $f^\dagger$ of the integral equation $g = Af$ is given by

$$f^\dagger(x) = \sum_{k=0}^{+\infty} \frac{1}{\alpha_k}(g, v_k)u_k(x) \tag{20}$$

where (g, v_k) denotes the usual scalar product of $L^2(-\infty, +\infty)$.

As we have already remarked in the Introduction, thanks to the scanning procedure we only need to compute $f^\dagger(0)$ for each scanning position. In the case of noise free data, i.e. $g = Af$, from the relation

$$(g, v_k) = (Af, v_k) = (f, A^* v_k) = \alpha_k (f, u_k) \tag{21}$$

we get

$$f^\dagger(0) = \int_{-\infty}^{+\infty} T(x') f(x') dx' \tag{22}$$

where

$$T(x) = \sum_{k=0}^{+\infty} u_k(0) u_k(x). \tag{23}$$

Finally, if we recall that to each step of the scanning procedure corresponds a translation $-x$ of the object and if we denote by ${f_x}^\dagger(0)$ the recovered value (22) corresponding to $f(x' + x)$, then the image $\tilde{f}(x) = f_x^\dagger(0)$ obtained in this way is just the convolution of the function $T(x)$ with the object $f(x)$. The function $T(x)$ is the impulse response function describing the total effect of imaging, data inversion and scanning.

In the coherent case one can show[5] that $\hat{T}(\omega) = 1$ when $|\omega| < 2\pi$ and zero elsewhere. As a consequence the Fourier transform of $\tilde{f}(x)$ coincides with the Fourier transform of $f(x)$ over $[-2\pi, +2\pi]$. On the other hand, in the incoherent case one can show[10] that $\hat{T}(\omega)$ is not constant but different from zero everywhere over the band $[-4\pi, 4\pi]$. As a consequence one can recover the Fourier transform of $f(x)$ over this band from the Fourier transform of $\tilde{f}(x)$.

The basic difficulty of the previous procedure is that the singular values α_k tend to zero when $k \to \infty$ and therefore the generalized solution (20) may not exist in the case of noisy data. The problem is again ill-posed.

In this case a very simple regularization algorithm consists in truncating the series (20), the regularization parameter being essentially the inverse of the number of terms in the truncated series expansion[11,12]. In order to correctly apply this method we must investigate in detail the singular systems of the integral operators (17) and (18).

As concerns the operator A_C, it has been proved[17] that its singular values and singular functions have very simple analytic expressions. No similar result holds in the case of the operator A_I. Therefore it can be useful to have numerical methods for computing the singular systems of both operators. The exact result known in the case of the operator A_C can be used as a test of the accuracy of the computations.

A rather good numerical method consists in discretizing the integral equations by means of the sampling expansions of the data of and of the solution[5,15]. In this way the integral equations are transformed into infinite-dimensional linear systems. Finally these can be truncated and the convergence of the results can verified by increasing the number of sampling points.

In the coherent case the noise free image has bandwidth π and the generalized solution has bandwidth 2π and zeros at the integer sampling points (except the origin). Thus we can put

$$g(x) = \sum_{n=-\infty}^{+\infty} g(n) sinc(x-n) \tag{24}$$

and

$$f^{\dagger}(x') = \sum_{m=-\infty}^{+\infty} f^{\dagger}(x'_m) sinc[2(x'-x'_m)] \tag{25}$$

where $x'_0 = 0$ and $x'_m = \pm(|m|-1/2)$ (the sign + corresponds to the positive values of m and the sign $-$ to negative values of m). If we substitute these expansions in the integral equation $g = A_C f$, we get the liner system[5]

$$g(n) = \sum_{m=-\infty}^{+\infty} (A_C)_{nm} f^{\dagger}(x'_m) \; ; \quad n = 0, \pm 1, \ldots \tag{26}$$

where

$$(A_C)_{nm} = \frac{1}{2} sinc(x'_m - n) sinc(x'_m). \tag{27}$$

An important property , which has been recently proved[16], of this infinite dimensional matrix is that its inverse has also a very simple analytic expression. However we do not give here this result because we are mainly interested in the use of eqs. (26),(27) for the approximate computation of the singular system of the operator A_C. In fact it is possible to show[5] that one can obtain very good approximations of the first singular values using a rather small number of sampling points, i.e. a rather small finite section of the infinite dimensional matrix.

The previous method can be easily extended to the operator A_I. In fact, in such a case the noise free image has bandwidth 2π and the generalized solution has bandwidth 4π and zeros at the integer sampling points (except the origin). Thus we can put

$$g(x) = \sum_{n=-\infty}^{+\infty} g(\frac{n}{2}) sinc[2(x - \frac{n}{2})] \tag{28}$$

and

$$f^{\dagger}(x') = \sum_{m=-\infty}^{+\infty} f^{\dagger}(x'_m) sinc[4(x' - x'_m)] \tag{29}$$

where $x'_0 = 0$ and $x'_m = \pm(k + j/4)$ if $m = \pm(3k + j)(k = 0, 1, 2, \ldots; j = 1, 2, 3)$. By substituting these expansions in the integral equation $g = A_I f$ we find again an infinite dimensional linear system

$$g(\frac{n}{2}) = \sum_{m=-\infty}^{+\infty} (A_I)_{nm} f^{\dagger}(x'_m); n = 0, \pm 1, \ldots \tag{30}$$

where

$$(A_I)_{nm} = \frac{1}{4} sinc^2(x'_m - \frac{n}{2}) sinc^2(x'_m). \tag{31}$$

This system can also be truncated in order to compute approximations of the singular values and singular functions.

In Table I we compare the singular values of the coherent and incoherent case. In both cases 25 sampling points for the image and 49 sampling points for the object have been used. We notice that the singular values of the incoherent case tend to zero more rapidly than the singular values of the incoherent one.

Table I. Singular values of the integral operators A_C and A_I.

k	Coherent	Incoherent
0	0.82190	0.60336
1	0.44286	0.25675
2	0.20642	0.10761
3	0.14761	0.04029
4	0.10984	0.02377
5	0.08856	0.01987
6	0.07420	0.01013
7	0.06325	0.00807
8	0.05591	0.00770
9	0.04918	0.00432

Since the singular functions are alternatively even and odd in both problems[5,15], only the singular functions with even value of the index contribute to the series (23). In the case of a truncated expansion the function $T(x)$, eq. (23), is replaced by

$$T_K(x) = \sum_{k=0}^{K-1} u_k(0)u_k(x). \tag{32}$$

In Fig. 3 we plot the Fourier transforms of the functions $T_K(x)$, with $K = 5$ and $K = 9$, both in the coherent and in the incoherent case. Notice that, when $K = 5$, the condition number α_0/α_{K-1} is 7.48 in the coherent case and 25.38 in the incoherent one. When $K = 9$, the condition number is 14.7 in the coherent case and 78.38 in the incoherent one. The last value is still acceptable when the signal-to-noise ratio is of the order of 100.

As follows from the figures, the functions $\hat{T}_K(\omega)$ have a rather complicated structure over the hand. In order to eliminate the effect of this modulation of the

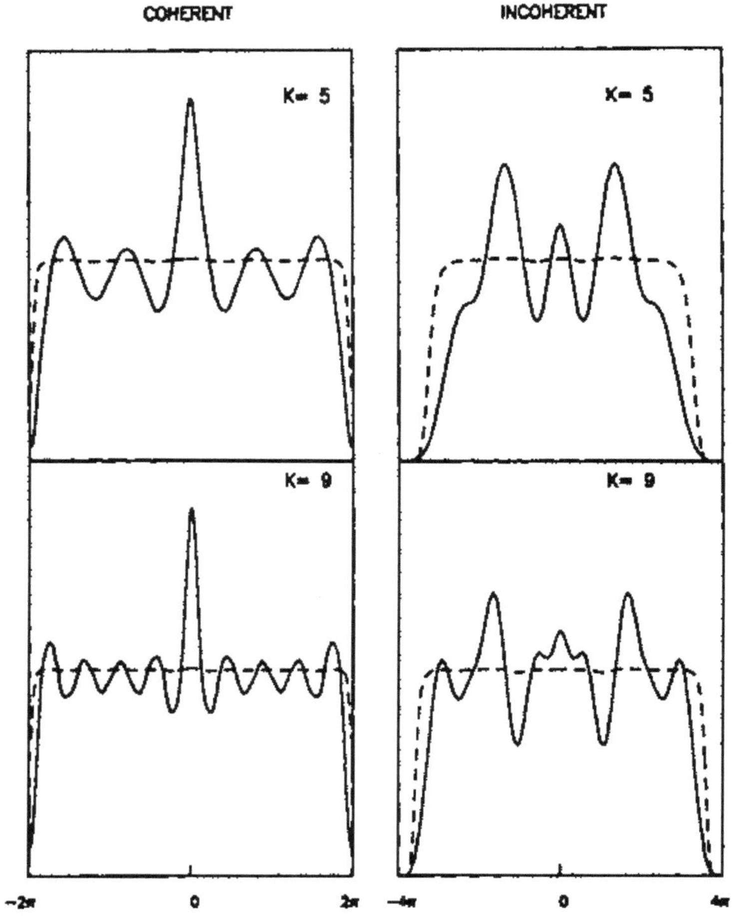

Fig. 3. Fourier transforms of the functions $T_K(x)$, eq. (32). The dotted lines represent the corresponding functions $\hat{T}_{\alpha,k}(\omega)$, eq. (33), with $\alpha = 10^{-2}$.

Fourier components of the object one can consider a regularized inversion of the image obtained by means of the convolution of $T_K(x)$ with $f(x)$. As discussed in the previous section this is equivalent to replace the function $\hat{T}_K(\omega)$ by the function

$$\tilde{T}_{\alpha,K}(\omega) = \frac{\left|\tilde{T}_K(\omega)\right|^2}{\left|\tilde{T}_K(\omega)\right|^2 + \alpha}. \tag{33}$$

These functions are also represented in Fig. 3 (dotted lines) in the case $\alpha = 10^{-2}$. If we compare the functions corresponding to the incoherent problem with the function $\tilde{K}_\alpha(\omega)(\alpha = 10^{-2})$ of Fig. 1 we see that the new microscope, equipped with an array of detectors, provides a better resolution than the microscope with only one detector.

4. EXTENSION TO TWO-DIMENSIONAL AND THREE-DIMENSIONAL PROBLEMS

The investigation of the super-resolution of the multidetector microscope has been completed only in the one-dimensional case. However the solution of the one-dimensional problems provides also the solution of the two dimensional problems (coherent and incoherent) with square pupils. In fact, in such a case we have

$$S_1(\vec{\varrho}) = S_2(\vec{\varrho}) = sinc(x)sinc(y) \tag{34}$$

and the singular systems of the integral operators obtained by replacing eq. (34) in eq. (1) or eq. (5) can be obtained from the singular systems respectively of the operators (17) and (18) as follows: the singular values of the 2D operators are all the possible products of the singular values of the corresponding 1D operators

$$\alpha_{i,k} = \alpha_i \alpha_k \; ; \quad i,k = 0,1,2,\ldots \tag{35}$$

and the associated singular functions are the tensor products of the corresponding 1D singular functions

$$u_{i,k}(\vec{\varrho}) = u_i(x)u_k(y) \ , \ v_{i,k}(\vec{\varrho}) = v_i(x)v_k(y). \tag{36}$$

It is obvious that the singular values with $i \neq k$ have multiplicity 2 while the singular values with $i = k$ have multiplicity 1.

The case of square pupils, however, is not very interesting in pratice since, in general, the pupils have a circular shape. Now the case of circular pupils is more difficult from the computational point of view even if it exibits some interesting theoretical properties.

If we assume again that the lenses are uniformly filled and that they have the same aperture, then the point spread functions are given by

$$S_1(\varrho) = S_2(\varrho) = 2\frac{J_1(\pi\varrho)}{\pi\varrho} \ , \ \varrho = |\vec{\varrho}| \tag{37}$$

(the normalization has been chosen in such a way that $0 \leq S(\varrho) \leq 1$) and the integral equations for the coherent and incoherent case are obtained by replacing eq. (37) in eq. (1) and eq. (5) respectively.

The two-dimensional integral equations can be reduced to an infinite set of uncoupled one-dimensional equations by means of Fourier series expansions of the image, of the object and of the kernel[4]. If we remember, however, that our problem is the restoration of the object at the origin, we recognize that this problem has circular symmetry and therefore only the circularly symmetric singular functions contribute to its solution. The same conclusion is reached by noticing that all the singular functions, which depend also on the angle, have nodal lines passing through the origin.

Thus, if we put

$$g_0(\varrho) = \frac{1}{2\pi}\int_0^{2\pi} g(\varrho,\phi)d\phi \ , \ f_0(\varrho') = \frac{1}{2\pi}\int_0^{2\pi} f(\varrho',\phi')d\phi' \tag{38}$$

(notice that $f_0(0) = f(\vec{0})$) and also

$$H_{C,0}(\varrho, \varrho') = \int_0^{2\pi} S\left(\sqrt{\varrho^2 + \varrho'^2 - 2\varrho\varrho' cos(\vartheta)}\right) d\vartheta \tag{39}$$

(where $\vartheta = \phi - \phi'$) for the coherent problem and

$$H_{I,0}(\varrho, \varrho') = \int_0^{2\pi} S^2\left(\sqrt{\varrho^2 + \varrho'^2 - 2\varrho\varrho' cos(\vartheta)}\right) d\vartheta, \tag{40}$$

in the incoherent case, from eq. (1) we obtain

$$g_0(\varrho) = \int_0^{+\infty} \varrho' H_{C,0}(\varrho, \varrho') S(\varrho') f_0(\varrho') d\varrho' \tag{41}$$

and from eq. (5) we also obtain

$$g_0(\varrho) = \int_0^{+\infty} \varrho' H_{I,0}(\varrho, \varrho') S^2(\varrho') f_0(\varrho') d\varrho'. \tag{42}$$

Recalling the relantionship between the Fourier transform and the Hankel transform[19)], it is easy to recognize that the Hankel transform of $g_0(\varrho)$ has the band $[0, \pi]$ in the coherent case and $[0, 2\pi]$ in the incoherent one. Analogously the Hankel transform of the generalised solution $f_0^\dagger(\varrho')$ has the hand $[0, 2\pi]$ in the coherent case and $[0, 4\pi]$ in the incoherent one (recall that the band of $S(\varrho)$ is $[0, \pi]$). Therefore in both cases the image and the generalized solution can be represented by means of a sampling expansion of the following kind[19)]

$$h(\varrho) = \sum_{n=1}^{\infty} \frac{\sqrt{2}}{cJ_1(x_n)} h\left(\frac{x_n}{c}\right) s_n(\varrho) \tag{43}$$

where c is the bandwidth of $h(\varrho)$, the x_n are the zeros of $J_0(x)$ and the sampling functions $s_n(\varrho)$ are given by

$$s_n(\varrho) = \sqrt{2} c x_n \frac{J_0(c\varrho)}{x_n^2 - (c\varrho)^2}. \tag{44}$$

These sampling functions $s_n(\varrho)$ form an orthonormal system in the following sense

$$\int_0^{+\infty} \varrho s_n(\varrho) s_m(\varrho) d\varrho = \delta_{nm} \tag{45}$$

and they have the following projection property

$$\int_0^{+\infty} \varrho h(\varrho) s_n(\varrho) d\varrho = \frac{\sqrt{2}}{c J_1(x_n)} h\left(\frac{x_n}{c}\right). \tag{46}$$

By replacing the appropriate sampling expansions of $g_0(\varrho)$ and $f_0^\dagger(\varrho')$ in eq. (41) or eq. (42) and using the projection property (46), these integral equations can be transformed into infinite dimensional linear systems, just as it was done in Sect. 3. These systems can be truncated in order to get approximations for the singular values and singular functions of the corresponding integral operators.

Preliminary results have been obtained in the coherent case. Using 20 sampling points both for the image and for the object we have obtained the following singular values

$$\alpha_0 = 0.869,\ \alpha_1 = 0.187,\ \alpha_2 = 0.080,\ \alpha_3 = 0.046,\ \alpha_4 = 0.030,\ \alpha_5 = 0.022.$$

The singular functions can be computed using the interpolation formula (43). Then one can define functions $T_K(\varrho)$, as in eq. (32), which have the meaning explained in Sect. 3. Using 4 singular values (the corrisponding condition number α_0/α_3 is 18.37), the Hankel transform of the corresponding $T_K(\varrho)$ is significantly different from zero over the full interval $[0, 2\pi]$ and this means an improvement of a factor of 2 with respect to the resolution of the conventional coherent microscope.

Computations are in progress in the case of the incoherent problem.

As we have remarked in the Introduction, in the case of a thick, weakly fluorescence object, the image of a confocal system, for a fixed scanning position, is given by eq. (7). If we consider again the case of uniformly filled and identical circular lenses, then we have[10)]

$$S_1(\varrho, z) = S_2(\varrho, z) = 2\int_0^1 tJ_0(\pi\varrho t)e^{-\frac{i\pi}{2}zt^2}dt\ , \quad \varrho = |\vec{\varrho}|. \tag{47}$$

Here the normalization has been chosen in such a way that $S(\varrho, 0)$ coincides with the point spread function (37). Since, for fixed z, the function (47) has circular symmetry, all the remarks made above for the two-dimensional problem apply also to this case. Therefore one can reduce the dimension of the problem by integrating over the angular variable in the transverse planes.

Preliminary computations have been performed in the case of a two-dimensional model with one lateral and one axial variable. In such a case the appropriate integral operator is the following one

$$(Af)(x) = \int\int_{-\infty}^{+\infty} |S(x-x', z')|^2\, |S(x', z')|^2\, f(x', z')dx'dz' \tag{48}$$

where

$$S(x, z) = \frac{1}{2}\int_{-1}^{1} e^{-i\pi xt - i\frac{\pi}{2}zt^2}dt. \tag{49}$$

Also in this case the normalization has been chosen in such a way that $S(x, 0) = sinc(x)$, the point spread function of the 1D problem. The preliminary computations performed for this problem show that the number of significant singular values of the operator (48),(49) is larger than that of the integral operator (18). For instance, in the case of a condition number of the order of 25, we can use 11 singular values instead of 5. Also in this case, however, only the singular functions with even index contribute, so that we have 6 terms (instead of 3) in the truncated singular function expansion. This result indicates that it must be possible to obtain an improvement in resolution also in the axial direction.

ACKNOWLEDGEMENTS

This work has been partly supported by EEC contract No. BAP-0293-NL(GDF) and by Ministero Pubblica Istruzione, Italy.

REFERENCES

1) Lemons, R.A. and Quate, C.F. Acoustic microscopy - scanning version. Applied Phys. Lett., 24, 163-165 (1974).

2) Sheppard, C.J.R. and Choudhury, A. Image formation in the scanning microscope. Optica Acta, 24, 1051-1073 (1977).

3) Brakenhoff, G.J., Blom, P. and Barends, P. Confocal scanning light microscopy with high aperture immersion lenses. J. Microsc., 117, 219-232 (1979).

4) Bertero, M. and Pike, E.R. Resolution in diffraction limited imaging, a singular value analysis. Optica Acta, 29, 727-746 (1982).

5) Bertero, M., Brianzi, P. and Pike, E.R. Super-resolution in confocal scanning microscopy. Inverse Problems, 3, 195-212 (1987).

6) McCutchen, C.W. Superresolution in microscopy and the Abbe resolution limit. J. Opt. Soc. Am., 56, 1190-1192 (1967).

7) Cox, I.J., Sheppard, C.J.R. and Wilson, T. Super-resolution by confocal fluorescent microscopy. Optik, 60, 391-396 (1982).

8) Wijnaends van Resandt, R.W., Marsman, H.J.B., Kaplan, R., Davoust, J., Stelzer, E.H.K. and Stricker, R. Optical fluorescence microscopy in three dimensions: microtomoscopy. J. Microsc., 138, 29-34 (1985).

9) Brakenhoff, G.J., van der Voort, H.T.M., van Sprosen, E.A. and Nanninga, N. Three-dimensional imaging by confocal scanning fluorescence microscopy. Annals N.Y. Acad. of Sci., 483, 405-415 (1986).

10) Born, M. and Wolf, E. Principles of Optics, pp. 434-448. Pergamon Press, London, (1959).

11) Tikhonov, A.N. and Arsenin, V.Y. Solutions of Ill-posed problems. Wiston & Sons, Washington, (1977).

12) Bertero, M., De Mol, C. and Viano, G.A. The stability of inverse problems, in Inverse Scattering Problems in Optics, ed. H.P. Baltes, Topics in Current Physics, vol. 20, pp. 161-214. Springer, Berlin, (1980).

13) van der Voort, H.T.M., Brakenhoff, G.J. and Janssen, G.C.A.M. Determination of the 3-dimensional optical properties of a confocal scanning laser microscope. Optik, 78,48-53 (1988).

14) van der Voort, H.T.M. and Brakenhoff, G.J. Modeling of 3-D confocal imaging at high N.A. in fluorescence. To appear in Proc. of the International Congress on Optical Science and Engineering, Hamburg, 19-23 September 1988, SPIE, (1988).

15) Bertero, M., Boccacci, P., Brianzi, P. and Pike, E.R. Inverse problems in confocal scanning microscopy, in Inverse Problems: An Interdisciplinary Study, ed. P.C. Sabatier, Adv. Electr. and Electron. Phys., Supplement 19, 225-239 (1987).

16) Bertero, M., Boccacci, P., Defrise, M., De Mol, C. and Pike, E.R. Super-resolution in confocal scanning microscopy: the incoherent case (in preparation) (1988).

17) Gori, F. and Guattari, G. Signal restoration for linear systems with weighted inputs. Singular value analysis for two cases of low-pass filtering. Inverse Problems, 1, 67-85 (1985).

18) Bertero, M., De Mol, C. and Pike, E.R. Analytic inversion formula for confocal scanning microscopy. J. Opt. Soc. Am. A, 4, 1748-1750 (1987).

19) Papoulis, A. Systems and Transforms with Applications to Optics, pp. 434-448. McGraw-Hill, New York, (1968).

LIST OF PARTICIPANTS

AUDET	J.	*Service Acoustique, Thomson-Sintra*	VALBONNE
AUDOLY	C.	*G.E.R.D.S.M.*	SIX FOURS
BARDOS	C.	*Labo. Mathématiques Appliquées, E.N.S.*	PARIS
BARIBAUD	M.	*LEMO-ENSERG*	GRENOBLE
BELLIARD	M.	*Labo. de Radioélectricité, Fac. St Jérôme*	MARSEILLE
BENTOSELA	F.	*Centre de Physique Théorique et Univ. Aix-Marseille II*	MARSEILLE
BERTERO	M.	*Dipartimento di Fisica*	GENOVA
BERTHON	A.	*Société AERO*	PARIS
BERTRAND	P.	*ONERA*	CHATILLON
BOELLE	J.L.	*IFREMER*	PLOUZANE
BONZOM	T.	*Electronique S. Dassault*	SAINT-CLOUD
BOURRELY	C.	*Centre de Physique Théorique*	MARSEILLE
BRIZE-TRAMONI	L.	*Labo. de Mécanique et d'Acoustique*	MARSEILLE
CHETY	L.	*D.R.E.T.*	PARIS
CHIAPPETTA	P.	*Centre de Physique Théorique*	MARSEILLE
COMBES	J.M.	*C.P.T. et Univ. de Toulon et du Var*	MARSEILLE
CONOIR	J.M.	*LCTAR*	VELIZY-VILLACOUBLAY
CRISTOL	X.	*Thomson-Sintra*	CAGNES/MER
DALMAS	J.	*Labo. de Radioélectricité, Fac. St Jérôme*	MARSEILLE
DELEUIL	R.	*Labo. de Radioélectricité, Fac. St Jérôme*	MARSEILLE
DENIAU	G.	*E.C.A.N.*	SAINT TROPEZ
DUCHENE	B.	*Labo. des Signaux, E.S.E.*	GIF SUR YVETTE
EMERIC	N.	*Dépt. de Phys. et Epistémologie, Univ.Provence*	MARSEILLE
GIBIAT	V.	*Groupe de Physique des Solides, ENS*	PARIS
GUILLOT	J.C.	*Centre Scient. et Polytech., Un.Paris Nord*	PARIS
HARDY	J.	*Laboratoire d'Acoustique*	LE MANS
JOLY	P.	*INRIA*	LE CHESNAY
LAMY	Ph.	*Laboratoire d'Astronomie Spatiale*	MARSEILLE
LE LOUARN	R.	*E.D.F., Dépt. acoustique*	CLAMART
LEFEBVRE	J.P.	*Labo. de Mécanique et d'Acoustique*	MARSEILLE
LEGRAND	O.	*Observatoire de Nice*	NICE
LEMAIRE	T.	*Centre de Physique Théorique*	MARSEILLE
LESSELIER	D.	*Laboratoire des Signaux, E.S.E.*	GIF SUR YVETTE
LEVIANDIER	L.	*Thomson Sintra*	ARCUEIL
LOMBARD	J.M.	*Avions Marcel Dassault*	SAINT CLOUD
LUZZATO	E.	*E.D.F., Dépt. acoustique*	CLAMART
MACON	L.	*Labo. Phys. de la Matière Condensée*	NICE
MAZE	G.	*L.E.A.M., Univ. du Havre*	LE HAVRE
MONTES	C.	*Labo. Phys. de la Matière Condensée*	NICE
MOTET	J.C.	*CELAR*	BRUZ
OLIVAIN	J.	*CEN Cadarache*	ST PAUL LEZ DURANCE
PERRIN	J.M.	*Labo. d'Astronomie Spatiale*	MARSEILLE
PHAM	M.H.	*Thomson Sintra*	ARCUEIL
POIREE	B.	*D.R.E.T.*	PARIS
POITEVIN	J.	*I.M.S.T.*	MARSEILLE
POTTIER	E.	*CELAR*	BRUZ
PRADA	C.	*Groupe de Physique des Solides, E.N.S.*	PARIS
RODRIGUEZ	R.	*Centre de Physique Théorique et Univ. Aix-Marseille II*	MARSEILLE
ROGER	A.	*SNE(A)*	PAU
ROUET	A.	*Science et Tec*	CROISSY SUR SEINE
ROUSSELOT	J.L.	*L.C.T.A.R.*	VELIZY VILLACOUBLAY
SABATIER	P.	*Labo. de Physique Mathématique*	MONTPELLIER
SAILLARD	J.	*Labo. de Radioélectricité*	RENNES
SALAZAR CARDOZO	R.E.	*ENSERG*	GRENOBLE
SARACCO	G.	*Labo. de Mécanique et d'Acoustique*	MARSEILLE

SESSAREGO	J.P.	*Labo. de Mécanique et d'Acoustique*	MARSEILLE
SORNETTE	D.	*Labo. Phys. de la Matière Condensée*	NICE
SOUILLARD	B.	*Centre de Physique Théorique*	PALAISEAU
TCHAMITCHIAN	Ph.	*Centre de Physique Théorique et Univ. Aix-Marseille III*	MARSEILLE
TERRASSON	G	*Avions Marcel Dassault*	SAINT CLOUD
TORRESANI	B.	*Centre de Physique Théorique*	MARSEILLE
TOURNOIS	P.	*Thomson-CSF*	BOULOGNE-BILLANCOURT
TRAN VAN NHIEU	M.	*Thomson-Sintra*	ARCUEIL
VAUCHER DE LA CROIX	D.	*METRAVIB*	ECULLY
VINCENT	P.	*CNET*	ISSY LES MOULINEAUX
ZAKHARIA	M.	*Labo. Traitement du Signal*	LYON